Cichlid Fishes

Cichlid Fishes

Behaviour, ecology and evolution

Edited by
Miles H. A. Keenleyside
Professor of Zoology,
The University of Western Ontario,
Canada

CHAPMAN & HALL
London · New York · Tokyo · Melbourne · Madras

UK	Chapman and Hall, 2–6 Boundary Row, London SE1 8HN
USA	Van Nostrand Reinhold, 115 5th Avenue, New York NY10003
JAPAN	Chapman and Hall Japan, Thomson Publishing Japan, Hirakawacho Nemoto Building, 7F, 1-7-11 Hirakawa-cho, Chiyoda-ku, Tokyo 102
AUSTRALIA	Chapman and Hall Australia, Thomas Nelson Australia, 480 La Trobe Street, PO Box 4725, Melbourne 3000
INDIA	Chapman and Hall India, R. Seshadri, 32 Second Main Road, CIT East, Madras 600 035

First edition 1991

Typeset in 10/12 Photina by Thomson Press (India) Ltd., New Delhi
Printed in Great Britain at the University Press, Cambridge

ISBN 0 412 32200 5   0 442 31157 5 (USA)

British Library Cataloguing in Publication Data
Cichlid fishes.
1. Cichlidae
I. Keenleyside, Miles A. II. Series
597.58

ISBN 0–412–32200–5

Library of Congress Cataloging-in-Publication Data available

Contents

Contributors

Gerard P. Baerends
Zoologisch Laboratorium, Rijksuniversiteit Groningen, The Netherlands

C. D. N. Barel
Research Group in Ecological Morphology, Zoologisch Laboratorium, Rijksuniversiteit Leiden, The Netherlands

G. W. Barlow
Department of Integrative Biology and Museum of Vertebrate Zoology, University of California, Berkeley, Cal., USA

T. Goldschmidt
Research Group in Ecological Morphology, Zoologisch Laboratorium, Rijksuniversiteit Leiden, The Netherlands

P. C. Goudswaard
Haplochromis Ecology Survey Team, Mwanza, Tanzania

P. H. Greenwood
Department of Zoology, British Museum (Natural History), London, England

M. H. A. Keenleyside
Department of Zoology, University of Western Ontario, London, Ontario, Canada

I. Kornfield
Department of Zoology, University of Maine, Orono, Maine, USA

K. F. Liem
Museum of Comparative Zoology, Harvard University, Cambridge, Mass., USA

W. Ligtvoet
Haplochromis Ecology Survey Team, Mwanza, Tanzania

R. H. Lowe-McConnell
Streat near Hassocks, Sussex, England

K. R. McKaye
Appalachian Environmental Laboratory, University of Maryland, Frostburg, Maryland, USA

M. H. J. Nelissen
Laboratorium Algemene Dierkunde, Rijksuniversitait Centrum Antwerpen, Belgium

D. L. G. Noakes
Institute of Icthyology, Department of Zoology, University of Guelph, Guelph, Ontario, Canada

R. S. V. Pullin
Aquaculture Program, ICLARM (International Center for Living Aquatic Resources Management), MC P.O. Box 1501, Makati, Metro Manila, Philippines

T. J. Ribbink
J. L. B. Smith Institute of Ichthyology, Grahamstown, Republic of South Africa

M. L. J. Stiassny
Department of Herpetology and Ichthyology, American Museum of Natural History, New York, USA

F. Witte
Research Group in Ecological Morphology, Zoologisch Laboratorium, Rijksuniversiteit Leiden, The Netherlands

K. Yamaoka
Faculty of Agriculture, Kochi University, Japan

Series Foreword

Among the fishes, a remarkably wide range of biological adaptations to diverse habitats has evolved. As well as living in the conventional habitats of lakes, ponds, rivers, rock pools and the open sea, fish have solved the problems of life in deserts, in the deep sea, in the cold Antarctic, and in warm waters of high alkalinity or of low oxygen. Along with these adaptations, we find the most impressive specializations of morphology, physiology and behaviour. For example we can marvel at the high-speed swimming of the marlins, sailfish and warm-blooded tunas, air-breathing in catfish and lungfish, parental care in the mouthbrooding cichlids and viviparity in many sharks and toothcarps.

Moreover, fish are of considerable importance to the survival of the human species in the form of nutritious and delicious food of numerous kinds. Rational exploitation and management of our global stocks of fishes must rely upon a detailed and precise insight of their biology.

The Chapman and Hall *Fish and Fisheries Series* aims to present timely volumes reviewing important aspects of fish biology. Most volumes will be of interest to research workers in biology, zoology, ecology and physiology, but an additional aim is for the books to be accessible to a wide spectrum of non-specialist readers ranging from undergraduates and postgraduates to those with an interest in industrial and commercial aspects of fish and fisheries.

This volume, the second in the *Fish and Fisheries Series*, presents a synoptic biology of the Cichlidae, by all accounts one of the most advanced teleost families and one which has not been summarized in book form since Fryer and Iles' famous monograph in 1972. Professor Miles Keenleyside has succeeded in assembling an impressive range of cichlid experts from all around the globe, and furthermore, we are most fortunate to have the foreword written by Gerard Baerends, a pioneer in cichlid ethology.

Cichlids exhibit a paradox between morphological similarity and ecological diversity. At first sight, these fish have been structurally conservative once their advanced design evolved, yet they occupy a prodigious range of ecological niches and their speciation in African Rift Valley lakes is justly described as explosive. Consequently, cichlid taxonomy is a challenging issue which often needs to deploy subtle biochemical methods, while cichlid behaviour, imbued with spectacular learning, parental care and courtship-segregation of species, is a rich and fascinating subject. While concentrating on functional morphology, ecology, behaviour and genetics, the book does not lose sight of the applied biology of cichlids. Augmenting their intrinsic evolutionary interest, cichlids have a most important role in tropical freshwater fisheries and aquaculture, and in some cases have raised significant conservation questions. This book will be an absorbing source of ideas and material to all who are interested in this group of fishes.

Dr Tony J. Pitcher
Editor, Chapman and Hall Fish and Fisheries Series
Reader in Biology, Bangor

CHAPMAN & HALL FISH AND FISHERIES SERIES

1. Ecology of Teleost fishes
   R. J. Wooton
2. Cichlid Fishes: Behaviour, ecology and evolution
   Edited by M. H. A. Keenleyside
3. Cyprinid Fishes: Sytematics, biology and exploitation
   Edited by I. J. Winfield and J. S. Nelson
4. Early Life History of Fish: An energetics approach
   E. Kamler
5. Fisheries Acoustics
   D. N. MacLennan and E. J. Simmonds

Forthcoming titles

Fish Chemoreception
Edited by T. Hara

Behaviour of Teleost Fishes, 2nd edn
Edited by T. J. Pitcher

Fish swimming
J. Videler

Sea Bass
G. Pickett and M. Pawson

Fisheries Ecology, 2nd edn
Edited by T. J. Pitcher and P. Hart

Foreword

In the last four decades, interest in fish of the family Cichlidae has shown a spectacular rise. However, mankind's relationships with cichlid fish date from ancient times. Specimens, possibly of *Oreochromis niloticus*, can be recognized on Egyptian frescos and reliefs. Further, at least two species, *Sarotherodon galilaeus* and *Tilapia zillii*, are likely to have been present in the catches St Peter and his colleagues took from Lake Galilee (Tiberias). Indeed, many native inland fisheries in the tropics, particularly in Africa, have long depended on stocks of cichlid fish. Since Linnaeus described *O. niloticus* and *S. galilaeus*, cichlids have been the subject of taxonomic considerations, but their classification in a separate family was only accomplished at the end of the last century by George Boulenger. As the exploration of tropical waters intensified, the taxonomy of this family became more and more difficult, particularly because clear morphological discontinuities in different, but apparently related forms often could not be established. In more modern times Regan, Trewavas and Poll have laid the basis for overcoming these difficulties.

Until the Second World War, cichlid fish were not popular with aquarists, mainly because many species tend to damage the vegetation and the substrate of aquaria, fight among themselves and suppress or even kill other fishes present. These 'bad habits' arise because many species can easily be induced to reproduce in aquaria and, because they all practise some form of parental care, they need a lot of free space. Within the family this reproductive behaviour varies from monogamous biparental care of eggs and fry to mouth-brooding by the female. The behaviour involved in pair formation, spawning and caring for the young is often spectacular and poses manifold scientific questions. This is why – as ethology emerged in the second half of the 1930s – some practitioners of this behavioural discipline chose cichlids as their subjects (Breder and Noble in the USA; Lorenz, Seitz, and Peters in Europe). It is a great advantage to ethologists to use for their observations and experiments animals that can in the laboratory relatively easily be provided with a sufficiently natural environment to allow them to develop their full behavioural repertoire.

My own interest in cichlids I owe to Konrad Lorenz who, during a visit to Niko Tinbergen's laboratory at the University of Leiden, strongly recommended these fish for use in practical courses of ethology, especially because they offer excellent opportunities for comparing behaviour among closely related species. In 1941 I was charged, as Tinbergen's assistant, with exploring the possibilities the cichlids could offer. My wife and I worked on the behaviour of the few species then available, until after a year's work at the university our research was brought to a halt by the Nazi occupation. We published the results in 1950, in a monograph with the rather presumptuous title *An Introduction to the Study of the Ethology of Cichlid Fishes* (*Behaviour*, Supp. 1). I hope that its influence may help in part to explain why it has now become very much out of date and badly needs replacement.

Results of behavioural research soon penetrated the literature of aquarists. This was one reason for the increasing popularity of cichlids in the second half of this century, in many countries clearly expressed in the formation of special societies of cichlidophiles. A variety of popular books on cichlids appeared and some firms in the ornamental fish trade specialized in this group. Another reason was that developments in aquarium techniques made it easier to cope with the 'bad habits' of these fish. The ornamental fish export business in the native countries of cichlids responded by sending more and more new species to Europe and North America. The multitude of species revealed in this way, and the great beauty of many, inspired the collecting urge in many cichlid lovers.

The scientific study of cichlids also benefited from the increased assortment of available species. The high number of closely related forms in some areas, particularly in Lakes Victoria, Tanganyika and Malawi, were meat and drink to evolutionary biologists. To assess the adaptive value of the various behavioural and morphological characteristics and to understand the selection pressures at work, data on the ecology of different cichlid species and on their feeding behaviour are essential. The aquarium studies had to be extended by field work. Conditions for such work had in the meantime improved; the organization of expeditions had become easier, and in several places in the distribution area of cichlids, local field stations were established. Scuba-diving and underwater photography had become important research tools.

For the study of the mechanisms of evolution, the value of the cichlid family now equals or has even surpassed that of the Galapagos Darwin's finches. It encompasses a multitude of biological disciplines such as taxonomy, genetics, functional anatomy, ecology and ethology. Within ethology the relatively new branch of sociobiology builds an extremely interesting link between behaviour and evolution. The present book deals with all these aspects. In addition a chapter is devoted to the application of the biological knowledge gained in the increasing aquaculture of cichlids.

Unfortunately, human interference in the presumed interest of fisheries has recently almost destroyed the cichlid ecosystem in Lake Victoria. One chapter deals with this dramatic event, which means on the one hand an irreparable loss to the study of evolution, and on the other an interesting, though unwanted, experiment.

The editor, Professor Miles Keenleyside, is an expert with a lifetime experience in fish behaviour. He has managed to bring together an excellent choice of authorities in the different fields concerned. Their chapters allow the reader to become informed about the present state of cichlid research and they give a good introduction to the literature. The chapters are written in a way that makes them valuable not only to the professional, but also to the many aquarists who wish to go more deeply into the biology of cichlid fishes. There was a clear need for this book and I strongly recommend it, not only to ichthyologists, but also to biologists who do not focus their interest on a particular taxonomic group, but who are looking for suitable experimental subjects to study basic problems in other branches of zoology.

Gerard P. Baerends
Groningen,
The Netherlands

Introduction

Two landmark publications on cichlid fishes have appeared since the end of the Second World War. The first was 'An introduction to the study of the ethology of cichlid fishes' by G. P. Baerends and J. M. Baerends-van Roon. This pioneering study appeared in 1950 as Supplement 1 of the journal *Behaviour.* It contained descriptions of the morphology and behaviour of several cichlid species, and some experimental analyses of social behaviour, especially the interactions among parents and their offspring. It has probably been cited by students of cichlid behaviour more often than any other single publication. The second influential book was *The Cichlid Fishes of the Great Lakes of Africa* by G. Fryer and T. D. Iles, published in 1972. In it the authors synthesized virtually all available information on the biology and evolution of the remarkable groups of cichlids found in Lakes Victoria, Tanganyika, Malawi and some smaller lakes of East Africa.

Since 1972 there has been a rapid increase in the number of publications on cichlid fishes, many dealing with behaviour, ecology, evolution or some combination of these. However, much of this work has been focused on specific problems and has appeared in special-interest journals. Now seems to be an opportune time to consolidate this recent material. Accordingly, this book presents a series of review articles, each written by a specialist with considerable experience in a particular area of cichlid biology. Each contributor was asked to prepare a review of recent developments and the present state of knowledge in one area. Luckily for the editor, everyone responded with enthusiasm, and the results of their efforts fill the pages of this book.

The opening chapter deals with an often vexing problem in cichlid research, The present state of classification and nomenclature within the family Cichlidae. Taxonomic revisions of certain groups appear at intervals (see p. xxi). Some are soon widely accepted; others meet more resistance. Melanie Stiassny discusses the major problems and presents a coherent view of the current state of knowledge of the phylogenetic relationships within the family. Her conclusions are appropriately cautious.

This raises the question of the nomenclature to be used in the book. Many cited studies used species that have since been given a different genus or species name; some have been changed more than once. I have rejected the suggestion that a standard nomenclature be used throughout. Any such attempt would be unsatisfactory for some readers because of the dynamic state of cichlid taxonomy. My solution, therefore, has been to retain each author's preferred nomenclature which, when combined with the original publication cited by the author, should engender little confusion about the species under discussion.

Chapters 2 and 3 discuss the relations between cichlids and their biotic and physical environments. Tony Ribbink considers the situation in the three Great Lakes of East Africa. He stresses the importance of philopatry and extended parental care of the young in shaping the within-lake distribution of species. The degree of philopatry is also related to the extent of speciation among the fishes of a single lake. Resource partitioning, based on physical, behavioural and ecological features, permits the coexistence of many species in each distinctive habitat. Rosemary Lowe-McConnell reviews the extensive material on distribution, feeding habits and behaviour of riverine cichlids in Africa and South America. A major difference between the habitats in rivers and lakes is the greater variation in water level and flow in the former. This variability causes fluctuations in physicochemical and biotic conditions which may influence both the feeding and reproductive activities of the fishes. In regions with wet and dry seasons these water-flow changes are seasonal, leading to more clearly defined breeding seasons among the cichlids (and other fishes) in rivers than in lakes.

In Chapter 4 Humphry Greenwood critically reviews the literature concerning speciation in the family Cichlidae. Most of this material is based on fishes of the African Great Lakes. Indeed, it was the discovery of cichlid species flocks in these lakes that led to a vigorous debate among evolutionists about the possibility of sympatric speciation. Much of the chapter deals with the relative importance of the mechanisms that may underlie the explosive speciation among the cichlids of these lakes. Special attention is paid to the relative merits of the 'biological species concept', with its emphasis on the origin of interspecific reproductive isolating mechanisms (Mayr, 1982), and the more recent 'recognition concept' which stresses the origin of the means by which potential conspecific mating partners recognize each other (Paterson, 1985). Greenwood suggests that several different speciation mechanisms may have been responsible for the spectacular evolutionary success of the African lake cichlids.

In Chapter 5 Irv Kornfield describes the contributions being made by geneticists to a better understanding of cichlid evolution. Basic studies in descriptive genetics are still needed, especially in the search for the most appropriate types of fish for aquaculture. Progress is being made by using monosexual populations, experimentally induced polyploids and hybrid crosses. The relative influence of genetic and environmental factors, especially during early ontogeny, in determining variation in morphology and in coloration is being examined by carefully controlled experimental crossings.

The bulk of the chapter deals with the application of problems in cichlid systematics of several recently developed molecular techniques. These are electrophoretic isozyme analysis, mitochondrial DNA restriction analysis and the comparative analysis of DNA sequences. Emphasis is on the assumptions behind the methods and some of the technical problems associated with each.

The next two chapters consider the feeding biology of cichlids from two different points of view. Both concentrate on work done on African Great Lakes species, and especially those of Lake Tanganyika. Diversity in the trophic apparatus, even among closely related species, has been recognized since Fryer and Iles' monograph and before. Karel Liem's approach (that of functional morphology) is to examine the form and function of the feeding apparatus in selected species by using the techniques of electromyography and kinematics. This allows him to demonstrate objectively the diversity in design of the trophic structure within the family. In addition, there is evidence from both field and laboratory studies that cichlids, like many other fishes, can be remarkably plastic in their diet. Liem argues that the trophic diversity within, for example, a single large lake such as Lake Tanganyika, is based on variability in both structure and behaviour, and is best explained as stemming from multiple ancestry rather than any form of monophyly.

Kosaku Yamaoka concentrates on differences and similarities in feeding behaviour as a way of understanding the apparent coexistence of different species. He first outlines the various forms of food acquisition, ranging from grazing on epilithic algae to various specialized forms of piscivory, including scale eating and the capture of eggs or young from the buccal cavity of mouthbrooding adults. A central feature of this chapter is the examination of interspecific associations during foraging. For example, two species which share a common diet (e.g. fish scales or eggs) may have greater success by feeding together than alone because they use slightly different tactics of attack. On the other hand, species with different diets may also benefit from commensal foraging because the feeding activities of one increase the capture possibilities for the other.

The next five chapters deal mainly with aspects of social behaviour; the first two discuss reproduction. The contrast between cichlids and most other fishes is nowhere more apparent than in the variety of their mating systems and the prolonged period of active parental care of the young. George Barlow describes the mating systems and the factors which appear to have shaped their evolution. The latter include the most fundamental feature of all sexually reproducing animals, anisogamy. This decrees that maximum reproductive output for males is attained by maximizing the number of eggs fertilized, using their quickly replaced sperm, whereas for females it requires the physiologically more costly process of developing smaller numbers of mature ova. Thus, even though monogamy appears to be the ancestral condition in the family, the male–female bond is frequently unstable, leading to polygyny, and ultimately, when combined with maternal mouthbrooding of eggs and young, to avian-like lek systems. These can be quite spectacular, as in the Lake Malawi mouthbrooding species, *Cyrtocara eucinostomus*, in which over 50 000 males congregate in a narrow breeding arena extending several km along the shore. Barlow also considers some environmental factors which can influence cichlid mating systems. These include the physical substrate, since many species spawn in excavations under rocks, inside empty mollusc shells, etc., and the abundance, distribution and variety of predators and of food. He stresses the need to consider the broad spectrum of ecological factors in trying to understand the evolution of cichlid mating systems.

In the next chapter I review parental care behaviour. The guarding of offspring for several weeks after hatching is rare among other fishes but universal among cichlids. This is a key feature (along with the attractive colours and small size of many species) in the great popularity of cichlids among aquarists. Descriptions of the various forms of parental care (e.g. uni- and biparental, substrate- and mouthbrooding) and of the behaviour patterns used by the guarding adults are followed by discussion of the extent of male–female parental role-sharing, the degree of pair-bond stability in biparental species, and the phenomenon of brood-mixing, both accidental and apparently deliberate under some conditions. Finally the evolution of parental care within the Cichlidae is considered with, one hopes, a minimum of uninformed speculation.

The chapter by David Noakes emphasizes the value of knowledge about the development of behaviour in young cichlids for a proper understanding of adult behaviour. There is a critical difference in the early post-hatching environment for the young of the two major parental types, guarders and bearers (equivalent to substrate- and mouthbrooders, respectively). When the young of bearer species become independent of their parents they are fully-formed juvenile fish, having spent their entire lives (except for brief periods of free swimming during which they feed) inside the parent's buccal cavity. In contrast, young guarders begin free swimming as incompletely differentiated larvae, and their subsequent development, behavioural as well as morphological, occurs while they are swimming in a school of siblings being guarded by one or both parents. This difference has important repercussions for the further development of juvenile and adult behaviour patterns. Noakes discusses the ontogeny of feeding behaviour and social behaviour, including interactions with siblings, parents, other conspecifics and heterospecifics. He pays special attention to the effects of early social experience on the behaviour of adults, including their social preferences, species recognition and mate choice.

Mark Nelissen next reviews selected areas of the voluminous literature on communication in cichlid fishes. The bulk of the chapter deals with the major senses used in sending and receiving signals. These include the visual, acoustic, chemical and tactile systems. Special emphasis is given to the work on changes in colour patterns as vehicles for information transfer. The final section considers the roles of these different mechanisms of communication in such social functions as species, family and individual recognition, dominance interactions and sexual behaviour.

Kenneth McKaye, like many other evolutionary biologists, is attracted by the challenges of understanding the forces driving the explosive speciation among African Great Lakes cichlids. In his chapter he describes experimental work he and his colleagues have done in Lake Malawi to explore the role that sexual selection, through female choice among potential mates, has played in this spectacular evolutionary scenario. A male character which females might use in selecting a mate is the form of the spawning platform, or bower, constructed by the males of haplochromine species. This structure is a good example of an 'extended phenotype' (Dawkins, 1982), and there is evidence that bower form is a character that can be used to distinguish species. These are lek-breeding, maternal mouthbrooders in which the male's role in reproduction is to establish a territory on the lek, build a bower and then attract, court and spawn with females in the bower. He plays no part in parental care. There is also variation within species in bower size, and data are presented supporting the proposition that breeding success among males is directly related to the size of their bower. This work is in its early stages but already clearly shows the value of carefully designed field experiments for testing the idea that sexual selection through female choice occurs among lek-breeding cichlids.

Kees Barel and four colleagues have contributed a chapter on the impact of the Nile perch, *Lates niloticus*, on the endemic haplochromine cichlids of Lake Victoria. The Nile perch was introduced in the early 1960s as a food source for the surrounding human populations. It is a piscivorous centropomid which can reach 2 m in length. It soon spread through the lake with devastating effects on the much smaller cichlid fishes. In providing a background to this deplorable situation, the authors describe the history of research on the Lake Victoria haplochromines. It has centred on two main problems, the taxonomy of the group (see Greenwood's chapter) and the role of trophic adaptations in the coexistence of several hundred closely related species. The lake has been a justly famous 'laboratory' for evolutionary biologists. The impact of the Nile perch on the haplochromines is then described and it is a gloomy scenario. Many of the species are considered close to extinction, and there is little reason to expect them to recover, no matter what happens to the predator population. Even though the commercial fishery of the lake, and therefore the consumers of the catch, have benefited from the great increase in the Nile perch population, the destruction of yet another natural ecosystem by human intervention is a depressing story.

In the final chapter Roger Pullin reviews the present status of cichlid fishes in aquaculture throughout the world. Many species are being cultured for human food but the preferred one, because of its rapid growth and versatile feeding habits, is *Oreochromis niloticus.* A major problem is that in ponds containing both sexes the fish mature early (at small size), produce many offspring, and soon overpopulate the pond with undersized fish. Among several methods to prevent this, the most promising is to feed the fry a diet containing androgenic steroids, which reverses the sex of females. The hormone is soon eliminated from the growing fish and the resulting all-male population develops without breeding to harvestable size. The chapter ends with a discussion of current trends in cichlid culture in the major producing regions of the world, together with recent statistics (mainly from the FAO) on production by individual countries.

In conclusion, I am grateful to all those who contributed to the planning and preparation of this book. Its conception occurred during a symposium on 'The Behaviour of Fishes' held in July 1986 in Bangor, Wales. Tony Pitcher, who was a co-organizer of that meeting, made useful suggestions at an early stage. Others who contributed to its development are Gerard Baerends, George Barlow, David Noakes and David M. Scott. Most chapters were written in 1988 and revisions were completed in early 1989. Thus, taxonomic revisions within the Cichlidae published since then are not reflected in the text.

I am especially thankful for the assistance I received from the Zoology Department, University of Western Ontario. In particular, Lori Dos Santos repeatedly and cheerfully processed the manuscript of my own chapter. Financial assistance for my editorial work came from the Natural Sciences and Engineering Research Council of Canada. Finally, no acknowledgements would be complete without recognizing the contribution made by that copy editor extraordinaire, Charles Hollingworth. His suggestions to improve clarity, consistency and logic have added immensely to the entire book.

Miles H. A. Keenleyside
London, Ontario, Canada.

Errata

p.112, line 9, and p.125, para 3, line 7, *Heros* should read *Cichlasoma*
p.124, line 5, (Fig. 5.7) should read (Fig. 5.8)
p.125, para 3, line 9, ignore reference to Fig. 5.8
p.126, line 6, Tanzania should read Lake Tanganyika
p.193, line 19, (page 20) should read (page 201)
p.279, line 1, Green Wood should read Greenwood

Cichlid Fishes: Behaviour, ecology and evolution

Chapman & Hall 0 412 32200 5
Van Nostrand Reinhold Inc. 0 442 31157 5

Chapter one

Phylogenetic intrarelationships of the family Cichlidae: an overview

M. L. J. Stiassny

1.1 INTRODUCTION

A brief review of any of the numerous popular texts for fish hobbyists (e.g. Goldstein, 1973; Staeck and Linke, 1984; Loiselle, 1985a) illustrates the extraordinary variety of forms constituting that entity we classify as the Cichlidae. Ufermann *et al.* (1987) have compiled a useful catalogue of species names, yet it remains difficult to give a definitive total for the family as many species have yet to be formally described. Reviewing recent revisional and faunistic works leads to an expectation of a total somewhere in excess of 1000 species, and if some of the less conservative estimates prove to be correct (e.g. Lewis *et al.*, 1986, judge there to be some 500–1000 species endemic to Lake Malawi alone!), then the final total may be considerably higher. Interestingly, while nearly all workers have been in agreement with Boulenger (1898) that this superficially disparate assemblage constitutes "a very natural group", it is only relatively recently that the monophyly of the family has been established by any stated and acceptable criteria (Stiassny, 1981a; Zihler, 1982; Gaemers, 1984).

Paradoxically, considering their diversity of shapes and sizes, cichlids are rather conservative in underlying anatomy. Although this is a subjective assessment it is true that in most cases, if morphological structures are regarded in terms of presence or absence, or composition, extremely few character differences are evident. When contrasted with that of other comparably speciose percomorph clades (e.g. labrids or gobies), the level of

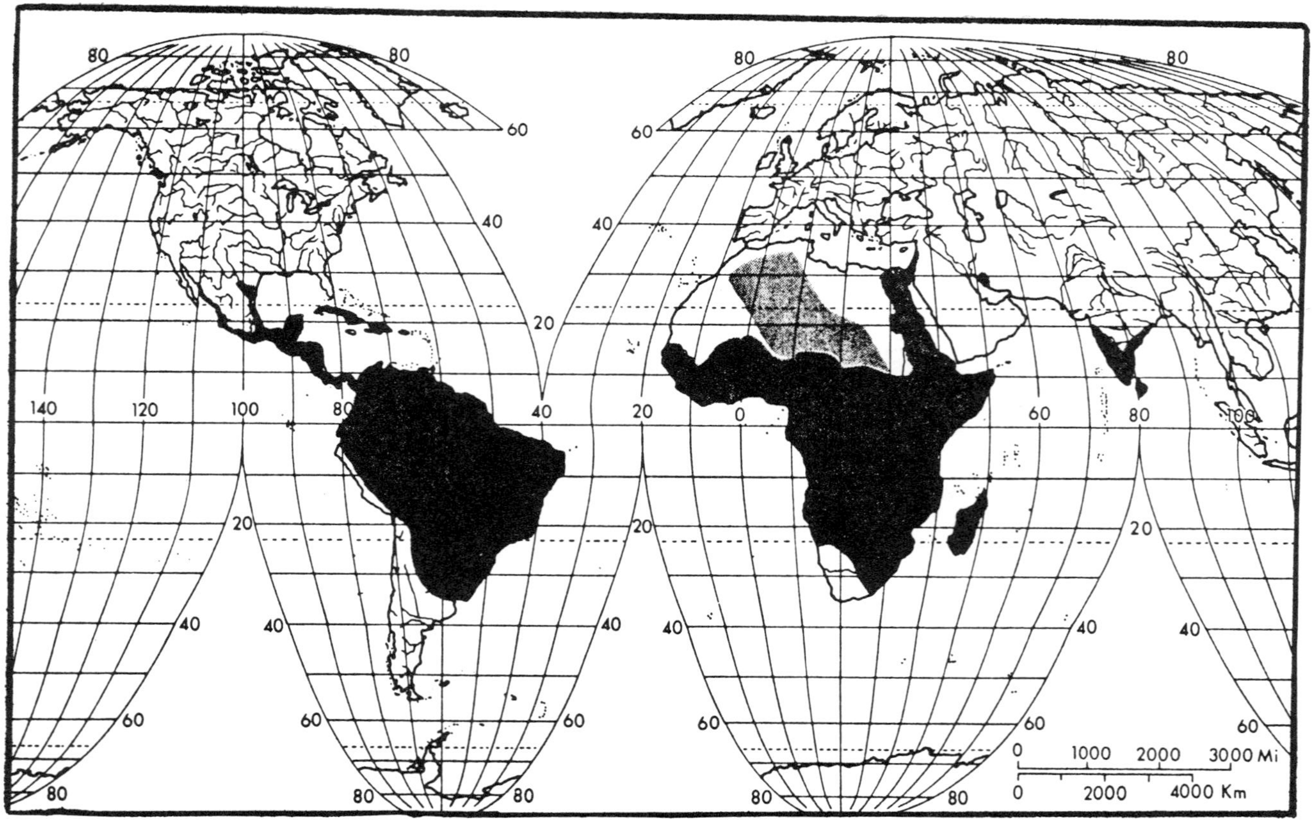

Fig. 1.1 Distribution map for the family Cichlidae (modified after Berra, 1981). Light grey patch across Sahara Desert indicates scattered distribution in isolated water-holes, etc.

morphological innovation within the Cichlidae is notably muted. The conclusion of Greenwood (1974a, 1984) and Strauss (1984) that much of the differentiation observed in Lake Victoria haplochromines is linear variation upon a basic 'bauplan', developed through small changes in relative growth rates, is generally true for the entire family. That is to say, much of the morphological differentiation associated with cichlid diversification (whether accompanying speciation, or accruing phyletically) appears to conform to a model of 'evolution via concertina' rather than by major anatomical reorganization and innovation. This view supports the claim that the functional and behavioural versatility of the cichlid 'bauplan' (e.g. Liem, 1973, 1980a; Liem and Osse, 1975) facilitates trophic diversification without major structural modification. Needless to say, such a situation, however fruitful for the cichlid clade, renders the problem of determining monophyletic lineages on the basis of shared derived morphological characters extremely vexing, and the lack of synapomorphic variation amongst cichlid clades has almost become a 'leitmotif' of analyses of cichlid interrelationships.

Cichlids are primarily fishes of the lowland tropics with a natural geographical distribution that conforms to an essentially Gondwanan pattern (Fig. 1.1). Currently the oldest known fossils appear in the Eocene of South America (Woodward, 1939; although the age of these deposits is questioned by van Couvering, 1982) and the Oligocene of Africa (van Couvering, 1982). These early cichlids have a distinctly modern facies and are highly specialized in a number of dental features, and it is most certainly the case that the origin of the group long predates its earliest fossil record. The Cichlidae (along with many other modern percoid and labroid families) probably arose sometime early in the Cretaceous, and taxonomic differentiation was well under way prior to the separation of the Gondwanan fragments (Stiassny, 1987). Interesting discussions of the ecological and historical biogeography of the family are to be found in Cichocki (1976) and Loiselle (1985a).

1.2 CICHLID PLACEMENT WITHIN THE PERCOMORPHA

The position of the Cichlidae in relation to other percomorph taxa has been a point of contention for many years and alternative classifications have been proposed (e.g. Müller, 1843; Jordan, 1905; Regan, 1913; Berg, 1940; Bertin and Arambourg, 1958; Greenwood *et al.*, 1966; Nelson, 1984). Whilst it is true that our understanding of the interrelationships of percomorph fishes remains rudimentary, recent years have seen a growing consensus amongst systematists and functional morphologists that Müller's (1843) original grouping of the families Pomacentridae (damselfishes)), Labridae (wrasses, odacids and parrotfishes), Embiotocidae (surf perches) and Cichlidae has phylogenetic integrity. This notion has been formalized by Kaufman and Liem (1982) with the assembly of those families into the suborder Labroidei.

The Labroidei is a huge assemblage comprising over 1800 species (5–10% of all living fishes), and much interest has centred on the evolutionary dynamics of these species-rich and adaptively multiradiate clades (e.g. Liem, 1973, 1986; Liem and Osse, 1975; Liem and Greenwood, 1981; Kaufman and Liem, 1982; Barel, 1983, 1984; Liem and Sanderson, 1986; Stiassny and Jensen, 1987). Recent research suggests that features of the labroid pharynx have important evolutionary consequences, not only for systematic studies, but for the diversification of the clade as well. Stiassny and Jensen (1987) discuss this phenomenon at length, and after examining possible reasons why a particular morphological structure or functional complex may play a disproportionately predominant role in the systematics of a group, conclude that the pharynx truly represents a significant locus of evolutionary change for the Labroidei.

Stiassny and Jensen (1987) also review the data bearing on the question of labroid interrelationships and present their analysis in the form of a cladogram (Fig. 1.2). This scheme differs from others previously proposed (e.g. Stiassny, 1980; Liem and Greenwood, 1981; Kaufman and Liem, 1982) in placing the Cichlidae as the sister group of the remaining labroids. However, as Stiassny and Jensen (1987) repeatedly stress, although theirs is the most parsimonious interpretation of the available data, the large number of crossbars superimposed onto their cladogram starkly illustrates that even this scheme requires the loss or independent gain of many derived characters. In this regard, perhaps the most significant observation is that morphological

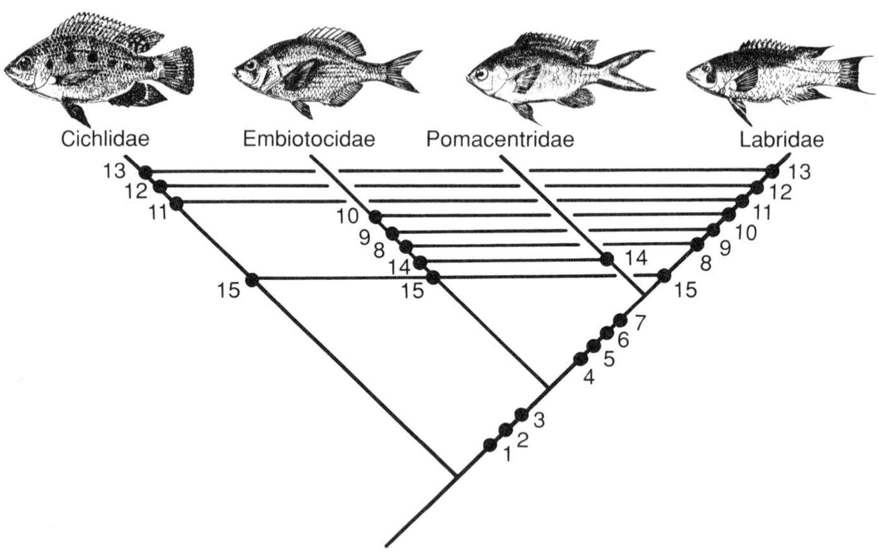

Fig. 1.2 Labroid interrelationships (modified after Stiassny and Jensen, 1987; readers are referred to that paper for a detailed discussion of the characters numbered here).

character transformations within the Labroidei display a disconcertingly large amount of homoplasy. No matter which scheme of relationships is ultimately chosen for the labroids, it must be accepted that character distributions within the Labroidei present a perplexing array of discord. It would be interesting to know whether the degree of homoplasy revealed by Stiassny and Jensen's study is a general phenomenon, or whether the magnitude of the problem is peculiar to the group, and is perhaps indicative of something specific about labroid evolutionary mechanisms.

1.3 A REVIEW OF CICHLID INTRAFAMILIAL RELATIONSHIPS

Despite almost a century of anatomical investigation aimed at unravelling the phylogenetic relationships of major groups of cichlid fishes (e.g. Pellegrin, 1904; Regan, 1906, 1920, 1922b; Vandewalle, 1971; Trewavas, 1973, 1983; Greenwood, 1974a, 1978, 1987a; Cichocki, 1976; Stiassny, 1981a, 1987; Kullander, 1983b, 1986; Oliver, 1984), we are still far from a

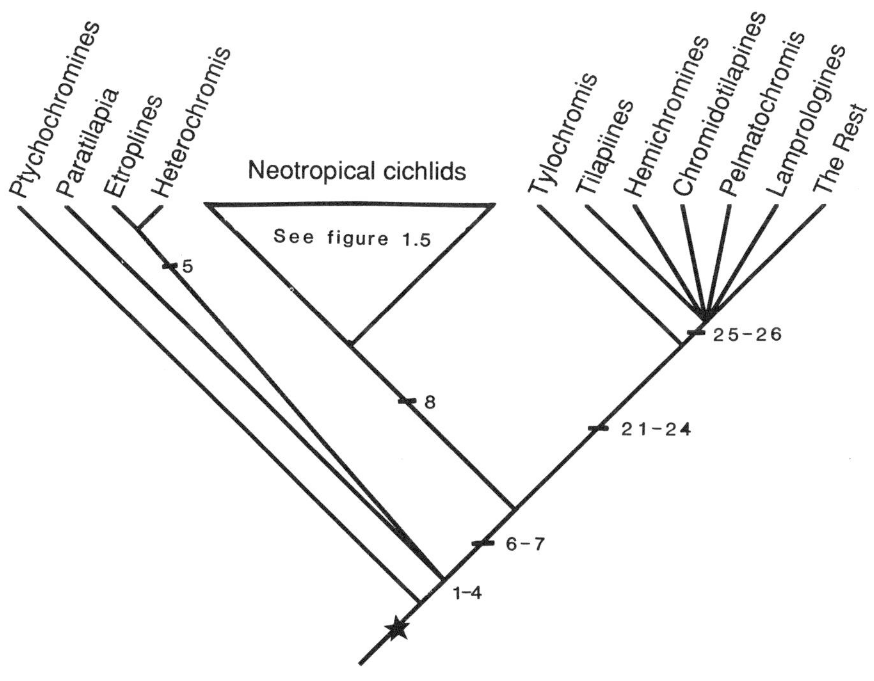

Fig. 1.3 Summary cladogram of cichlid intrafamilial relationships. Character descriptions are listed on pages 33–5 (for a list of the characters diagnosing the family see Stiassny, 1981a; Zihler, 1982; Gaemers, 1984).

comprehensive understanding of high-level familial intrarelationships. Although a number of schemes exist (Ufermann *et al.*, 1987), the lack of resolution evident in the summary cladogram (Fig. 1.3) renders the recommendation of a formal classification for the Cichlidae premature. However, some vocabulary is necessary. In this chapter I shall adopt, without the implication of a formal subfamilial or tribal classification, an increasingly popular device employed for reference to suprageneric groups of cichlids: the informal usage of the ending 'ine' (e.g. tilapiine, geophagine, cichlasomine, etc.). To avoid ambiguity and for ease of reference, the genera included in each of the putatively monophyletic terminal groupings on my cladogram are listed, along with pertinent comments and literature citations, on page 31–2.

Greenwood (1986) used Oliver's (1984) cladogram of cichlid intrarelationships as a reference diagram (Fig. 1.4). Oliver's scheme is based primarily upon an earlier analysis of Cichocki (1976), to which it provides a useful complement with the inclusion of data from many more African taxa, and in particular the little known Zairean genus, *Heterochromis.* My summary cladogram (Fig. 1.3) differs from the Oliver/Cichocki scheme in a number of

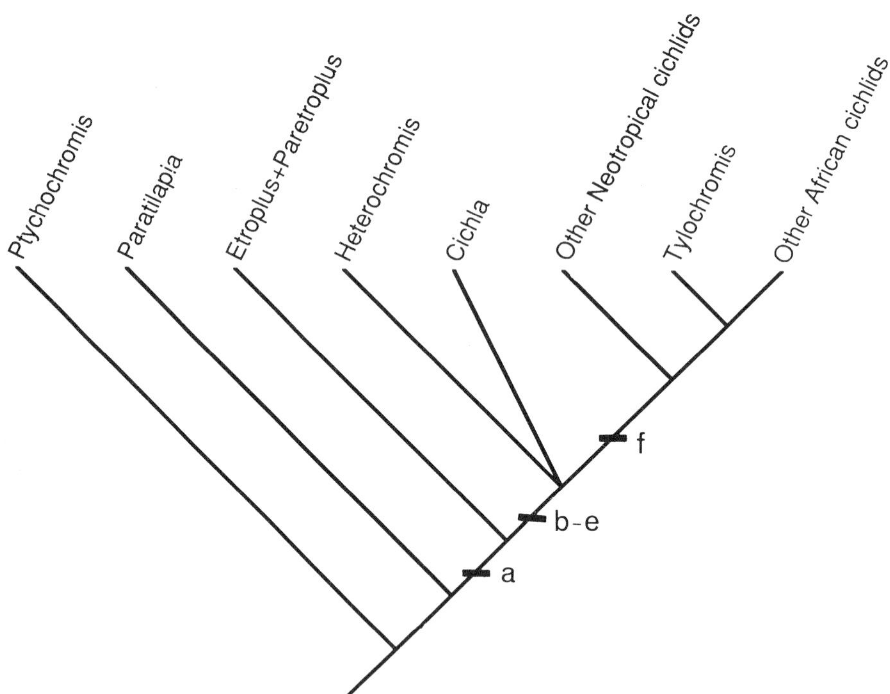

Fig. 1.4 Summary cladogram of Oliver (1984). Characters a–f are discussed in the text.

details, in particular with regard to the degree of resolution of the Madagascan and Indian taxa, and in the placement of the Neotropical genus *Cichla*.

In this analysis I have used J. S. Farris' new phylogenetic program for MS-DOS microcomputers, HENNIG86. All characters have been derived either from a reanalysis of data available in the literature, or are described here for the first time. Throughout the study, character polarity has been assessed by the application of the Outgroup Method (Watrous and Wheeler, 1981; Maddison *et al.*, 1984) using a generalized percomorph outgroup (e.g. Stiassny, 1981a, 1987), and all data have been coded as two-state characters of equal weight (see page 34 for the data matrix). From this matrix 54 trees of equal (and minimum) length were obtained (tree length 34), and the Nelson consensus tree (in which only those groups present in all 54 trees are represented) is presented with the data matrix (Fig. 1.20). Following Farris' successive weighting procedure (Farris, 1969; Carpenter, 1988), a final total of three trees was derived from the matrix. These differ only in the extent to which they

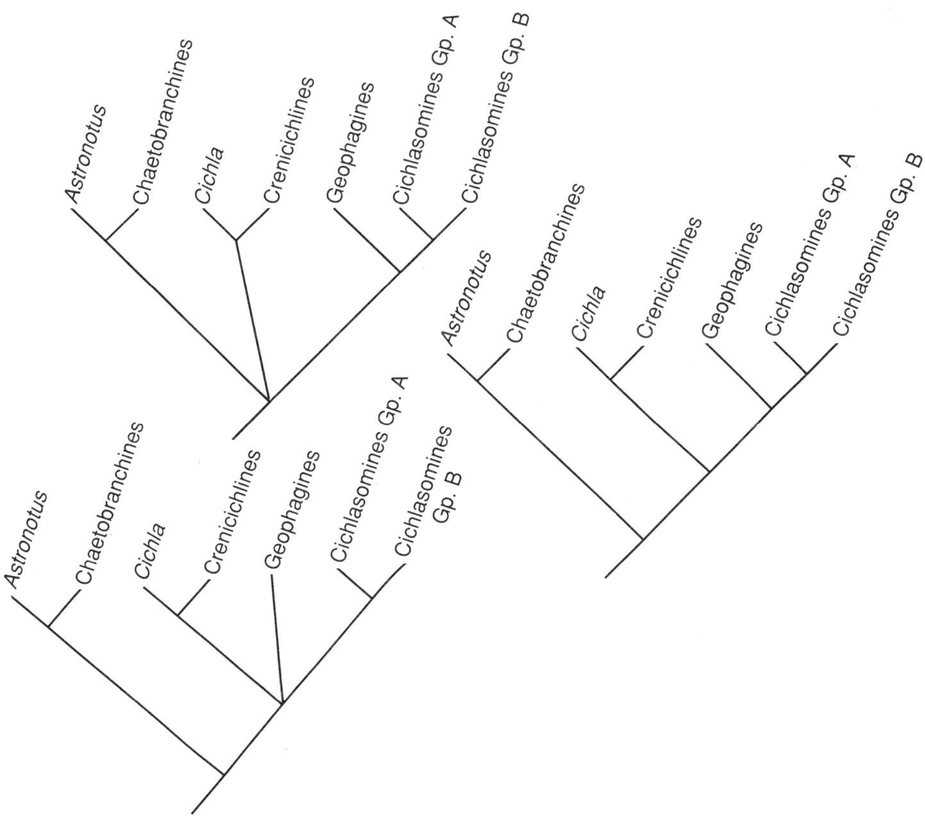

Fig. 1.5 Alternative schemes for the interrelationship of the Neotropical lineages resulting from Farris' successive weighting procedure.

resolve the interrelationships of major lineages of Neotropical cichlids (Fig. 1.5).

From the foregoing it should be clear that the scheme of relationships presented in this chapter is extremely tentative and that considerably more data will need to be integrated into the analysis before a stable classification may be proposed for the family. In many ways this study represents an interesting testament to how little we still can claim to know about the precise interrelationships of these enigmatic fishes.

The Madagascan/Asian cichlids and the Zairean genus, *Heterochromis*

Perhaps the most significant result of Cichocki's (1976) study is the notion of the Madagascan (Kiener and Mauge, 1966) and Asian (Day, 1987) cichlids as a basal grade within the family. Similarly, Oliver's (1984) recognition of the Zairean genus, *Heterochromis*, as a taxon phylogenetically removed from the remaining African assemblage also represents a major advance in our understanding of high-level intrarelationships.

Although both Cichocki's and Oliver's studies of these taxa were based upon single specimens of rather few species, their basic conclusions have been borne out by my own investigation of the assemblage (Stiassny, 1987; Stiassny and Reinthal, in prep.). In addition to examining many specimens of most taxa, I have been able to include in the study representatives of the monotypic genus, *Oxylapia*, a taxon not available to either Cichocki (1976) or Oliver (1984), and have been led to reinterpret a number of the characters and character states utilized by them in their analyses.

The ptychochromines form the sister group of the rest of the family, and all non-ptychochromine cichlids are united by the possession of three pharyngeal features assessed to be phylogenetically derived in relation to the ptychochromine condition (Cichocki, 1976):

1. Primitively amongst perciforms the uncinate process of the first epibranchial bone is distinctly shorter than the anterior arm (Rosen, 1973; Travers, 1981). Uniquely amongst cichlids the ptychochromines retain the primitive condition of this element with the anterior arm of the first epibranchial somewhat longer than the uncinate process (Fig. 1.6(d)). In other cichlid taxa the two arms are either sub-equal in length (e.g. Fig. 1.6(b)), or the uncinate process is considerably longer than the anterior arm (e.g. Fig. 1.17(b));
2. The presence of a well-developed toothplate overlying (but not fused to) the proximal region of the second epibranchial bone is not uncommon amongst percomorphs (see Cichocki, 1976 for a distributional review; Johnson, 1980). The ptychochromines retain a second epibranchial toothplate (Fig. 1.6(d)) whilst in all other cichlids a separate toothplate is lacking (e.g. Fig. 1.6(b));

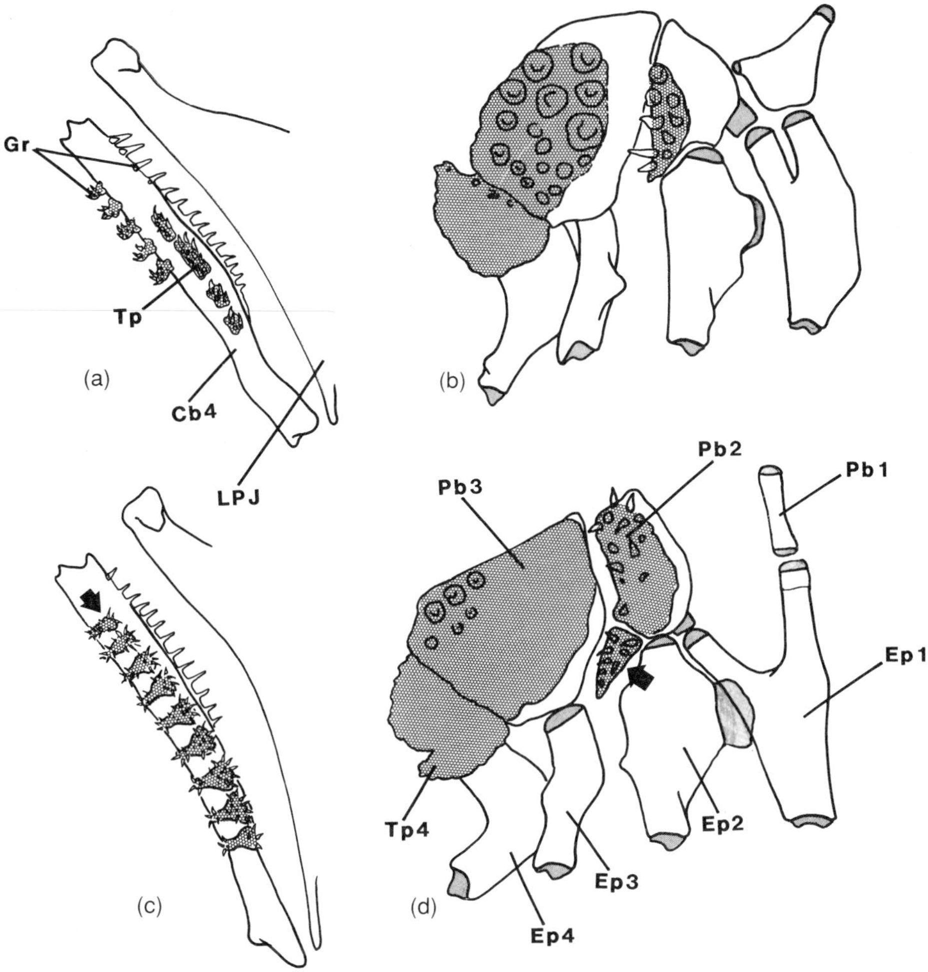

Fig. 1.6 (a) and (b) *Oxylapia*: (a) fourth ceratobranchial element: (b) isolated left upper pharyngeal jaw elements. (c) and (d) *Ptychochromis*; (c) fourth ceratobranchial element, arrow indicates fusion of median toothplates and the outer gill rakers; (d) isolated left upper pharyngeal jaw elements, arrow indicates separate Ep2 toothplate. Abbreviations used in Figs. 1.6 to 1.18: AAP, adductor arcus palatini muscle; BB3, basibranchial 3; Cart.BB4, cartilaginous basibranchial 4; Cb4, fourth ceratobranchial; Cent.lig, central ligament; Ect., ectopterygoid; End., endopterygoid; EP1–4, epibranchial 1–4; For.Mag., foramen magnum; FR, frontal bone; Gr. gill rakers; HBr3, third hypobranchial; LPJ, lower pharyngeal jaw; NLFO, neurocranial lateral line foramen 0; Occ.For., occipital foramen; Op. spt, opercular spot; Pal., palatine; Pal.mx.lig, anterior palato-maxillary ligament; Pb1–3, pharyngobranchial 1–3; Phcl.int, pharyngocleithralis muscle; Pred, predorsal; Preop.spt, preopercular spot; Psp, parasphenoid; SOC, supraoccipital bone; Tp, toothplates; Tp4, fourth upper toothplate; Vom., vomer.

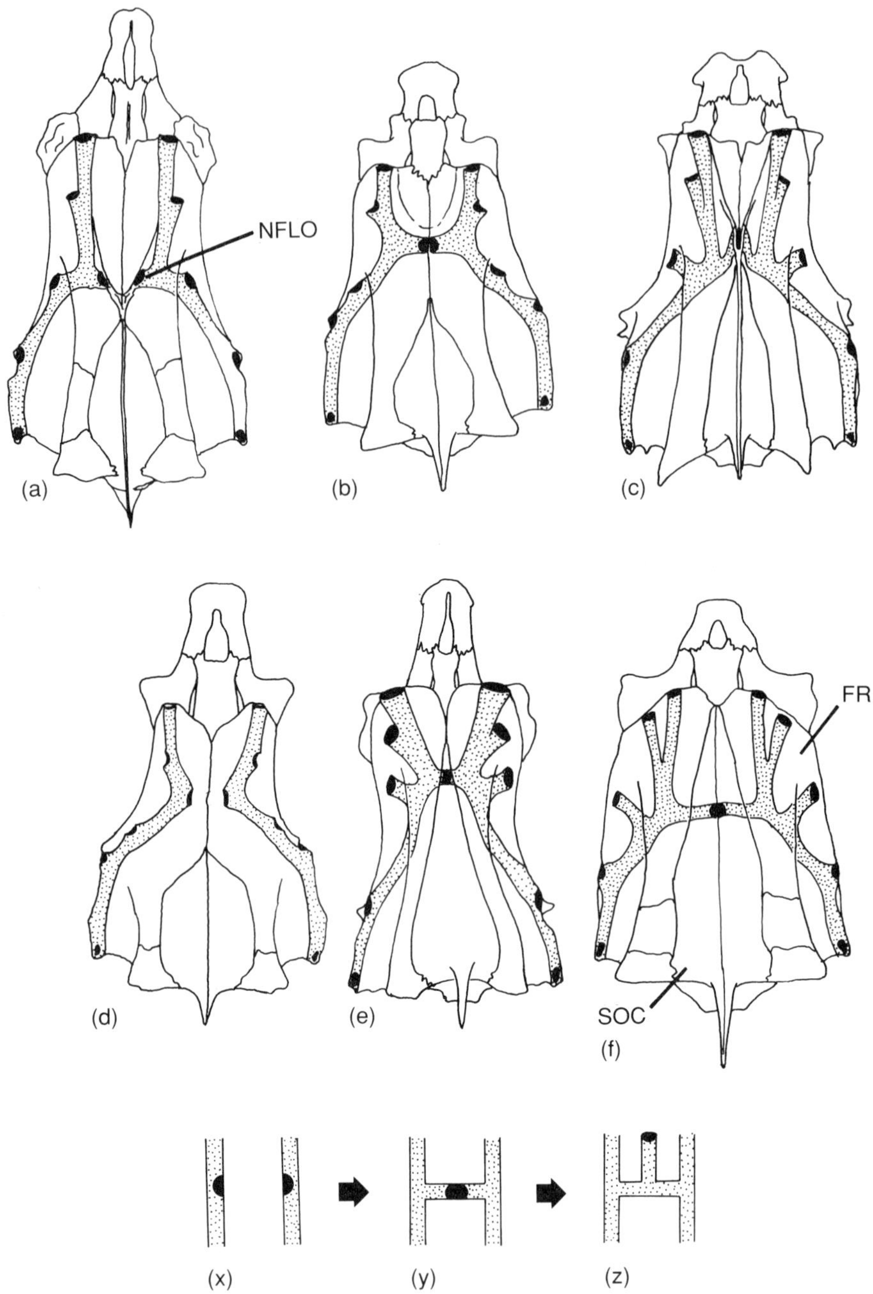

Fig. 1.7 (a) to (f) Dorsal view of the neurocrania of (a) *Ptychochromis*; (b) *Paratilapia*; (c) *Cichla*; (d) *Oxylapia*; (e) *Heterochromis*; (f) *Etroplus*. (x) to (z) Schematic representation of the NLFO canal and pore canal configuration in (a) to (c), respectively. Abbreviations as in Fig. 1.6.

3. The fourth ceratobranchial toothplates of ptychochromines are confluent with the outer row gill rakers of the arch (Fig. 1.6(c)), whilst in other cichlids with fourth ceratobranchial toothplates, those toothplates are separate from the gill rakers (e.g. Fig. 1.6(a)). Cichocki's review of this features indicates that the ptychochromine condition is probably best interpreted as plesiomorphic relative to that exhibited by other cichlids.

Cichocki (1976) recognizes a complex array of character states involving the neurocranial lateral line system and its supporting bones, and utilizes this character complex extensively in his resolution of the Madagascan and Indian cichlid interrelationship. My own observations of this character complex led me to a somewhat different interpretation from Cichocki's and this is briefly discussed here.

Primitively amongst perciforms the median frontal pores (NLFO of Barel *et al.*, 1976) are more or less widely separated from one another (Branson and Moore, 1962; Cichocki, 1976). This is the condition found in the ptychochromines (Fig. 1.7(a)), and in the etropline, *Oxylapia* (Fig. 1.7(d)). Contrasted with this is the condition expressed in *Paratilapia* (Fig. 1.7(b)), *Etroplus* (Fig. 1.7(f)), its sister genus *Paretroplus*, and in the Zairean genus *Heterochromis* (Fig. 1.7(e)). In these taxa the NLFO pores are borne at the end of short, medially-directed tubes and are partially (or fully) coalesced in the midline. Figures 1.7(x) and (y) are a schematic representation of these two states. The sequence x→y also represents the developmental sequence for taxa in which state y is the definitive condition. The NLFO configuration in the remaining Cichlidae is discussed on page 15.

That *Oxylapia* retains the plesiomorphic condition of NLFO is of interest as it shares with *Etroplus* and *Paretroplus* a number of unambiguously derived features of the suspensorium, ethmovomer and hyoid linkage system, indicating a sister-group relationship with that pair (Stiassny and Reinthal, in prep.), and as such would be expected to share the derived NLFO configuration. Given a monophyletic etropline group that includes *Oxylapia*, I interpret the lack of central canal development and apposition of the NLFO foramina in *Oxylapia* as a phylogenetic reversal, possibly resulting from a simple developmental truncation. Following this interpretation, the non-ptychochromine cichlids are further united by:

4. The presence of median frontal canals bearing NLFO foramina more or less coalesced in the midline (with a phylogenetic reversal in *Oxylapia*).

Etroplus, *Paretroplus* and *Heterochromis* share an additional derived feature of the NLFO system. In these taxa the centrally apposed NLFO pores are overlain by the supraoccipital bone (Fig. 1.7(e) and (f)) and the supraoccipital crest must be peeled back to reveal the median canal in the frontal bone beneath it. In all other taxa the anterior margin of the supraoccipital borders the frontal posterior to the NLFO foramina (e.g. Fig. 1.7(a), (b), (c)). The former configuration is undoubtedly a derived condition within the family and once again, the absence of the feature in *Oxylapia* (Fig. 1.7(d)) is problematical; two readings of this anomalous distribution are considered here. In the first it is assumed that the absence of the derived supraoccipital morphology in *Oxylapia* is another phylogenetic reversal, possibly also the result of developmental truncation. This interpretation argues for a monophyletic group consisting of *Etroplus*, *Paretroplus*, *Oxylapia* (reversal) and *Heterochromis*, diagnosed on the basis of the supraoccipital–frontal relations. An alternative interpretation would be to view supraoccipital modification in the *Etroplus*/*Paretroplus* pair as being independently derived from that of *Heterochromis*, and therefore not homologous (*sensu* Patterson, 1982) with it.

The former interpretation, which requires a reversal in *Oxylapia* (the alternative posits a dual origin for the supraoccipital modification in two independent clades), is favoured in the successive weighting analysis and *Heterochromis* is placed as the sister group to the etroplines. This is a highly tentative placement and considerably more data are necessary before this phylogenetic assertion may be made with any degree of confidence.

The scheme of relationships that I have presented so far differs from that of Oliver and Cichocki, unfortunately most notably in a reduction of resolution. However, the five characters (or character states) presented in support of their hypothesis cannot be upheld:

1. Character a (Fig. 1.4), separating the remaining Cichlidae from the ptychochromines and *Paratilapia*, stated by Cichocki (1976: 148) as "coronal lateralis canals coalesced medially producing a single median canal shared by both frontals. . . ." I am unable to distinguish between this feature (i.e. median canal development) and the central apposition or coalescence of the NLFO pores (combined in my character 4, Fig. 1.3);
2. Character b (Fig. 1.4), separating the remaining Cichlidae from the ptychochromines, *Paratilapia* and the etroplines, stated by Cichocki (1976: 149) as "epaxial cranial roof musculature extending medially and laterally to or anterior to the level of the coronal aperture". Contrary to Cichocki I find lateral epaxial migration to the level of the NLFO pores and a little beyond in large specimens (> 15 cm SL) of *Ptychochromis* and *Paratilapia* (> 10 cm SL). In *Etroplus* and *Paretroplus* the epaxial musculature extends the length of the supraoccipital to a level well in advance of the NLFO pores and median canal. It is only in *Oxylapia* that the epaxial musculature is restricted to the supraoccipital, epiotic and parietal regions at all sizes of fish examined.
3. Character c (Fig. 1.4), separating the remaining Cichlidae from the ptychochromines, *Paratilapia* and the etroplines, stated by Oliver (1984: 12) as "There are diverging frontal ridges anterior to the single median coronal canal and pore (NLFO)." Diverging frontal ridges are variously developed in Neotropical and African cichlids (Cichocki, 1976; pers. obs.), and are present in the larger (> 15 cm SL) specimens of *Ptychochromis* (Fig. 1.7(a)). The phylogenetic significance of diverging frontal ridge development within the

Cichlidae is unclear. The presence of a single median coronal canal and pore is of more interest and is discussed further on page 15;

4. Character d (Fig. 1.4), separating the remaining Cichlidae from the ptychochromines, *Paratilapia* and the etroplines, stated by Cichocki (1976: 149) as "pseudobranch not free and gill-like but glandular and completely sunken beneath the bucco-pharyngeal mucosa". There is much variation in pseudobranch morphology within the Cichlidae, and while there is clearly a considerable range in the degree to which the element is gill-like rather than glandular, and exposed within the buccopharyngeal cavity rather than deeply embedded in fatty connective tissues, I have been unable to locate a clear-cut distinction of the type suggested by Cichocki (1976);
5. Character e (Fig. 1.4), separating the remaining Cichlidae from the ptychochromines, *Paratilapia* and the etroplines, stated by Cichocki (1976: 149) as "swimbladder extending at most only slightly anterior to the epipleural ribs or with secondarily-developed, narrow, short and thickened diverticula". Marathe and Khosla (1958) described a peculiar otophysic linkage in *Etroplus*, and Cichocki (1976: 168) notes the presence of a similar arrangement in *Paretroplus.* In both of these genera elaborate swimbladder extensions are embedded in complex exoccipital foramina (Fig. 1.8(b)). *Oxylapia* (Fig. 1.8(d)) lacks all trace of exoccipital foramina (other than the vagal foramina, which are also present and unmodified in the other Madagascan and Indian cichlids), and the swimbladder does not extend forward beyond the level of the first pleural rib. In the ptychochromines (e.g. Fig. 1.8(a)) the exoccipitals bear large foramina similar to those of *Paretroplus*, although somewhat smaller and less complex. Exoccipital foramina are also present in *Paratilapia*, although they are smaller still (Fig. 1.8(c)). In these genera the swimbladder is bifurcate anteriorly, and thin-walled extensions may reach the level of the first pleural rib, although they do not contact the neurocranium and have no structural association with the exoccipital foramina.

   Cichocki (1976) reviews the occurrence of anterior swimbladder extensions amongst percomorphs and concludes that, although the evidence is equivocal, the presence of anterior extensions approaching the neurocranium is probably primitive for cichlids. However, in view of considerable intraspecific variation in this feature in Malagasy taxa, as well as in certain African and Neotropical forms, it is doubtful that the character, as stated by Cichocki (1976), serves to diagnose the remaining Cichlidae.

Of more interest is the peculiar excavation of the exoccipital bones in the Malagasy and Asian cichlids. Similarly-situated foramina are very rarely (if ever) encountered in other percomorphs, and are lacking in other cichlid taxa. The presence of this, undoubtedly derived, exoccipital morphology in the Malagasy and Asian cichlids (once again, *Oxylapia* is anomalous in lacking the

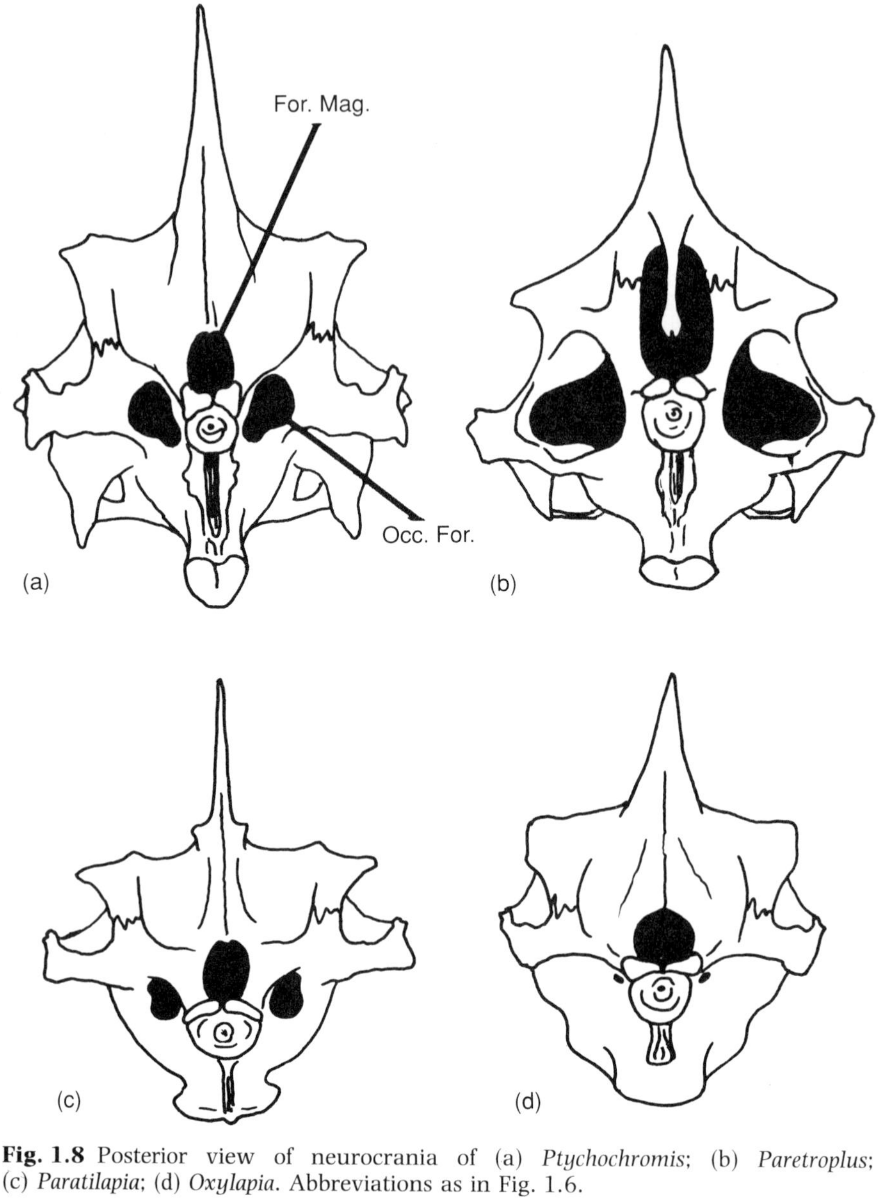

Fig. 1.8 Posterior view of neurocrania of (a) *Ptychochromis*; (b) *Paretroplus*; (c) *Paratilapia*; (d) *Oxylapia*. Abbreviations as in Fig. 1.6.

feature) strongly argues for a monophyletic origin of these fishes. However, when this character was entered into the data matrix (character 28) it was effectively removed from final tree construction during the successive weighting routine. With the data currently available, the argument for a monophyletic Malagasy/Asian assemblage, based upon one (albeit extremely 'impressive') morphological character, is considered weaker than that arguing for the scheme I am presenting. Nevertheless the possibility of a monophyletic Malagasy/Asian radiation is clearly posed and is a proportion that bears further investigation (Stiassny and Reinthal, in prep.).

The Afro-neotropical radiation

The remaining Cichlidae are represented here (Fig. 1.3) as constituting a monophyletic lineage that includes all of the Neotropical and African taxa (except the Zairean genus *Heterochromis*). The monophyly of this Afro-Neotropical lineage is attested to by the presence in all included taxa of two derived characters:

1. The first of these (character 5 in Fig. 1.3) is a feature of the ventral branchial apparatus which is discussed in detail by Stiassny (1987). Primitively amongst cichlids, the central ligament (= ligamentum centralis of Anker, 1978) is bifurcate and passes from the rostral tip of the LPJ to insert onto the contralateral third hypobranchial elements (e.g. Fig. 1.9(b)). This is the condition found in all of the Madagascan and Asian cichlids, and in the Zairean genus *Heterochromis* (Stiassny, 1987). In all other cichlids the central ligament is formed of a single band that passes directly from the LPJ to insert onto the median fourth basibranchial element (e.g. Fig. 1.9(a)). Stiassny (1987) discusses the distribution of these two configurations amongst labroid and other percomorph taxa and concludes that the bifurcate LPJ hypobranchial linkage represents the primitive configuration. The single ligament/LPJ basibranchial linkage is interpreted as the derived condition and serves to diagnose the main Afro-Neotropical lineage;
2. The second morphological feature of interest at this level (character 6 in Fig. 1.3) involves a further elaboration of the acoustico-lateralis system in the frontal bones of the neurocranium. In the advanced state (diagnosing the members of the Afro-Neotropical lineage), the paired frontal canals are united in the midline and extend rostrodorsally as a single, central tube (Fig. 1.7(z)), elevated in the median frontal ridge (e.g. Fig. 1.10(c)), and terminating in a single NLFO pore. Essentially this is character c in Oliver and Cichocki's scheme (Fig. 1.4), but cannot be upheld at that level as *Heterochromis* lacks this derived configuration (Fig. 1.10(b)).

Monophyly of the Neotropical Cichlidae, and the phylogenetic placement of the genus **Cichla**

Although considerably less speciose than its African counterpart, the Neotropical Cichlidae is nonetheless an extremely varied and ecologically diverse assemblage (e.g. Staeck and Linke, 1984, 1985; Lowe-McConnell, this volume). Once again, it is difficult to give a precise number, but an estimated

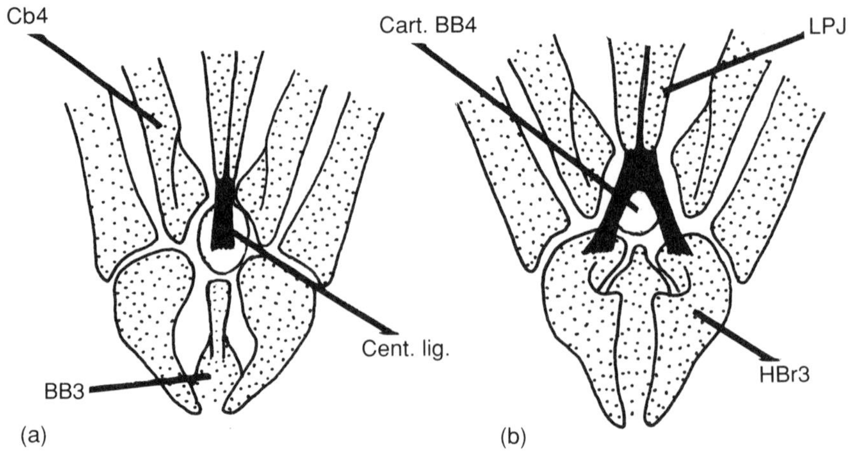

Fig. 1.9 Median elements of the ventral branchial apparatus of (a) *Cichla*; (b) *Oxylapia*. Abbreviations as in Fig. 1.6.

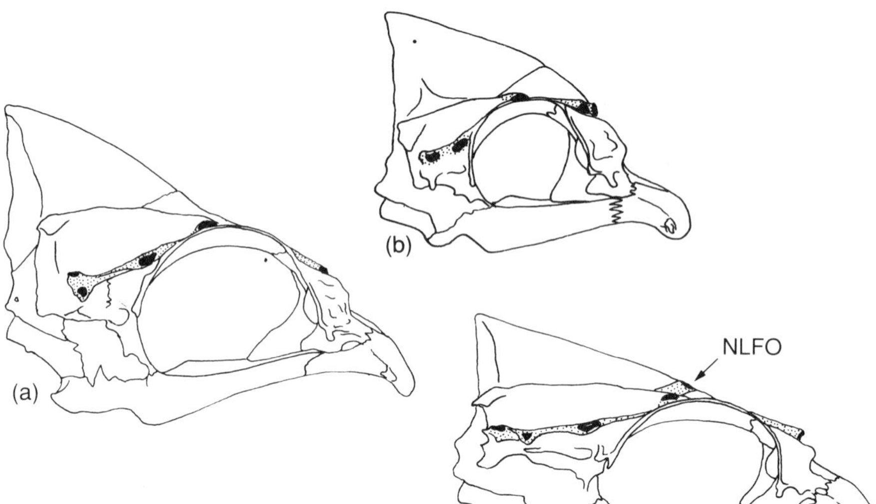

Fig. 1.10 Lateral view of neurocrania of (a) *Ptychochromis*; (b) *Heterochromis*; (c) *Cichla*. Abbreviations as in Fig. 1.6.

total somewhere in the region of 300 to 350 species is probably reasonable (e.g. Miller, 1976; Kullander, 1986; Kullander and Nijssen, 1989).

Regan, in a series of papers published in 1905 to 1908, provides the first comprehensive review of the entire Neotropical cichlid fauna. He summarized his ideas regarding their interrelationships (Regan, 1906) in the form of a

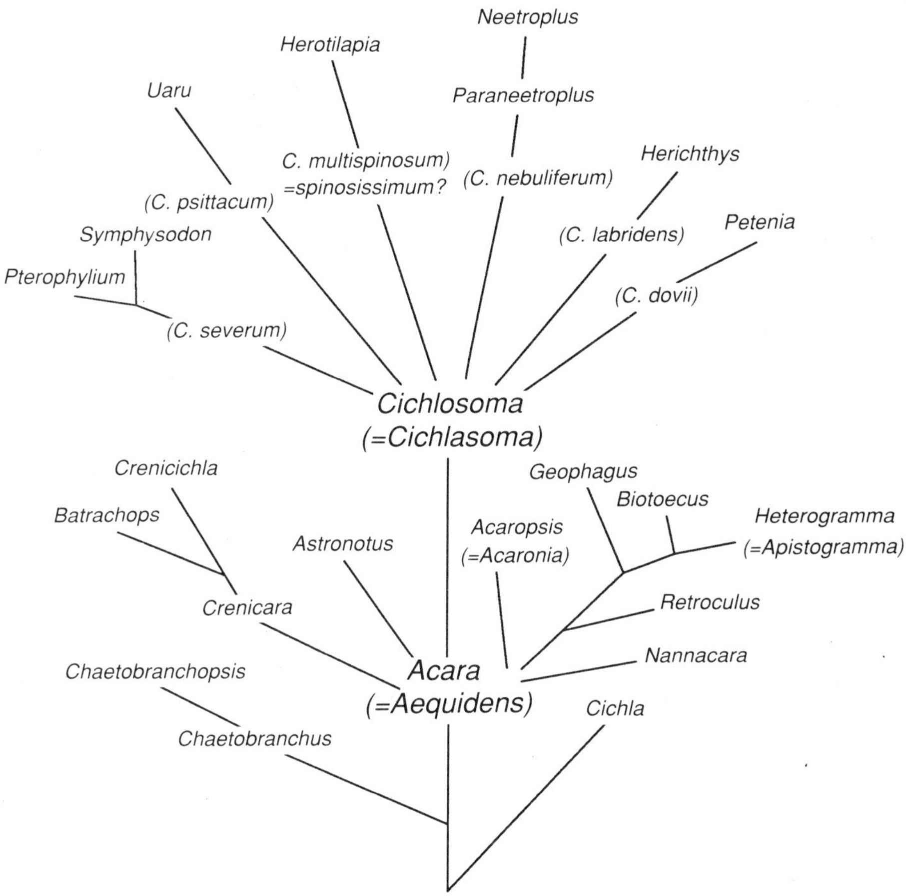

Fig. 1.11 The Neotropical Cichlidae (after Regan, 1906).

branching diagram (Fig. 1.11). Although he was not explicit as to the anatomical basis for his scheme, it is clearly influenced by his belief in the importance of anal spine number as a primary grouping character (see Kullander, 1983a, who argues convincingly for the rejection of Regan's three-spine versus 'more-than-three' spine dichotomy). Regan (1906) isolates the genus *Cichla* and places it as the sister group to the remaining Neotropical assemblage. However, in a later paper he strongly intimated a close link between *Cichla* and the haplochromine cichlids of Africa (Regan, 1920). Whilst the phylogenetic integrity of a subgrouping of the Cichlidae based upon apophyseal structure (Regan's character to unite *Cichla* with the haplochromines) has been seriously disputed (e.g. Greenwood, 1978, 1987a, b), the possibility persists that the Neotropical Cichlidae is not monophyletic (Cichocki, 1976; Kullander, 1983b; Oliver, 1984), and that

Cichla may be phylogenetically allied with the African Cichlidae (Cichocki, 1976). The summary cladogram presented here (Fig. 1.3) challenges both of these notions, and in the following section I shall attempt to outline briefly my argument for so doing.

In my 1987 study I was unable to find any positive evidence for Neotropical monophyly, but neither was I able to locate data negating such a grouping (Stiassny, 1987). However, readdressing this question I have recently located a feature of the ethmovomerine region of the neurocranium which may serve to diagnose the Neotropical assemblage (character 8, Fig. 1.3):

> In most acanthomorph taxa the stalk of the vomer is 'slotted' into the parasphenoid bar and forms a simple straight suture with it (e.g. Fig. 1.12(a); see also Stiassny, 1986). The degree of vomerine insertion, and the angle of inclination of the suture varies (e.g. Fig. 1.10(a)), but in very few instances is the suture between these two bones strongly interdigitated. This is also the case amongst the Malagasy, Asian and African taxa investigated (representatives of all terminal taxa on the cladograms in this chapter have been examined). The condition differs markedly in the Neotropical Cichlidae, where the vomerine wing and the parasphenoid bar are very firmly united in a strongly interdigitating suture (e.g. Figs 1.10(c), 1.12(b)). Interestingly, *Heterochromis* (Fig. 1.10(b)), alone amongst the non-Neotropical taxa examined, has a strongly interdigitated vomerine–parasphenoid suture. On the basis of its widespread distribution amongst percomorphs, a more-or-less simple sutural union between the vomer and the parasphenoid bar is interpreted as the primitive configuration, and within the Cichlidae the occurrence of the strongly interdigitated suture is derived.

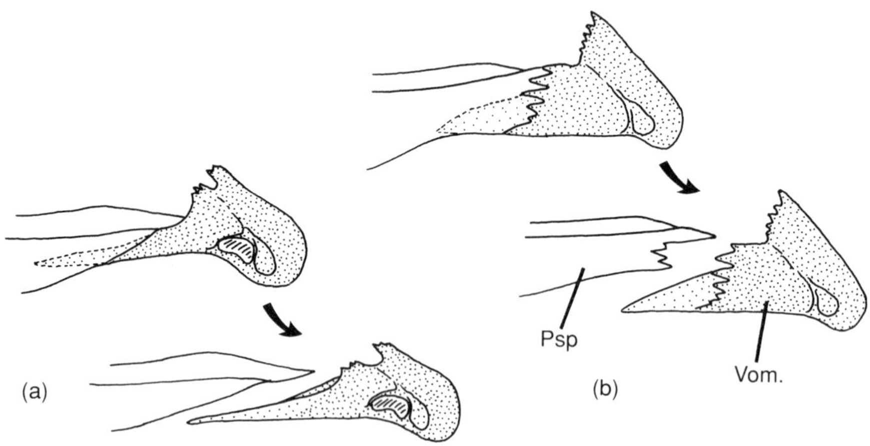

Fig. 1.12 The parasphenoid/vomerine complex (articulated and disarticulated) of (a) *Tylochromis*; (b) *Cichlasoma*. Abbreviations as in Fig. 1.6.

That *Heterochromis* has a complex interdigitated suture is problematical inasmuch as it clearly casts doubt on the reliability of this character as an indicator of Neotropical monophyly. Nonetheless, the Nelson consensus tree (Fig. 1.22) and the three trees obtained utilizing Farris' successive weighting procedure agree in favouring an interpretation of an independent origin of this derived morphology in *Heterochromis* on the one hand, and the remaining Neotropical Cichlidae on the other. However, it is clear that more evidence corroborating the phylogenetic integrity of the Neotropical assemblage is required before my claim of group monophyly may be considered to be substantiated.

Stiassny (1982) sought to resolve the question of the phylogenetic placement of *Cichla* within the family and concluded that the immediate relationships of *Cichla* lie with the Neotropical pike-cichlids of the genus *Crenicichla* (currently including *Batrachops*). *Cichla*, which is unquestionably a morphologically aberrant taxon, nonetheless shares uniquely with the crenicichlines a suite of derived features that strongly argue for group monophyly. The characters uniting the (*Cichla* + (*Crenicichla* + *Teleocichla*)) clade may be summarized as follows (for, detailed analysis of these features see Stiassny, 1987):

1. Pharyngocleithralis muscle originates from the lateral face of the cleithrum;
2. Urohyal bone bears a rostrally directed spine (reduced in *Teleocichla*);
3. An elaboration of the obliquus inferioris muscle is associated with the first and second postcleithra;
4. There is an increased number of vertebrae;
5. Presence of a characteristically notched vomer (reduced in *Teleocichla*);
6. The sphenotic foramen and canal housing otic branch of the facial nerve (*Teleocichla*?) are enlarged.

Kullander (1983b, 1986, 1988) listed a series of characters shared by *Crenicichla* and the majority of Neotropical cichlids, but not with *Cichla*. One of these (character 17 in this analysis) was considered by Stiassny (1987) to be in conflict with the integrity of a *Cichla* + Crenicichline clade. That character is the presence of six rather than seven lateralis canal foramina in the preoperculum. This is a feature of all Neotropical cichlids with the exception of *Cichla*, *Astronotus*, and the chaetobranchines (amongst geophagines only *Retroculus* has seven foramina). Percomorph outgroup data are equivocal regarding the polarity of the seven- versus six-pore pattern (Stiassny, 1987), but as all Malagasy and Asian cichlids, and *Heterochromis* (in addition to all African taxa) exhibit the seven-pore pattern, it seems reasonable to interpret the six-pore pattern as the derived condition within the family.

An additional character in apparent conflict with a crenicichline–*Cichla* alignment is a feature of the infraorbital series first described by Cichocki

(1976), and entered in the data matrix here as character 27. Cichocki (1976), and later Oliver (1984), considered that all Neotropical and African taxa (excluding *Cichla* and *Heterochromis*, but including the crenicichlines, e.g. Fig. 1.13(a)) have a derived condition of the infraorbital series. In these

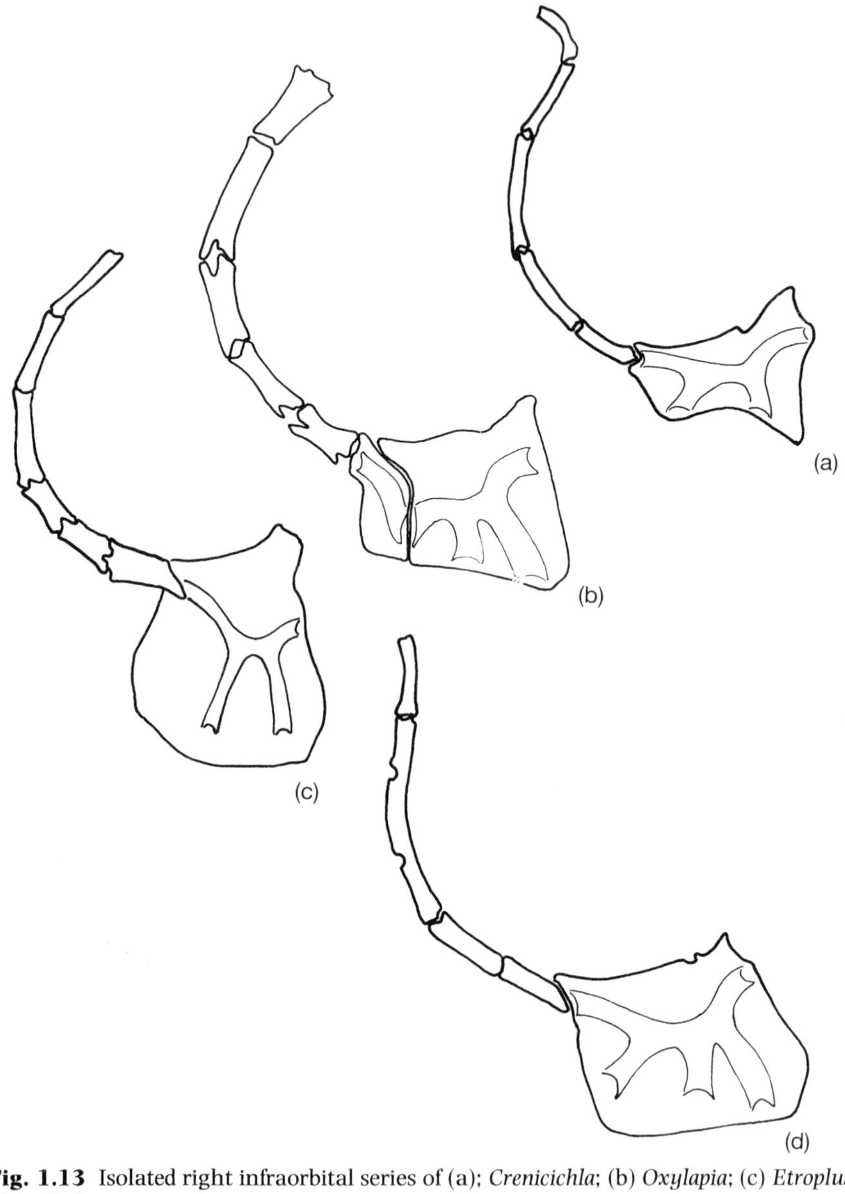

Fig. 1.13 Isolated right infraorbital series of (a); *Crenicichla*; (b) *Oxylapia*; (c) *Etroplus*; (d) *Tylochromis*.

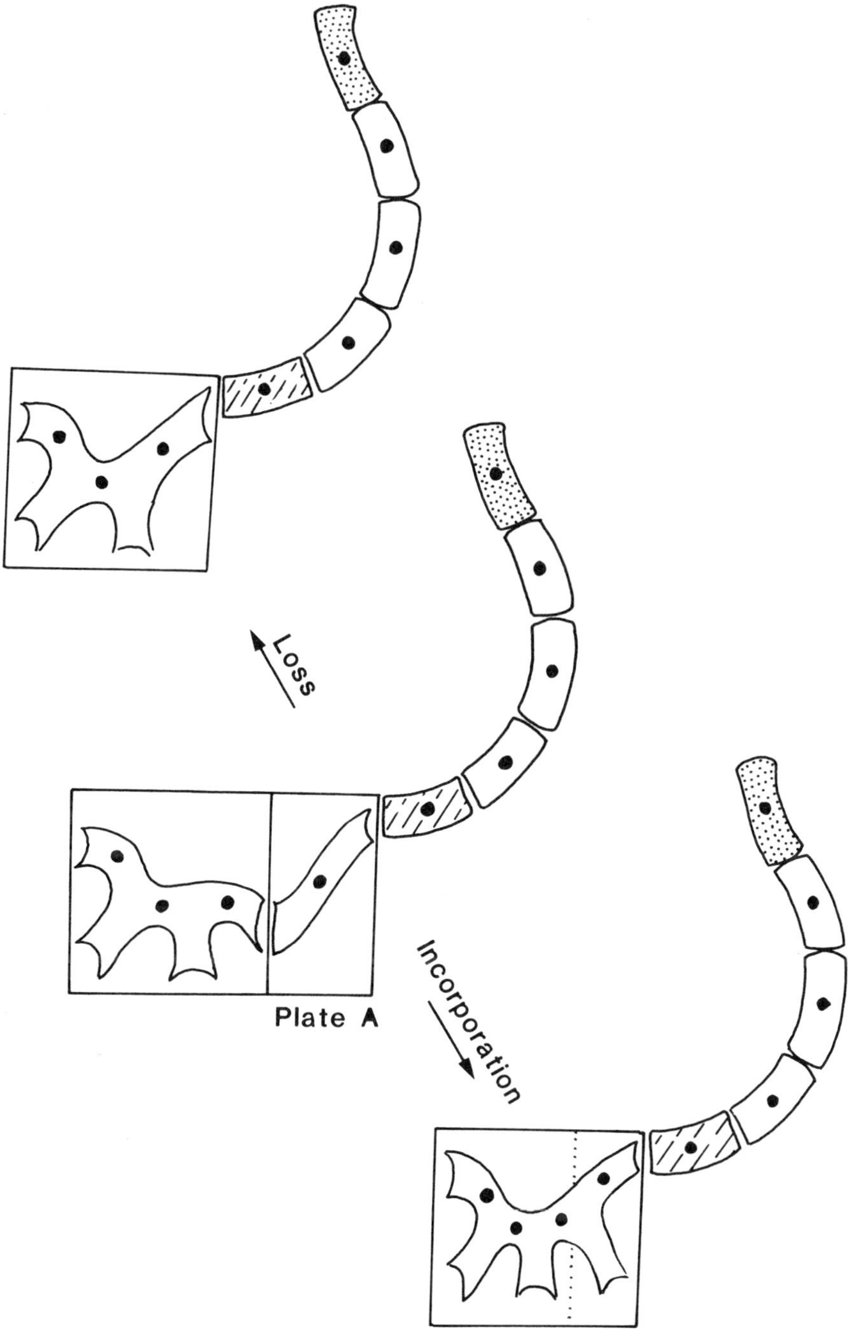

Fig. 1.14 Schematic representation of the possible derivation of the alternative cichlid infraorbital patterns.

fishes the lachrymal is a single element, and the next infraorbital bone is tubular and narrowly contiguous with the lachrymal (character f in Fig. 1.4), and is typically followed by infraorbitals 3–5 and the dermosphenotic. Primitively amongst cichlids the lachrymal, which bears three neuromasts and four marginal canals, appears to be subdivided such that a second canal-bearing bone is broadly contiguous with it (plate A in Fig. 1.14; Cichocki refers to this element as the 'primitive second infraorbital'). The second infraorbital bone (= third infraorbital of Cichocki, 1976 and Oliver, 1984) is tubular and is typically followed by three infraorbitals and a dermosphenotic. This condition is represented in all of the Malagasy cichlids (e.g. Fig. 1.13(b)), *Heterochromis* (Oliver, 1984) and in *Cichla* (Cichocki, 1976; Kullander, 1986), as well as in *Retroculus* and *Astronotus* (Cichocki, 1976; Kullander, 1986). The condition in *Etroplus* (Fig. 1.15(c)) is somewhat anomalous, as the plate-like subdivision of the lachrymal is lacking, and the lachrymal is followed by six tubular infraorbitals rather than five.

The lachrymal of many of the African cichlids appears to represent another derivation from the primitive system. In these taxa the lachrymal usually bears four neuromasts and five marginal canals (e.g. Fig. 1.13(d)), and is typically followed by a tubular second infraorbital, infraorbitals 3–5 and a dermosphenotic. I have represented this configuration with a specimen of *Tylochromis*; however, it should be noted that *Tylochromis* species (Stiassny, 1989) are frequently polymorphic for this feature, and many individuals exhibit the three-

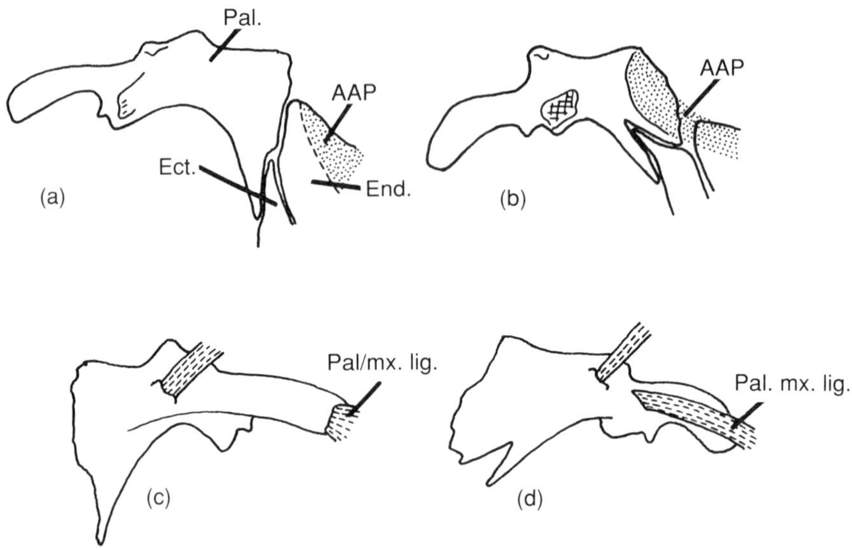

Fig. 1.15 (a) and (b) Right palatine bone in medial view: (a) *Ptychochromis*; (b) *Tylochromis*. (c) and (d) Right palatine bone in lateral view: (c) *Ptychochromis*; (d) *Tylochromis*. Abbreviations as in Fig. 1.6.

neuromast, four-marginal-canal pattern (see also Trewavas, 1983 for a discussion of the distribution of this feature amongst other African lineages).

Despite these two somewhat troubling characters I believe that the evidence currently available argues in favour of a *Cichla*–crenicichline clade embedded within the Neotropical assemblage (Fig. 1.3). Perhaps one of the most interesting aspects of such a placement is the striking morphological parallel between *Cichla* and many of its distant percoid relatives. As Stiassny (1987: 1327) points out,

> "It is not suprising that Kullander (1983b) has been led to speculate that *Cichla* represents an extremely ancient (protocichlid?) lineage whose large size and predatory lifestyle has shielded it from morphological change, but such a view of *Cichla* is no longer tenable. If my assessment of the relationships of *Cichla* are correct, then the interest attached to this taxon revolves not around a phylogenetically primitive placement within the cichlid clade, but in its presentation of complex, high-order morphological atavisms alongside typically cichlid synapomorphies and a suite of interesting autapomorphies (Stiassny, 1982)."

The other major Neotropical lineages depicted here (Fig. 1.5) are based primarily on data culled from Kullander's ongoing revisional studies of the South American Cichlidae (e.g. Kullander, 1980, 1983a, b, 1986; Kullander and Nijssen, 1989; see also character descriptions on page 33–5). Cichocki (1976) presents a rather different scheme and an impressive array of additional data on the anatomy of Neotropical cichlids which may prove to be of value, particularly at less inclusive levels.

Cichlasoma *and the cichlasomine radiation*

Before discussing the remaining African taxa I would like briefly to address a question of nomenclature that has resulted from a recent revisional study. Kullander (1983a) undertook a taxonomic revision of the genus *Cichlasoma*, a heterogeneous assemblage of some 100 or so species in South, Central and southern North America, and the Greater Antilles. He restricted *Cichlasoma* to the type species of the genus, *Labrus bimaculatus* Linneaus, and 11 others which were newly described (*C. amazonarum, C. araguaiense, C. boliviense, C. orientale C. orinocense, C. paranaese, C. pusillum,* and *C. sanctifranciscense*); a re-validated from synonomy (*C. taenia*); or previously assigned to the genus *Aequidens* (*C. dimerus, C. portalegrense*). The genus *Aequidens* was further dismembered, and *Aequidens sensu stricto* was restricted to include only *Ae. tetramerus, Ae. diadema, Ae. pallidus, Ae. viridis, Ae. metae, Ae. stollei, Ae. chimantanus,* and *Ae. uniocellatus*. Four additional lineages are recognized amongst the '*Aequidens*' residue (some of which have received generic assignations, e.g. *Bujurquina* Kullander, 1986; *Laetacara* Kullander, 1986; see also Kullander and Nijssen, 1989).

Kullander (1983a, 1986) is of the opinion that *Cichlasoma s. str., Aequidens s.*

str., Nannacara, Tahuantinsuyoa (Kullander, 1986), *Laetacara, Bujurquina* and the remaining '*Aequidens*' residue are monophyletic (i.e. *Cichlasoma* Gp. B in Fig. 1.5). Two characters may be cited in support of that hypothesis:

1. The first is the presence in these taxa of an exclusively uni- or tri-serial predorsal scale patterning. This pattern is assessed by Kullander (1983a: 264–5) to be derived relative to the alternative stochastic array exhibited by other cichlid taxa;
2. The second character suggestive of group B monophyly is a derived behavioural attribute of the group. According to Loiselle (pers. comm.), all of the group B cichlasomine genera for which relevant behavioural data are available lay their eggs in a characteristically tight circular plaque; this is unlike the more usual behaviour amongst substrate-spawning cichlids, which lay loose, or open (non-circular) egg plaques.

Kullander (1983a) assigns the South American species excluded from *Cichlasoma s. str.* to the genera *Caquetaia, Heros, Hoplarchus* and *Mesonauta* (see also Kullander, 1986), but most of the remaining '*Cichlasoma*' species of Central America, southern North America and the Greater Antilles are left without formal generic designations. This is clearly an unsatisfactory state of affairs, and has already caused considerable confusion in both the scientific and popular literature. However, until phylogenetic analyses of the interrelationships of these Mesoamerican taxa are available (e.g. Miller and Cichocki, in prep.), any generic assignments would represent purely formal nomenclatural actions and would not result in a stable system of names. As an interim measure to minimize nomenclatural confusion, I suggest that the convention proposed by Rosen and Patterson (1977) be adopted here. Species formerly included in the genus *Cichlasoma* should be referred to by that name cited between quotation marks. An additional aid when referring to these fishes is available with the use of Regan's (1906, 1906–08) informal sectional names. For example, '*Cichlasoma*' *spilurus* becomes '*Cichlasoma* (*Archocentrus*)' *spilurus* and '*C.*' *trimaculatus* becomes '*C.* (*Amphilophus*)' *trimaculatus*, etc. (reference should be made to Miller, 1976; Miller and Nelson, 1961 for the allocation of species to sections).

An analogous situation is encountered within the African assemblage where Greenwood's revisional studies of the genus *Haplochromis* (Greenwood, 1979, 1980) have resulted in similar nomenclatural problems. In attempting to resolve the speciose and clearly polyphyletic genus *Haplochromis* into monophyletic lineages, he restricted that genus to five species from Lakes Victoria, Kyoga, Edward, George, Nabugabo and Kivu (Greenwood, 1980). For the remaining 300 or more species he has designated numerous, putatively monophyletic genera (e.g. Greenwood, 1979, 1980) and has suggested, as a temporary expedient, the adoption of '*Haplochromis*' for reference to species previously referred to *Haplochromis sensu lato* and for which no alternative generic name is available.

The African Cichlidae

The remaining Cichlidae constitute the outstandingly speciose African radiation of the family. Once again, estimates vary widely, but perhaps as many as 80% of all cichlid species are to be found in African fresh waters, and of these the great majority are represented amongst the major lacustrine radiations (Fryer and Iles, 1972; Ribbink, this volume). Given the marked preponderance of lacustrine taxa amongst the African radiation and the absence of comparable lacustrine biotopes in the Neotropics, perhaps a more informative comparison between the Old and New World assemblages would be between the African fluviatile species and their Neotropical counterparts.

Despite the superficially disparate array of morphological and ecological diversity represented by the African Cichlidae, the monophyly of this spectacular radiation (excluding the Zairean genus *Heterochromis*) is strongly corroborated by a suite of derived morphological features (characters 21–24 in Fig. 1.3). These features are discussed in detail elsewhere (Cichocki, 1976; Oliver, 1984; Stiassny, in press) and will only be described briefly here:

1. (character 21, Fig. 1.3) As was first noted by Cichocki (1976), the AAP muscle of African cichlids extends anteriorly to insert onto the palatine bone in the region of the palatine fossa (e.g. Fig. 1.15(b)). A similar fossa is lacking in *Heterochromis*, and in all Malagasy, Asian and Neotropical cichlids, where the anteriormost extent of the AAP is limited to the pterygoid series (e.g. Fig. 1.15(a); see also discussion in Greenwood, 1985a). Anteriad extension of the AAP is uncommon amongst percomorphs, and in no other taxon does the AAP insert onto the palatine in the manner of the African cichlid lineage;
2. (character 22, Fig. 1.3) The anterior palato-maxillary ligament of African cichlids is well developed and strap-like, and typically extends much of the length of the palatine prong (e.g. Fig. 1.15(d)). This is not the case in other cichlids, nor in most acanthomorphs (Stiassny, 1986), where the palatine prong is attached to the maxillary saddle by a short ligament (e.g. Fig. 1.15(c)). *Hemichromis* lacks a well-developed strap-like anterior palato-maxillary ligament (Greenwood, 1985a) and Cichocki (1976) interpreted this absence as an indication of a phylogenetically primitive position of *Hemichromis* in the African clade. However, as its sister group, *Anomalochromis*, has the ligament well developed (Greenwood, 1985b), the absence of the ligament in *Hemichromis* is probably better interpreted as a secondary loss functionally associated with the extensive remodelling of the whole palatine–ethmoid region (Greenwood, 1985a);
3. (character 23, Fig. 1.3) Primitively amongst percomorphs, the pharyngo-cleithralis internus muscles (Phcl. Int.) insert either musculously or via broad tendinous bands along the fifth ceratobranchial elements (see e.g. figures in Stiassny, 1987; Stiassny and Jensen, 1987). Amongst cichlids there is considerable variation in the mode of association of the

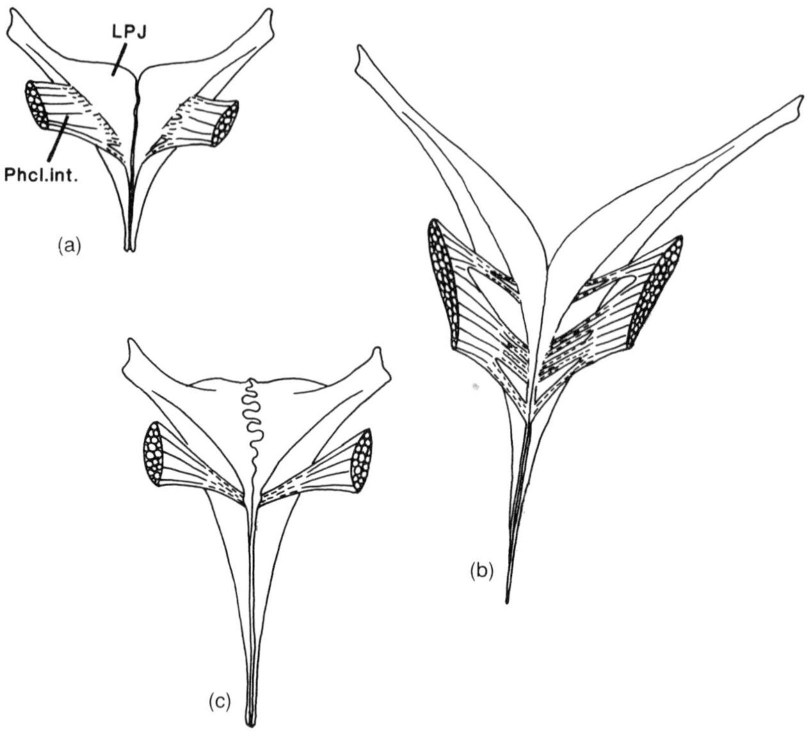

Fig. 1.16 Ventral view of LPJ and isolated pharyngocleithralis internus muscles of (a) *Oxylapia*; (b) *Chaetobranchus*; (c) *Tylochromis*. Abbreviations as in Fig. 1.6.

phcl. internus with the LPJ (Fig. 1.16). In *Heterochromis* and non-African taxa the muscle is rather broadly associated with the LPJ, usually via a combination of direct muscular and multiple tendinous insertions (e.g. Fig. 1.16(a)). In some taxa the tendinous associations are particularly elaborate. For example, in *Chaetobranchus* (Fig. 1.16(b)), and its sister taxon, *Astronotus*, the tendinous associations of the muscle with the LPJ are peculiarly complex. In members of the African clade, the Phcl. internus inserts onto the LPJ via a single, cord-like tendon (e.g. Fig. 1.16(c)). In view of its restricted distribution amongst percomorphs, this latter association is interpreted as being the derived condition within the Cichlidae;

4. (character 24, Fig. 1.3) In members of the African lineage, the uncinate process of the first epibranchial (Ep1) is orientated caudad and is considerably longer than the anterior arm of the element (e.g. Fig. 1.17(b)). This is not usually the case in other cichlid taxa where (with the exception of ptychochromines, page 8) the two arms of Ep 1 are typically sub-equal in length (e.g. Fig. 1.17(a)).

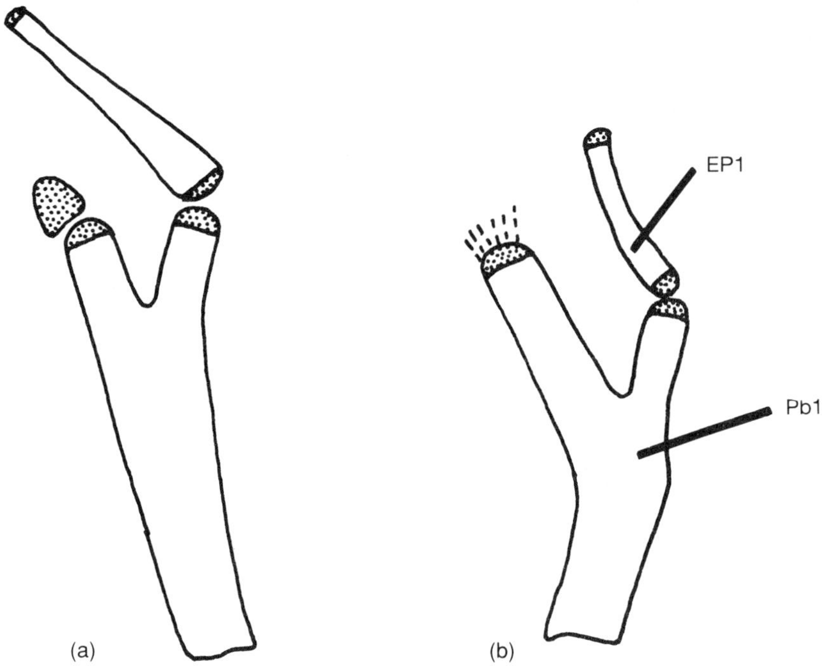

Fig. 1.17 Isolated left first epibranchial bones of (a) *Cichlasoma*; (b) *Tylochromis*. Abbreviations as in Fig. 1.6.

Regan's classic papers (1920, 1922b) provide the first, and to date the only critical analysis of all African cichlid genera. In dismantling Boulenger's three central genera (*Tilapia*, *Paratilapia*, and *Pelmatochromis*) Regan essentially created two great cichlid assemblages – the *Haplochromis* and the *Tilapia* lineages – and in so doing defined a conceptual framework that was to dominate views of cichlid interrelationships for the next 50 years. Since the early 1960s the phylogenetic validity of these two groupings has been repeatedly challenged (e.g. Wickler, 1963; Fryer and Iles, 1972; Liem, 1981a; Greenwood, 1978, 1987a, b), and it is evident that the concept of a simple haplochromine–tilapiine dichotomy of the African Cichlidae is no longer acceptable. However, an alternative scheme has yet to be worked out and as can be seen from Fig. 1.3, our understanding of the interrelationships of the various named lineages is minimal.

There is some indication that the virtually pan-African genus *Tylochromis* (Stiassny, 1989) represents the sister group of the remaining African lineage. The evidence for this assessment is twofold:

1. (character 25, Fig. 1.3). *Tylochromis* is unique amongst African cichlids in having two predorsal bones associated with the first neural spine

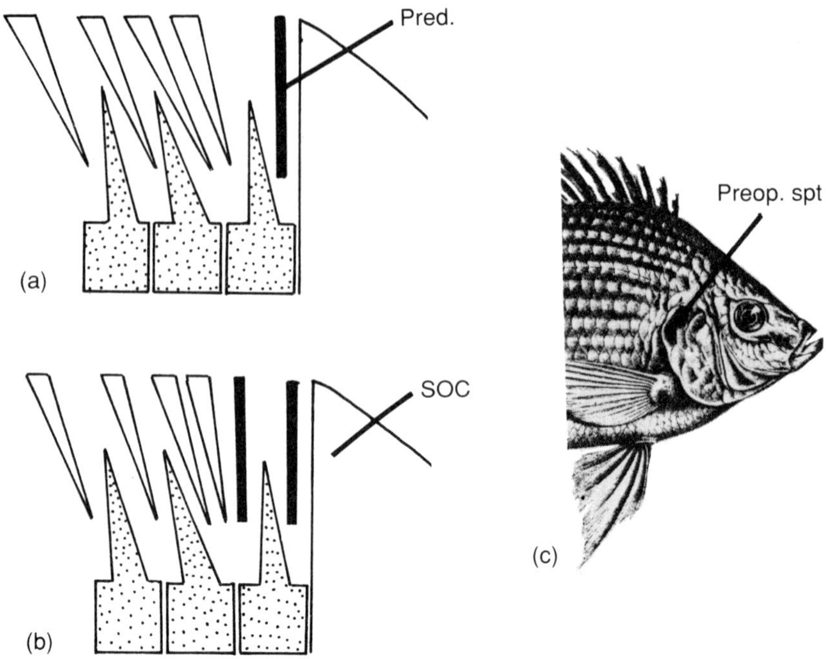

Fig. 1.18 (a) and (b) Schematic representation of the predorsal bones and associated structures in (a) most African cichlids; (b) *Tylochromis*. (c) Head of *Oreochromis galilaeus* (after Boulenger, 1907) illustrating the opercular spot typical of most African lineages. Abbreviations as in Fig. 1.6.

(Fig. 1.18(b)), whilst all others (excluding *Heterochromis*, which also has two) have apparently lost the posterior predorsal and retain only a single anterior element (e.g. Fig. 1.18(a)). Predorsal number, both within the Cichlidae and amongst percomorphs generally, was investigated by Cichocki (1976), who concluded that the presence of two predorsals is primitive for cichlids. The reduction to a single element (a similar reductional trend occurs in parallel in various Neotropical lineages) is a character separating the rest of the African lineage from *Tylochromis*, which lacks this derived feature;

2. (character 26, Fig. 1.3). *Tylochromis* lacks the well-developed and strongly pigmented opercular spot that is so characteristic of other African lineages (e.g. Fig. 1.18(c); Stiassny, 1989). Kullander (1983b: 307) was the first to note the presence of the dorsally situated opercular spot in African lineages and its general absence in the Neotropical assemblage (see also Trewavas, 1983 for a description of the opercular spot in tilapiines). The presence of the opercular spot characterizes virtually all lineages of African cichlid fishes (see, e.g. figures in Boulenger, 1915; Loiselle, 1985a). Within some lineages, e.g. the Mbuna of Lake Malawi (Ribbink *et al.*, 1983a; Lewis *et al.*,

1986), individual species may lack the opercular spot, but in all cases there are species included with typical opercular spot development. A similarly situated opercular spot is not developed in any of the Malagasy, Asian or Neotropical cichlids, nor is one present in *Heterochromis* (see Boulenger, 1915, Fig. 277; pers. obs.); amongst percomorphs, it is only in the family Centrarchidae that anything similar is developed.

With regard to the remaining African lineages it is clear that little more may be said with respect to their interrelationships until more data are available for analysis. At present I have nothing to add in the way of anatomical data bearing on these important questions. However, in the course of this study I

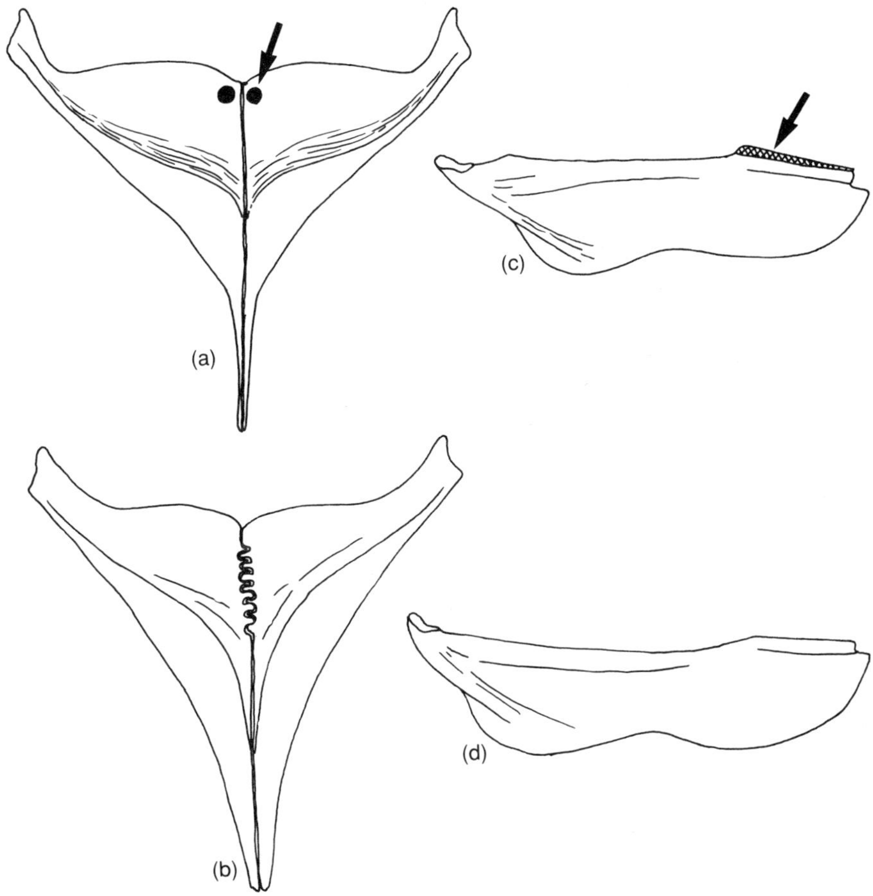

Fig. 1.19 Lower pharyngeal jaw of *Tilapia sparrmani* in (a) ventral view; (c) lateral view. Lower pharyngeal jaw of *Astatotilapia calliptera* in (b) ventral view; (d) lateral view. Arrows indicate position of foramina and median ridge on the pharyngeal keel.

have located two features of potential interest with respect to the redefinition of one of the major named lineages, the tilapiines:

1. The LPJ of tilapiines bears an interesting feature on its ventral surface. In these fishes there are usually two prominent foramina situated one on either side of the median suture (Fig. 1.19(a)). As far as can be ascertained from gross dissection and electron microscopy, no blood vessels or nerves issue from these foramina, which in life are covered by a connective-tissue elaboration of the sphincter oesophagii muscle. In some individuals only a single foramen is present instead of the more usual paired configuration. I have been unable to locate similarly positioned foramina in the LPJs of other cichlid (or percomorph) taxa (e.g. Fig. 1.19(b)). LPJ foramina are present in the following genera: *Danakilia*, *Iranocichla*, *Konia*, *Myaka*, *Oreochromis*, *Pungo*, *Sarotherodon*, *Stomatepia*, *Tilapia* and *Tristramella*. Amongst the *Tilapia* species examined, *T. busumana* is apparently polymorphic for the feature, with some individuals lacking LPJ foramina and others having either single or paired foramina on the jaw (Greenwood, pers. comm.). This polymorphism in *T. busumana* is of particular interest in view of Greenwood's recent suggestion that this taxon may be a primitive tilapiine (Greenwood, 1987a; see also Thys van den Audenaerde, 1968b; Trewavas, 1973);
2. The LPJ of tilapiines differs from that of other cichlids in the development of a median ridge of bone on the dorsal aspect of the keel (Fig. 1.19(c)). The ridge is variously developed in tilapiines, being most prominent in *Sarotherodon* and *Oreochromis* and least well developed in *Tilapia*. A similar ridge is absent in other cichlid (and percomorph) taxa (e.g. Fig. 1.19(d)).

Trewavas (1983) tentatively defined a tilapiine assemblage and in addition to the taxa included here, Trewavas' tilapiines include *Pelmatochromis*, *Pterochromis*, *Steatocranus* and *Gobiocichla*. However, in none of these taxa is there a median ridge present or are foramina developed on the ventral face of the LPJ, nor do they exhibit the characters discussed by Greenwood (1987a). Whilst I suspect that future analyses may bear out Trewavas' contention that these taxa are related to the tilapiines (of this paper), sound evidence supporting such an alignment has yet to be presented. As I have attempted only to include putatively monophyletic assemblages as terminal taxa in this analysis I favour the more restricted tilapiine lineage as defined here.

1.4 CONCLUDING REMARKS

I began this review by suggesting that my study may be viewed as a testament to how little we still may claim to know about the interrelationships of the major lineages of cichlid fishes. In conclusion I can only reiterate that rather negative claim. After almost a century of the study of cichlid morphological variation, virtually all that may reasonably be said (at these inclusive levels) is summarized in Fig. 1.3. Given the fact of such morphological conservatism, perhaps it is time to 'broaden the net' to incorporate the growing body of molecular data into cladistic phylogeny reconstruction (see Kornfield, this volume). At the less inclusive levels, the use of behavioural data has yet to be fully explored and it is to be hoped that behavioural analyses may yield much phylogenetically informative data.

The problems of phylogeny reconstruction in the face of muted morphological variation are manifold, yet without a sound phylogenetic framework our understanding of the behaviour, ecology and evolution of these fascinating fishes is greatly diminished. There is clearly great incentive to continue with phylogenetic studies of the family Cichlidae.

1.5 TAXONOMIC APPENDIX

Putatively monophyletic subgroupings discussed in the text

Ptychochromines. Ptychochromoides and *Ptychochromis* (Kiener and Mauge, 1966; Cichocki, 1976; Stiassny and Reinthal, in prep.).

Etroplines. Etroplus, Oxylapia, and *Paretroplus* (Kiener and Mauge, 1966; Cichocki, 1976; Stiassny and Reinthal, in prep.).

Crenicichlines. Crenicichla (including *Batrachops*) and *Teleocichla* (Kullander, 1983a, 1986, 1988; Stiassny, 1987).

Chaetobranchines. Chaetobranchopsis, Chaetobranchus and *Astronotus* (Kullander, 1986; Stiassny, this paper).

Geophagines. Acarichthys, Apistogramma, Apistogrammoides, Biotoecus, Crenicara, Geophagus, Gymnogeophagus, Papiliochromis, Retroculus, Satanoperca, Taeniacara (Gosse, 1975; Cichocki, 1976; Kullander, 1980, 1983b, 1986; Kullander and Nijssen, 1989).

Cichlasomine gp. A. Acaronia, Caquetaia, 'Cichlasoma Amphilophus', 'C. Archocentrus', 'C'. insertae sedis, 'C. Herichthys', 'C. Nandopsis', 'C. Paraneetroplus', 'C. Theraps', 'C. Thorichthys', Hoplarchus, Heros, Herotilapia, Hypselecara, Mesonauta, Neetroplus, Pterophyllum, Symphysodon, and *Uaru* (Miller and Nelson, 1961; Cichocki, 1976; Kullander, 1983b, 1986; Kullander and Nijssen, 1989). I have included *Acaronia* amongst the gp. A cichlasomines. Kullander (1983b, 1986) has expressed some doubt as to the correct placement of this genus, primarily because of the fact that "*Acaronia* is very different from cichlasomines", particularly with respect to its primitive jaw and gill-raker structure. However, retention (or re-expression) of a primitive morphology does not exclude *Acaronia* from the cichlasomine radiation, with which it shares characters 17 and 18.

Cichlasomine gp. B. *Aequidens s. str.*, *'Aequidens'*, *Bujurquina*, *Cichlasoma s. str.*, *Laetacara*, *Nannacara* and *Tahuantinsuyoa* (see Cichocki, 1976; Kullander, 1983a, 1986; Kullander and Nijssen, 1989, where additional generic designations and descriptions for components of the '*Aequidens*' residue are provided).

Tilapiines. *Danakilia*, *Iranocichla*, *Konia*, *Myaka*, *Oreochromis*, *Pungu*, *Sarotherodon*, *Stomatepia*, *Tilapia* and *Tristramella* (Thys van den Audenaerde, 1964, 1968a, b; Trewavas, 1973, 1983; Greenwood, 1987a; Stiassny, this paper).

Hemichromines. *Hemichromis* and *Anomalochromis* (Loiselle, 1979; Greenwood, 1985a, b).

Chromidotilapiines. *Chromidotilapia*, *Limbochromis*, *Nanochromis*, *Parananochromis*, *Pelvicachromis*, and *Thysochromis* (Thys van den Audenaerde, 1968a; Loiselle and Welcomme, 1972; Trewavas, 1973; Greenwood, 1987a). Following Daget (1988), the name *Thysochromis* is adopted as a replacement name for *Thysia* (Loiselle and Welcomme, 1972).

Lamprologines. *Chalinochromis*, *Julidochromis*, *Lamprologus*, *Lepidolamprologus*, *Neolamprologus*, *Teleogramma* and *Telmatochromis* (Poll, 1978, 1986; Colombe and Allgayer, 1985). Poll (1986), I believe quite correctly, questions the phylogenetic validity of Colombe and Allgayer's (1985) subdivision of *Lamprologus sensu lato*, and proposes an alternative scheme. I have included *Teleogramma* amongst the lamprologines as these fishes share uniquely with other lamprologines a highly derived and characteristic buccal dentition (Roberts and Stewart, 1976; Stiassny, in prep.).

The rest. This 'category' includes all the remaining African genera not yet included in the groups outlined above. This residue is most certainly not monophyletic and is simply appended here as a convenience (for included genera see, e.g. Poll, 1957, 1986; Trewavas, 1935, 1983; Greenwood, 1979, 1980; Ufermann *et al.*, 1987). Numerous additional lineages have been identified within this residue, e.g. the Serranochromines (Greenwood, 1979), the *Opthalmotilapia* lineage (Liem, 1981a; Greenwood, 1983b), the *Hemibates–Trematacara* lineage (Stiassny, 1981b), but virtually nothing is known of their interrelationships.

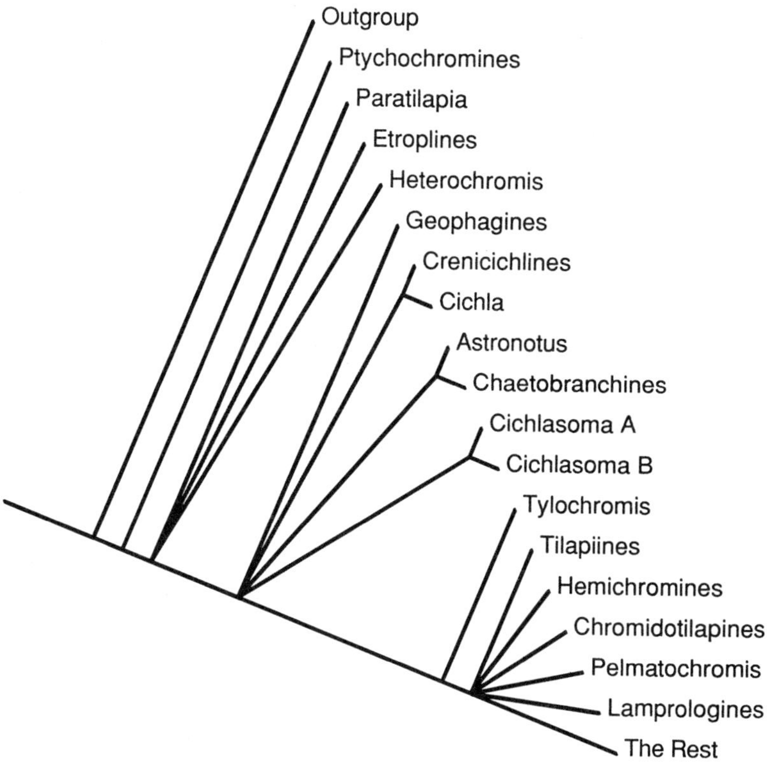

Fig. 1.20 The Nelson ancensus tree for the 54 minimum-length trees retrieved by HENNIG86.

1. The two arms of the first epibranchial bones are sub-equal in length (page 8);
2. Absence of separate second epibranchial toothplates (page 8);
3. Fourth ceratobranchial toothplates separate from gill rakers (page 11);
4. NLFO foramina are more-or-less coalesced in the midline (page 11);
5. NLFO foramina are overlain by the supraoccipital bone (pages 11–12);
6. Single, central ligament inserts onto the median fourth basibranchial (page 15);
7. Paired frontal canals extend rostrodorsally as a single central tube elevated in the frontal ridge, and terminating in a single central NLFO pore (page 15);
8. Presence of a strongly interdigitating suture between the vomerine wing and the parasphenoid bar (page 18);

Data matrix and characters

Taxon	Character Number																											
	1	2	3	4	5	6	7	8	9	10	11	12	13	14	15	16	17	18	19	20	21	22	23	24	25	26	27	28
Ptychochromines	0	0	0	0	0	0	0	0	0	0	0	0	0	0	0	0	0	0	0	0	0	0	0	0	0	0	0	1
Paratilapia	1	1	1	1	0	0	0	0	0	0	0	0	0	0	0	0	0	0	0	0	0	0	0	0	0	0	0	1
Etroplines	1	1	1	1	1	0	0	0	0	0	0	0	0	0	0	0	0	0	0	0	0	0	0	0	0	0	0	1
Heterochromis	1	1	1	1	1	0	0	1	0	0	0	0	0	0	0	0	0	0	0	0	0	0	0	0	0	0	0	1
Crenicichlines	1	1	1	1	0	1	1	1	1	1	1	1	1	1	0	0	1	0	0	0	0	0	0	0	0	0	1	0
Cichla	1	1	1	1	0	1	1	1	1	1	1	1	1	1	0	0	0	0	0	0	0	0	0	0	0	0	0	0
Astronotus	1	1	1	1	0	1	1	1	0	0	0	0	0	0	1	1	0	0	0	0	0	0	0	0	0	0	0	0
Chaetobranchines	1	1	1	1	0	1	1	1	0	0	0	0	0	0	1	1	0	0	0	0	0	0	0	0	0	0	1	0
Geophagines	1	1	1	1	0	1	1	1	0	0	0	0	0	0	0	0	1	0	0	0	0	0	0	0	0	0	1	0
Cichlasomine gp.A	1	1	1	1	0	1	1	1	0	0	0	0	0	0	0	0	1	1	0	0	0	0	0	0	0	0	1	0
Cichlasomine gp.B	1	1	1	1	0	1	1	1	0	0	0	0	0	0	0	0	1	1	1	1	0	0	0	0	0	0	1	0
Tylochromis	1	1	1	1	0	1	1	0	0	0	0	0	0	0	0	0	0	0	0	0	1	1	1	1	0	0	1	0
Tilapiines	1	1	1	1	0	1	1	0	0	0	0	0	0	0	0	0	0	0	0	0	1	1	1	1	1	1	1	0
Hemichromines	1	1	1	1	0	1	1	0	0	0	0	0	0	0	0	0	0	0	0	0	1	1	1	1	1	1	1	0
Chromidotilapiines	1	1	1	1	0	1	1	0	0	0	0	0	0	0	0	0	0	0	0	0	1	1	1	1	1	1	1	0
Pelmatochromis	1	1	1	1	0	1	1	0	0	0	0	0	0	0	0	0	0	0	0	0	1	1	1	1	1	1	1	0
Lamprologines	1	1	1	1	0	1	1	0	0	0	0	0	0	0	0	0	0	0	0	0	1	1	1	1	1	1	1	0
The rest	1	1	1	1	0	1	1	0	0	0	0	0	0	0	0	0	0	0	0	0	1	1	1	1	1	1	1	0

9. Pharyngocleithralis muscle originating from the lateral face of the cleithrum (Stiassny, 1982, 1987);
10. Urohyal bone bears a rostrally directed spine (Stiassny, 1982, 1987);
11. Increased number of vertebrae (Stiassny, 1982, 1987);
12. Elaboration of the obliquus inferioris muscle associated with the first and second postcleithra (Stiassny, 1982, 1987);
13. Characteristically notched vomer (Stiassny, 1987);
14. Enlarged sphenotic foramina (Stiassny, 1987);
15. Characteristic microbranchiospine form (Kullander, 1986: 61);
16. Complex tendon system of the pharyngocleithralis internus muscle (page 26);
17. Six preopercular pores (page 19);
18. Four pores in the dentary (Kullander, 1983a: 265, 1986: 200);
19. Uni- or tri-serial predorsal scale pattern (page 24);
20. Eggs laid in a tight circular plaque (page 24);
21. The adductor arcus palatini muscle inserts onto the palatine bone and into a palatine fossa (page 25);
22. Anterior palato-mixillary ligament is strap-like and extends much of the length of the palatine prong before inserting onto the mixillary head (page 25);
23. The pharyngocleithralis internus muscle inserts onto the LPJ via a single cord-like tendon (pages 25–6);
24. The uncinate process of the first epibranchial bone is considerably longer than the anterior arm (page 26);
25. A single predorsal bone is present (pages 27–8);
26. A strongly pigmented opercular spot is developed (pages 28–9);
27. The lachrymal is a single element, and the second infraorbital is tubular and is typical followed by three infraorbitals and a dermosphenotic (pages 20–3);
28. Large foramina in the exoccipital bones of the neurocranium (pages 13–14);

Chapter two

Distribution and ecology of the cichlids of the African Great Lakes

A. J. Ribbink

2.1 INTRODUCTION

Two intriguing questions regarding the natural history of the cichlid species flocks of the African Great Lakes are (1) why are there so many species and (2) how can so many species coexist within each lake? These are not new questions. They have been exercising the minds of systematists since the turn of the century. It was in the early 1950s, however, that ecological and zoogeographic contributions supplemented taxonomic and anatomical studies to indicate how resource utilization occurred and intralacustrine allopatric speciation took place. Numerous publications on these subjects appeared, but real milestones were the contributions of Brooks (1950), Lowe (1952), Poll (1956), Fryer (1959) and a series of papers by P. H. Greenwood on the Lake Victoria haplochromines. Outstanding reviews of the cichlids of the African Great Lakes are those of Fryer and Iles (1972), Greenwood (1974a), Lowe-McConnell (1975, 1987) and Coulter (1991).

This account builds on the excellent foundations provided by previous work and focuses on the advances in understanding which were made in the last decade. It demonstrates that many species are restricted geographically to small areas within each lake. Such intralacustrine endemicity is attributed to philopatry, which is imposed by adult stenotopy and by parental care of offspring, ensuring that the entire life history of individuals takes place within a single habitat. Philopatry can promote speciation, but only if stenotopic populations are divided and geographically isolated. As philopatric populations are poor dispersers, vicariant events are probably necessary to separate narrowly stenotopic populations. Eurytopic fishes are more widely distributed geographically and are not as speciose as the stenotopic species flocks.

Therefore information on the distribution and ecology of populations helps to answer questions regarding speciation.

Within the principal habitats of each lake there are multispecies communities whose members have specific ecological requirements as well as anatomical and behavioural adaptations which enable them to partition the resources more finely than was realized previously. Such fine partitioning accounts for the species richness within each of these communities. Collectively, these species-rich communities contribute to the total number of species inhabiting each lake. A consequence of intralacustrine endemicity of cichlids is that geographically isolated habitat patches have species that are not found anywhere else. Therefore, a tentative answer to the question regarding the accommodation of so many species in each lake is that fine resource partitioning enables numerous species to coexist in each habitat and that the community in each geographically isolated habitat patch is unique, comprising endemic and more widespread species.

The discussion presented here has been confined to cichlid interactions and does not consider the extremely important limnological factors that influence the ecology of the lakes. Similarly, almost no reference is made to energy flow and productivity. Readers interested in these aspects of ecology are referred to Livingstone and Melack (1984), Coulter *et al.* (1986) and Coulter (1991).

2.2 GEOGRAPHIC DISTRIBUTION

Endemicity

The most striking feature of cichlid distribution in Lakes Victoria, Tanganyika and Malawi is the high incidence of endemicity. Each lake has its own unique fauna with a cichlid endemicity of 99% or more. Not one species is shared by these lakes. In addition, regional endemicity within each lake is marked, as demonstrated by the many species which occur naturally in certain, often very small, areas and nowhere else.

Interlacustrine endemicity is readily explained by the formation of each lake at widely different geological times and in far apart sites. Furthermore, available data suggest that the riverine ancestral fauna which gave rise to the various cichlid flocks was different in each case (Trewavas, 1935; Fryer and Iles, 1972; Greenwood, 1974a; Lewis, 1981; Coulter *et al.*, 1986). Lakes Tanganyika and Malawi were formed as long, deep troughs when different sections of the western Rift Valley filled with water. The pioneer species which colonized Lake Tanganyika gave rise to *c.* 60 species of substrate spawner and 110 mouthbrooding species (Coulter, 1991). All of the Lake Malawi endemic species whose behaviour is known are maternal mouthbrooders. The riverine haplochromines, which colonized this lake originally, were probably similar to the extant *Astatotilapia calliptera* (Trewavas, 1949; Fryer and Iles, 1972) and perhaps also to other riverine ancestors.

Lake Victoria is not a Rift Valley lake, but was formed when tectonic events caused a mountain ridge to rise and block the western passage of several rivers draining the eastern highlands of Kenya (Greenwood, 1974a). The reversal of river flow and a back-ponding flooded the river courses and the surrounding low-lying area to produce an expansive lake (69 000 km^2, Greenwood, 1974a). The endemic haplochromines of Lake Victoria are similar to those of Lake Malawi in many respects, including the fact that they are all maternal mouthbrooders, as are the two endemic tilapiines. An ancestral-like form for the haplochromines of Lake Victoria is *Astatotilapia bloyeti* (Greenwood, 1974a).

Interpretations of intralacustrine endemicity invoke physical environmental events which occurred during the history of each lake (Livingstone and Melack, 1984), but more important perhaps is that the distribution of the contemporary species within the lakes is dictated by characteristics of the fishes themselves. This was not recognized by earlier workers (e.g. Plate, 1913 in Greenwood, 1974a), who were impressed by the increasing numbers of species that were being discovered in each lake. It was apparent that the cichlids had undergone spectacular speciation and formed large species flocks, but this posed enigmatic questions as it was difficult to perceive how fishes, which are vagile and are potentially capable of swimming to any part of a lake, could speciate in allopatry within a single lake. It is now known that most cichlids are not evenly or widely distributed within each system (Poll, 1952, 1956; Fryer and Iles, 1972; Lewis, 1982; van Oijen, 1982; Ribbink *et al.*, 1983a; Witte, 1984a; Eccles, 1986).

Distribution in Lake Malawi

Although the classic examples of intralacustrine allopatry are those of the cichlids of Lake Tanganyika (Fryer and Iles, 1972), the most detailed documented patterns of geographic distribution of species within the African Great Lakes are those exhibited by cichlids which are associated with rocky habitats in Lake Malawi. These speciose lithophilous communities are dominated by a monophyletic group of genera, collectively referred to by their Chitonga name, Mbuna (Trewavas, 1935; Fryer, 1959; Fryer and Iles, 1972; Ribbink *et al.*, 1983a; Reinthal, 1987). Mbuna are small fishes, most species being less than 100 mm standard length (SL) as adults, although some may grow to more than 150 mm SL and others are no larger than 40 mm SL (Ribbink *et al.*, 1983a). The majority of Mbuna are colourful and are characterized by an affinity for rocky habitats; some species are very closely tied to these habitats while others have much weaker links to rocky zones (Fryer, 1959; Ribbink *et al.*, 1983a; McKaye and Gray, 1984). The Mbuna probably comprise more than two hundred species which have been assigned to a number of genera, species complexes (monophyletic assemblages) and species groups (polyphyletic assemblages) (Ribbink *et al.*, 1983a). The taxonomy of the Mbuna is currently under revision and many of the informal groups depicted by Ribbink *et al.*, (1983a) are being described as separate genera (Lewis, in prep.).

Although the Mbuna dominate rocky habitats numerically and taxonomically, they are by no means the only fishes to reside there. Numerous other cichlids, as well as members of other families of fish, share these environments, either permanently or temporarily, with the Mbuna (Fryer, 1959; Fryer and Iles, 1972; Ribbink *et al.*, 1983a).

In a preliminary survey of fishes of rocky habitats, at 121 stations in 14 regions throughout the Malawian waters of the lake, 196 species were identified, their distribution was plotted and their numerical density was assessed in underwater transects (Ribbink *et al.*, 1983a). Only two strictly lithophilous species, *Genyochromis mento* and *Labeotropheus fuelleborni*, occur in every study region and even then these species were absent from some stations. Of the Mbuna of intermediate habitats, where rocks are interspersed with aquatic macrophytes and sandy substrates, only *Cyathochromis obliquidens* is widely distributed. All other Mbuna have narrower natural distributions and many species are restricted to small parts of larger rocky shores or to small rocky outcrops. Indeed, the total distribution of some species is no more than several thousand square metres. A consequence of such narrow distributions is that the species assemblage in each area is unique, with contributions coming from species which are endemic to the site as well as from those which are widespread. The endemicity within each region is remarkably high (Fig. 2.1). Distance between stations has clear effects on species composition. Invariably, the number of shared species decreases as the distance between stations increases, even when stations have similar habitats. Frequently, too, when species are common to several adjacent stations they differ in relative abundance, being rare at one, common at another and abundant at yet another. This variability in numerical abundance of a species and in the species composition of cichlid communities throughout the lake means that widespread species are subjected to the interactions of a different, unique collection of fishes at each locality.

Small rocky islands or pinnacles of rock rising from sandy substrates, often in deep water, have fewer species than larger islands or stretches of rocky mainland shore. For example, the rocky pinnacle, Zimbawe Islet, supports only nine Mbuna species. The slightly larger islets of Chinyamwezi and Chinyankwazi support 12 and 14 species respectively, whereas the even larger islands Thumbi West, Maleri and Mbenji all have 29 or more species. The largest island, Likoma, has 49 species. Large areas of rocky habitat probably have the greatest species richness because their habitat diversity is broader and better able to accommodate a wide variety of species than can that of small, stenotopic areas. Rocky pinnacles, for example, comprise large rocks and boulders and provide habitats for only those few fishes which can cope with such an environment. Larger islands have a broad spectrum of habitats with a corresponding species diversity. The disjunct distribution of some species

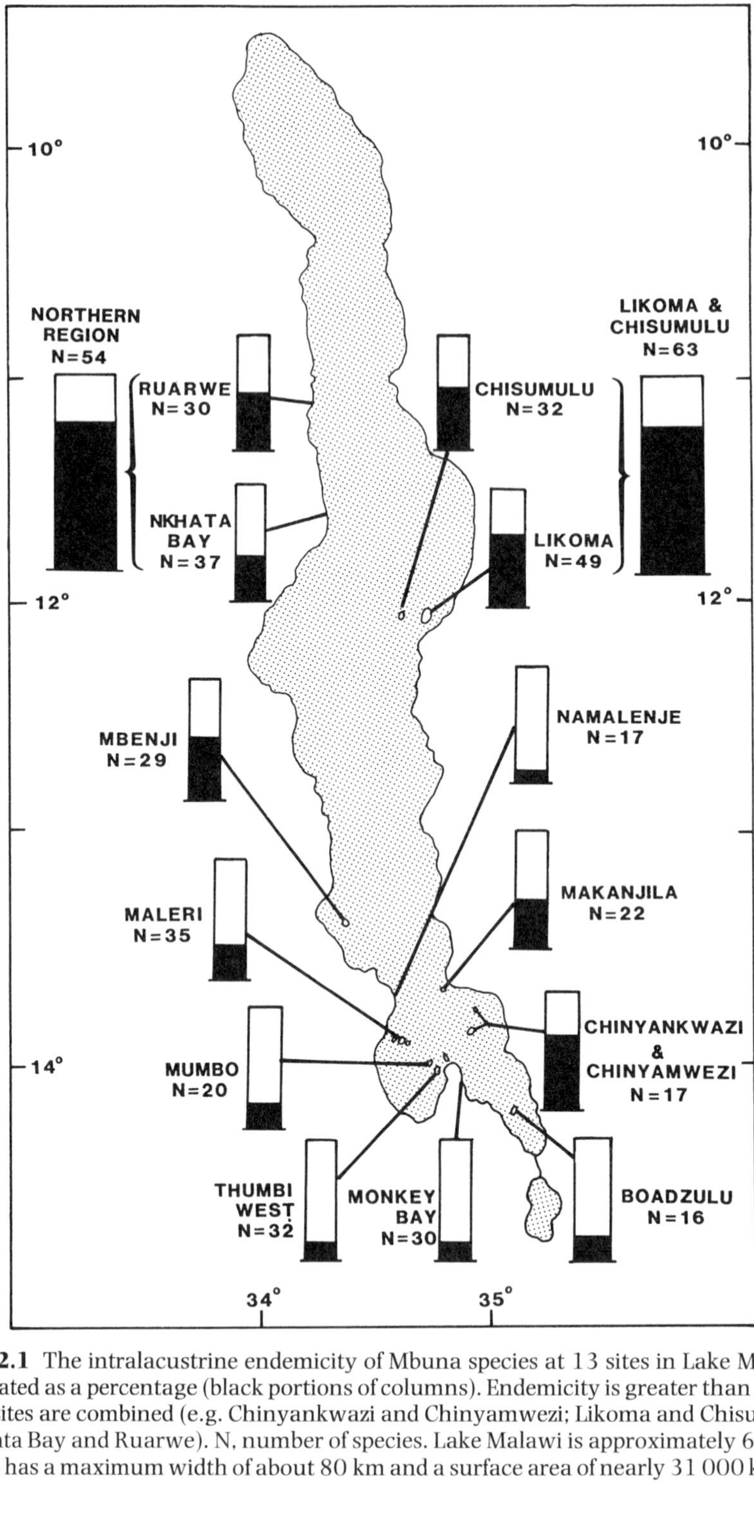

Fig. 2.1 The intralacustrine endemicity of Mbuna species at 13 sites in Lake Malawi, indicated as a percentage (black portions of columns). Endemicity is greater than 50% if two sites are combined (e.g. Chinyankwazi and Chinyamwezi; Likoma and Chisumulu; Nkhata Bay and Ruarwe). N, number of species. Lake Malawi is approximately 600 km long, has a maximum width of about 80 km and a surface area of nearly 31 000 km².

may be explained by the presence or absence of suitable habitat patches. *Cyathochromis obliquidens*, for example, has a wide distribution, occurring wherever intermediate habitats are suitable, but is absent from purely rocky regions. The disjunct distribution of other species is more difficult to explain. *Labeotropheus trewavasae* occurs along the north-western rocky zones of the lake, is absent from the islands of Likoma, Chisumulu and Mbenji in the central regions, but does occur at the Maleri Islands, Namalenje, Mumbo, Thumbi West and Zimbawe (near to Thumbi West, Fig. 2.1). It is absent from Domwe, the Nankumba Peninsula, Thumbi East (Monkey Bay), the remainder of the south-western coast, Chinyamwezi and Chinyankwazi, but reappears at Boadzulu Island in the south-eastern arm of the lake and at Makanjila on the south-eastern shore (Ribbink *et al.*, 1983a, b). Similarly, *Petrotilapia tridentiger* occurs along the north-western rocky habitats from Chilumba to Nkhata Bay, is absent from the central regions of the lake and reappears in the south-western habitats (Marsh, 1981, Ribbink *et al.*, 1983a). Why this species should be present in the northern and southern part of the lake, yet not in the central portion, is unknown.

There are also unexplained distributional gaps where populations of a species occupy areas that are close to one another, but sites within the distributional range, which appear to have suitable habitats, are unoccupied (Ribbink *et al.*, 1983a). Another distributional anomaly is the restricted distribution of a species to part of a rocky shore when the congruent areas of a continuous, apparently suitable habitat are not inhabited by the species (Ribbink *et al.*, 1983a).

Individual variation within Mbuna populations at any particular station is small, but with increasing distribution along a rocky shore or across a series of habitat islands, overt changes in coloration and size are discernible. In some cases, clinal gradations may be traced around a single island, along stretches of coast or even along a series of islands (Ribbink *et al.*, 1983a). Clearly, for geographic races to occur over relatively short distances, gene flow must be impeded.

On a broader scale, every Mbuna species in the lake which has a wide distribution exists as a series of disjunct populations, each of which is uniquely coloured (Ribbink *et al.*, 1983a, b). Indeed, it might be argued that several of the geographically isolated colour forms of *Genyochromis mento*, *Labeotropheus fuelleborni* and *Labeotropheus trewavasae* are so different from one another that they might be distinct species. The real issue, however, is that the existence of well-defined geographic races in widely distributed species, which may be considered to have better powers of dispersal than those with narrow distribution, indicates once again that gene flow between Mbuna populations is weak.

The distribution of cichlids in open water and over soft substrates (sand, mud and ooze) is not as clearly understood as that of the fishes that occupy rocky habitats. This is because habitat patches are relatively well defined in rocky habitats and boundaries between categories such as large, medium-sized or small rocks are readily recognized. In contrast, the boundaries between the different types of soft substrates and in pelagic zones gradually merge into one another. Moreover, Mbuna communities have been studied primarily by direct observation and by experimental manipulation by SCUBA divers, whereas most of the information regarding the distribution of non-Mbuna is gleaned from catch data of the trawl and traditional fisheries. The latter techniques cannot be as comprehensive as the former because they do not sample all the species in these habitats and they provide virtually no information about behavioural interactions and resource use. In addition, the cichlid fishes of the African Great Lakes are notoriously difficult to identify. This is particularly true of open-water species flocks ('Ncheni' = piscivores of the genus *Rhamphochromis*; 'Utaka' = planktivorous *Cyrtocara*) and demersal fishes (e.g. *Lethrinops* flocks, Eccles and Lewis, 1977, 1978, 1979; Eccles, 1986). Such taxonomic difficulties mask geographic limits as a series of sibling species with different but overlapping distributions might be considered to be a single, broadly distributed species.

Nevertheless, there is no doubt that while the ratio of narrowly distributed species to those with a lakewide distribution may be smaller than that of the lithophilous fishes, many 'Utaka', 'Ncheni', and numerous sand-dwelling species are restricted to specific geographic localities (Trewavas, 1935; Eccles and Lewis, 1977, 1978, 1979; Lowe-McConnell, 1987; Turner, 1977a, b; Eccles, 1986).

Distribution in Lake Tanganyika

The work of Poll (1952, 1956), Marlier (1959), Poll and Matthes (1962), Matthes (1962) and Brichard (1978) provides an excellent basis for an appreciation of the geographic distribution of cichlids in Lake Tanganyika. The similarities with Lake Malawi are striking; geographic differences between communities are marked, with each community comprising endemic species, local races and widespread species. The most clearly defined examples of geographic variation are those of the lithophilous communities, with *Tropheus moorii* providing the outstanding example. *T. moorii* has lakewide distribution and each population differs in coloration from the next, sometimes vividly. At each locality it exhibits identical behaviour and habitat specificity, which suggests that the colour forms represent geographic races and not sibling species. The poorest examples of geographic variation in Lake Tanganyika are provided by the fishes that inhabit deep water. Eccles (1986) suggests that fishes in this community are extremely eurytopic and are therefore too mobile to develop geographic isolates.

The painstaking work of the Japanese team at Lake Tanganyika is demonstrating that there are more lithophilous species than previously envisaged, that there is a species-rich community in the sand –rock interface zone (it may have more member species than similar zones in Lake Malawi) and that the community members are highly specific with regard to habitat requirements (Kawabata and Mihigo, 1982; Hori *et al.*, 1983; Yamaoka, 1983b; Kuwamura, 1987).

The strong evidence of geographic isolation of species, subspecies and colour forms within the two lakes, plus the indications that gene flow between allopatric populations is very low or non-existent, provides a sound basis for the conclusion that intralacustrine allopatric speciation gave rise to many of the extant endemic species in Lakes Tanganyika and Malawi. These data are particularly convincing for the lithophilous communities. However, seismic surveys of sediments have shown that during the geological history of the Rift Valley lakes, Lake Tanganyika was divided into at least three separate palaeolakes (Scholtz and Rosendahl, 1988). These subdivisions would have provided the isolation necessary for allopatric speciation and may account for at least some of the biotic diversity of the lake. The seismic surveys showed that Lake Malawi was a single basin and did not offer the same opportunities for geographic speciation. Nevertheless, McKaye and Gray (1984) suggested that isolation of populations in temporary ponds fringing Lake Malawi may have contributed to the divergence of populations.

Distribution in Lake Victoria

The pioneering studies in Lake Victoria suggested that the cichlids were distributed fairly uniformly throughout the lake. Initially there was no evidence of isolated populations which were recognizably distinct from those in different parts of the lake, and this apparent absence of geographic restriction was contrasted with the situation in Lakes Malawi and Tanganyika (Fryer and Iles, 1972). More recently, Greenwood (1974a) noted that the major habitats of Lake Victoria were not as clearly delimited as in Lakes Malawi and Tanganyika and that, unlike the two Rift Valley lakes, there was no indication of any intralacustrine differentiation of populations, but instead a genetic continuity within species existed throughout the lake. The deep-water species were considered to be a possible exception. The earlier publications of the *Haplochromis* Ecology Survey Team (HEST) also indicated that in sharp contrast to Lakes Malawi and Tanganyika, all Lake Victoria species were distributed lakewide over their specific habitats (van Oijen *et al.*, 1981). Lewis (1981) predicted, however, that a more extensive survey of Lake Victoria would reveal that many species had geographically restricted distributions. This proved to be true (Witte, 1984a). It is now apparent that there are clear geographic limits to the distribution of numerous cichlid species (Witte, pers. comm.).

Intralacustrine speciation in Lake Victoria may account for fewer species than does the fragmentation of this lake into numerous smaller lakes which were isolated from one another during interpluvial periods. This isolation of subpopulations in smaller lakes was apparently long enough on several occasions during the history of the lake for speciation to occur and ultimately to contribute to the species-rich cichlid faunas of Lake Victoria (Greenwood, 1974a).

2.3 ECOLOGY

Introduction

The history of ecological studies in these lakes has been characterized by increasingly detailed investigations elucidating more precisely the particular adaptations of individual species within various communities. The work of ecologists usually follows that of taxonomists, but in the African Great Lakes, where as many as 50% of cichlids may still be undescribed, some ecological studies are being made on undescribed taxa. Ecological understanding of these poorly known communities is in its infancy, but research effort on all three lakes is greater now than ever before.

The geographic restriction of species within a lake explains, in part, how these systems are able to accommodate so many species. It is evident, however, that geographic distribution is inextricably linked to the ecological requirements of most cichlids, and that to stenotopic species a foreign habitat is inimical. In general, therefore, stenotopes of sandy, rocky, vegetated and pelagic zones are unlikely, as adults, to frequent other habitats, except perhaps to find shelter for fry which are deposited there (see page 54). Eurytopes, on the other hand, may move among various habitats and are more likely to have broader distributions and to show less geographic variation. In this section, the foundation for these statements is examined by studies of the manner in which the habitat is partitioned (habitat partitioning is used herein as an 'effect', *sensu* Williams, 1966) by speciose communities, each of which may be represented by numerous individuals (e.g. 38 species and 7000 individuals in a 20 m × 20 m quadrat in Lake Tanganyika, Hori *et al.*, 1983; and similar findings at several study sites in Lake Malawi, Ribbink *et al.*, 1983a).

The fishes of all three lakes are distributed according to water depth and the nature of the substrate. Virtually all species for which there are data appear to have a particular vertical depth range, which may be extremely narrow (less than 5 m for some littoral, lithophilous species, Lewis, 1981; Ribbink *et al.*, 1983a; Kuwamura, 1987; Fig. 2.2) or broad (more than 100 m, Coulter, 1967a; Eccles, 1986; Fig. 2.3). Similarly, habitat preferences of species range from those which are highly specific to those which are extremely broad. Most researchers have reported on habitat and depth distribution of adult fishes. In many cases the habitat is the same for both adults and fry (e.g. some members of lithophilous and shell-bed communities), but the need for pelagic and semi-pelagic fishes and those of open sandy plains to find shelter for their young requires a migration from their normal habitat to benthic nursery areas. At various stages in their life history, therefore, many species do occupy different habitats, and this factor makes such species more eurytopic than those which remain within a single habitat throughout their ontogeny.

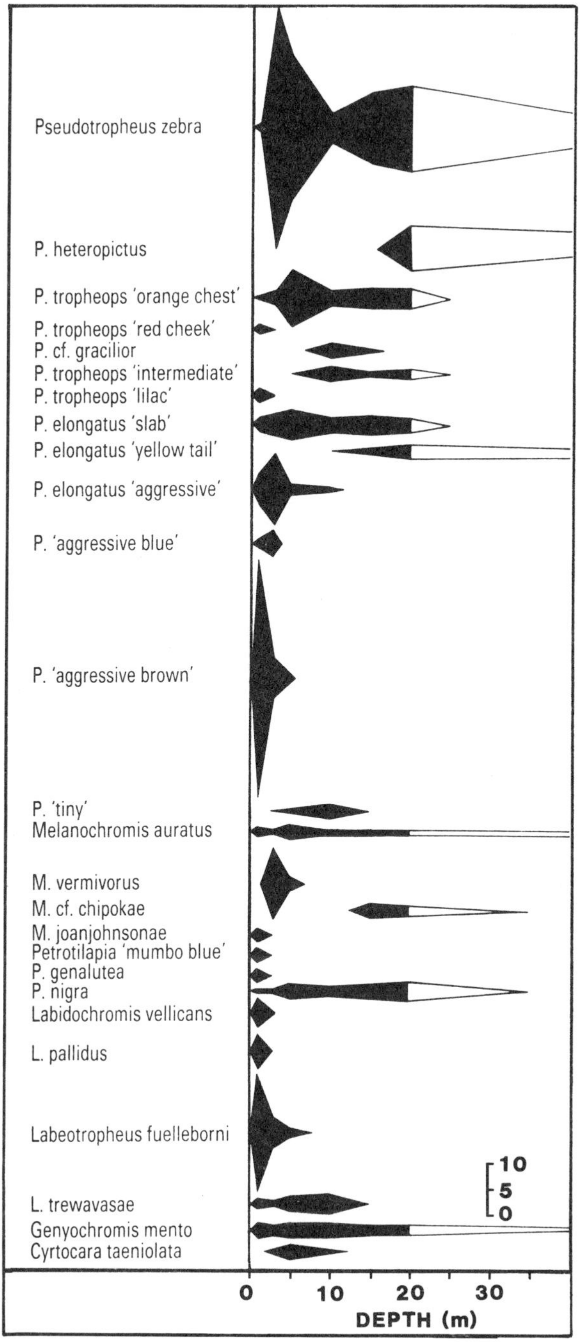

Fig. 2.2 Depth distribution of Mbuna at Thumbi West in Lake Malawi. Solid polygons represent quantitative counts of territorial individuals in a 50 m² transect, and the open sections represent estimates of numbers at point transects (after Ribbink *et al.*, 1983a).

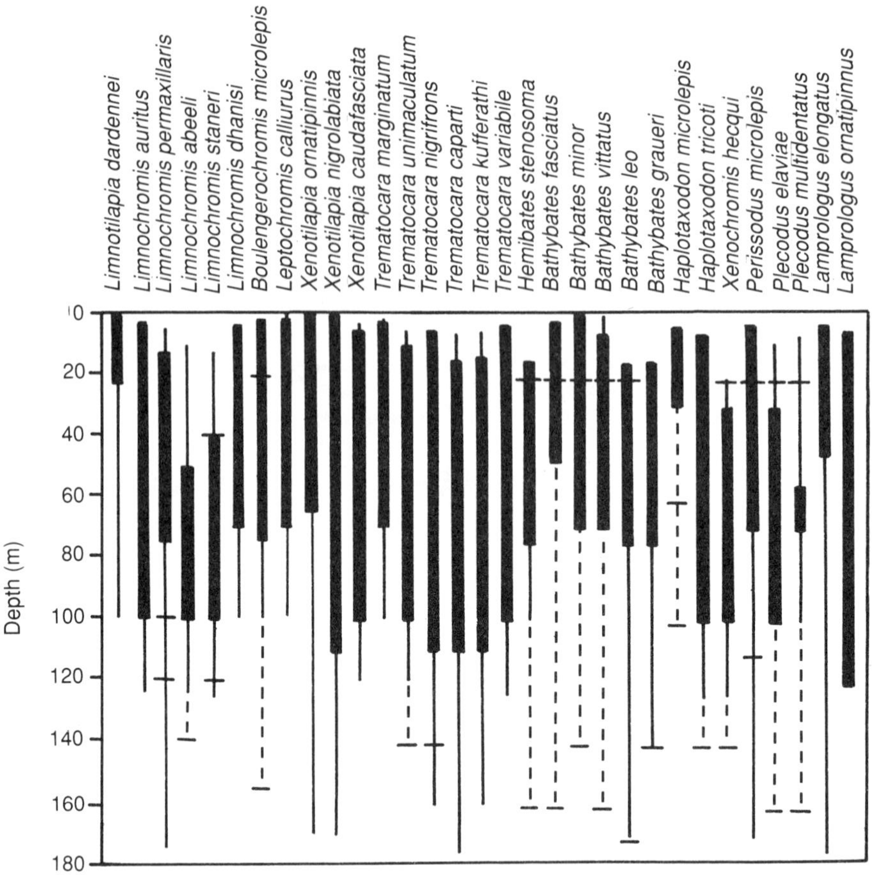

Fig. 2.3 The depth distribution of cichlid species known to occur below 80 m depth in Lake Tanganyika. Trawl samples are indicated by continuous lines. Thin lines indicate full depth range covered by trawls. Thick lines indicate minimum range between bottom of shallowest haul and top of deepest haul (data from Poll, 1956). Crossbars indicate range recorded in gill nets set in 20 m depth zones between 20 and 200 m (after Eccles, 1986). For a fuller presentation see Coulter (1991).

Lithophilous communities in Lake Malawi

Cichlids associated with rocky habitats show the most clearly defined habitat preferences and depth distributions (Poll, 1952, 1956; Fryer, 1959; Brichard, 1978; Lewis, 1981; van Oijen *et al.*, 1981; Hori *et al.*, 1983; Ribbink *et al.*, 1983a; Witte, 1984a; Kuwamura, 1987; Reinthal, 1987).

As the ecology of the fishes of rocky habitats in Lake Malawi has been more closely studied than that of other lakes, the principal descriptions given here will be based on the Malawi work (e.g. Fryer, 1959; Fryer and Iles, 1972; Holzberg, 1978; Lewis, 1981, 1982; Marsh, 1981; Marsh *et al.*, 1981; Sharp, 1981; Ribbink *et al.*, 1983a, b; Reinthal, 1987), followed by comparison with similar communities in the other lakes.

Many species have specific substrate requirements, inhabiting areas composed only of large, small or medium-sized rocks. Others occupy zones in which rock is interspersed with sand, and for certain species (e.g. *Cyathochromis obliquidens*) the presence of macrophytes among rocks is a necessary habitat constituent. If part of a rocky habitat, say of medium-sized rocks, is interrupted by belts or patches of substantially larger or smaller rocks, then the distribution of fishes reflects such habitat discontinuities (e.g. in Fig. 2.2, the narrow waist at 10 m depth for *P. zebra* indicates the presence of a suboptimal substrate). To stenotopic species both depth and substrate type are limiting factors. *Labeotropheus fuelleborni*, however, penetrates to greater depths at islets and rocky pinnacles, which are exposed to wave action and currents, than in sheltered bays. This suggests that *L. fuelleborni* avoids sediment-covered rocks. Therefore, in sheltered bays, where the sediment layer approaches the surface, this species is restricted to a narrow depth range, but at islets where sediment is deeper, so too is the depth penetration of *L. fuelleborni* (Ribbink *et al.*, 1983b). In contrast, an undescribed species, *Pseudotropheus tropheops* 'intermediate', becomes more common as the sediment layer thickens (Sharp, 1981; Reinthal, 1987). Apparently, in addition to the nature of the rocks themselves and depth *per se*, thickness of the organic sediment mantle affects the distribution of at least some Mbuna species.

The vast majority of Mbuna species occur in a depth range of 3–10 m (Ribbink *et al.*, 1983a), where the water is well oxygenated and algal productivity is high (Sharp, 1981; Reinthal, 1987). At least 26 species penetrate to 40 m depth. However, this figure is likely to be an underestimate as the time spent sampling at 40 m by SCUBA divers breathing compressed air is more limited than that in shallower water.

Explanations for the distribution pattern of members of the lithophilous community in the rocky habitats are emerging. Firstly, rock-frequenting species use the openings among rocks for cover from predators and for privacy, so that they might spawn without intrusions from sneakers (non-territorial conspecific males which attempt to steal fertilizations, see Gross, 1984) and egg-stealers. The openings among clusters of large rocks are bigger than those among small stones, so the permanent tenancy of a particular part of the habitat is dictated by the size of the rocks. All fishes that are too large would be physically excluded from these parts of the habitat. Secondly, the territorial requirements are species-specific (Holzberg, 1978; Marsh, 1981; Sharp, 1981; Ribbink *et al.*, 1983a; Marsh and Ribbink, 1985). The species-specific nature of territories was particularly well demonstrated by Sharp (1981), who carefully mapped the territories of Mbuna within a 36 m^2 quadrat which was traversed by grid lines. He then removed territorial individuals and found that each was replaced by a conspecific which, if removed, was itself replaced by a conspecific. From these experiments and from several hundred hours of behavioural observation, he defined precisely the microhabitat where each species was most likely to hold its territory.

The vigour of territorial defence ranges from that of fishes which are so aggressive that all intruders are excluded from the territories, to that of fishes which hold temporary, weakly defended territories (Sharp, 1981; Ribbink *et al.*, 1983a). The algal mat within the territories of highly aggressive species grows relatively unchecked as it is not subjected to the same heavy grazing as in surrounding areas. In the case of *Pseudotropheus elongatus* 'aggressive' (informal name of this undescribed species is from Ribbink *et al.*, 1983a) there is 2.25 times as much algae as in the surrounding areas (Sharp, 1981). These areas of thick algae, which mark the territories of highly aggressive species, are called 'algal gardens' (Sharp, 1981; Ribbink *et al.*, 1983a). However, as most species with algal gardens are medium-sized or small Mbuna, they could be robbed by larger fishes determined to enter the garden (Sharp, 1981; Marsh and Ribbink, 1985). Another way of 'stealing' from the garden is for individuals to band together to form a feeding school which then swamps the tenant (Marsh and Ribbink, 1986).

In addition to the influence of bottom topography and depth on the distribution of territories, interactions with other members of the community also play a role. Intraspecific competition for breeding sites is intense, with only a proportion of the sexually active fish winning territories. Those without territories constitute a supernumerary reserve and become territorial when an opportunity arises. Territories of conspecifics may abut with minimal or no overlap (Holzberg, 1978; Marsh, 1981; Sharp, 1981). In contrast, territories of heterospecifics do overlap considerably, and first impressions suggest that territories are formed with reference to the presence and activities of conspecifics, but without regard for other species. The following example indicates that this is not so. Marsh (1981) found that the largest Mbuna territories in the Monkey Bay rocky habitats were those of *Petrotilapia tridentiger*, being on average *c.* 22 m^2 in area. Within these territories were those of other, smaller fishes, including sibling species of *Petrotilapia* and at least two species with algal gardens. *P. tridentiger* dominated the social hierarchy within their territories as all other fishes avoided them. Although interspecific overlap of territories was high, overlap of feeding sites amongst territorial males was very low. Of the total feeding space utilized by congeneric males, only 2.5% was shared and simultaneous use of feeding sites was never recorded. This was attributed to the social dominance of *P. tridentiger* over its congeners. *P. tridentiger* fed exclusively from the algal gardens of the biggest *Pseudotropheus elongatus* 'aggressive' individuals. These gardens were the largest and had the densest algal growth. Clearly, there were benefits to *P. tridentiger* in having ready access to gardens that were maintained by other species. Equally, there were costs to those fishes which defended gardens. One might speculate that there were benefits also, perhaps in the form of protection afforded by the *P. tridentiger* which kept out other solitary intruders, though they appeared to have little effect on feeding schools.

Space for territories is limited and most of the area which is unsuitable for territories is used by non-territorial fishes, principally as feeding sites. The proportion of the substrate used for territories has not been calculated for any locality, but there is no doubt that it is variable. Areas dominated by large rocky slabs and boulders are unsuitable for stable territories for most species as they provide no shelter. These open areas are occupied almost exclusively by numerous feeding cichlids during the day, but are deserted at night. In contrast, regions of natural rubble comprising rocks of 60–90 cm diameter are dominated by active territorial fish during the day. At night these fishes remain in their territories, resting on or among the rocks.

The third aspect of resource partitioning, and that which has attracted most attention, is that associated with the trophic adaptations of cichlids. From the time cichlids were first described it became apparent that the most overt differences between species were those associated with feeding: dentition, jaw structure and size, head shape and lips. Ecological inferences were made on the basis of these anatomical differences, emphasizing the evolution of anatomical trophic specializations that may have led to resource partitioning (Bertram *et al.*, 1942; Trewavas, 1949). The classical work of Fryer (1959) awakened a real interest in the ecology of these fishes. He related dentition and jaw structure to feeding behaviour and demonstrated habitat stenotopy and specialized feeding mechanisms. However, stomach content analyses of the Mbuna he studied were enigmatic. Despite the anatomical and behavioural specializations for feeding, many species consumed identical filamentous algae (Aufwuchs) and its epiphytes (loose Aufwuchs). This paradox of dietary overlap was partially explained by the observation that several species fed upon the same components of the algal mat, but by virtue of their trophic specializations they harvested algae from different parts of the substrate. For example, fishes with broad mouths and coarse teeth were unable to feed from narrow cracks in the rock surface which were used by smaller species with thin, pointed mouths and long teeth (Fryer, 1959). However, numerous other species, e.g. *Petrotilapia tridentiger* and *Pseudotropheus zebra*, fed in the same way, from essentially the same surfaces, and consumed identical components of the algal film. Fryer (1959) considered the overlap of resource use by these and a variety of other species to be a direct contradiction of Gause's competitive exclusion principle. He concluded that this was possible as the epilithic algae were produced more rapidly under tropical conditions than they were harvested, and were therefore always present in superabundance. Fryer and Iles (1972) argued that the population densities of fishes in these habitats did not grow to a level at which they would overexploit the algae, because they were kept in check by predators and by a self-regulatory mechanism. Such self-regulation invoked theories of group selection and led to the suggestion that community members lived in a peaceful condominium (*sensu* Wynne-Edwards, 1962) rather than in competition with one another (Fryer and Iles, 1972).

Underwater observation of fishes reveals a much finer partitioning of habitat than was envisaged by Fryer (1959) and Fryer and Iles (1972). Although most Mbuna do rely on epilithic algae as their primary food source and many feed on essentially the same components, they collect their food from different parts of the habitat. Different species favour specific depths or particular rock surfaces (e.g. upper horizontal, vertical or under surfaces); some feed predominantly in holes or crevices, some in sediment-covered areas and others in wave-washed zones; algal gardens are cultivated by certain species and yet others are eurytopic (Marsh, 1981; Sharp, 1981; Ribbink *et al.*, 1983a, b; Marsh and Ribbink, 1985). These observations suggest that a partitioning of the habitat occurs and that Gause's exclusion principle is not contradicted by the Mbuna. Furthermore, behavioural field observations, experiments and analyses of condition factors suggest that epilithic food resources are scarce for much of the year and may never be superabundant (Sharp, 1981; Ribbink *et al.*, 1983a).

A broad categorization of food types tends to mask the trophic partitioning that is dependent upon the differential use of subclasses. Fryer (1959) recognized only two algal forms (filamentous and unicellular) contributing to an apparently homogeneous epilithic algal mat. He found an enormous amount of dietary overlap among the Mbuna. Sharp (1981) and Ribbink *et al.* (1983a, b) recognized several more algal categories and noted that the filamentous algae were distributed patchily, according to rock surface inclination and depth. Although their resolution of dietary resource partitioning was finer than that of Fryer (1959) and showed species-specific trends, they still found a great deal of interspecific overlap in the constituents of stomach contents. A real advance in an understanding of Mbuna herbivory was made by Reinthal (1987), who recognized 41 different categories of algae and painstakingly mapped the distribution of each attached form in the rocky habitats at Thumbi West (Fig. 2.1) and the nerest rocky zone of the mainland, Otter Point in the south-western part of Lake Malawi. He confirmed that the algae are distributed in distinct patterns and that the spatial variation corresponded to depth and substrate inclination. As in previous studies, there was a great deal of dietary overlap in gut contents, but using discriminant analysis he was able to classify the fishes from their gut contents and to demonstrate that species could be identified from their diets. Reinthal showed experimentally that herbivorous Mbuna are able to distinguish between particular epilithic algal components. This explains the tendency for Mbuna to feed from particular parts of the habitat.

In support of Sharp (1981) and Ribbink *et al.* (1983a), Reinthal (1987) concluded that algae are not produced in superabundance and that competition for resources (see also Marsh and Ribbink, 1985) characterized Mbuna communities.

All studies of herbivorous Mbuna revealed a high interspecific dietary overlap that needs explanation in view of the anatomical and behavioural specializations of these fishes to feed upon particular components of the epilithic algae. The various components of the algal mat are mixed in such a manner that it is inevitable that while feeding on one form, others would be ingested, despite the trophic specializations of the species. Furthermore, Liem's (1978, 1979, 1980a) kinematic and electromyographic profiles demonstrated that in specializing, the Mbuna had not lost earlier (pleisiomorphic) feeding repertoires, but had added to them. Highly specialized species, therefore, have retained the ability to feed upon resources other than those for which they are best adapted. Mbuna are opportunists that will consume a wide spectrum of foods (Fryer, 1959; Sharp, 1981; Ribbink *et al.*, 1983a), and several species may switch from one food substance to another (McKaye and Marsh, 1983). Reinthal (1987) cautions, however, that reports of switching which are based on behavioural observations need verification. Fishes which feed upon phytoplankton that has settled on the substrate or is in the water column are not switching from one type of food to another, they merely collect it from different places.

Lithophilous communities in Lake Tanganyika

This rock-dwelling cichlids of Lake Tanganyika parallel those of Lake Malawi in many respects: they are speciose and numerous; several genera have feeding mechanisms which are virtually identical to those of their Malawi counterparts (e.g. *Petrochromis* and *Petrotilapia*; *Tropheus* and *Pseudotropheus tropheops* species-complex; *Labidochromis* and *Tanganicodus*); the communities are distributed according to depth and bottom physiography in much the same way and they show similar general trends in resource partitioning (Fryer and Iles, 1972; Kawanabe, 1981; Kawabata and Mihigo, 1982; Yamaoka, 1982; Hori *et al.*, 1983; Takamura, 1983; Kuwamura, 1987). However, the taxa are completely different, as are many other biotic features of the lakes. Although the data are not available for a direct comparison, it appears that Lake Tanganyika has a much more diverse algal and invertebrate biota than Lake Malawi. Coulter (1991) records very high invertebrate species numbers for a lacustrine fauna, many of which (e.g. 74 of the 85 ostracods and 46 of the 75 molluscs) are endemic. The cichlids appear to have developed feeding specializations to utilize these resources, particularly the invertebrates. Accordingly, the number of predatory species in the lithophilic littoral communities is higher in Lake Tanganyika than in Lake Malawi. Prevalent amongst the predatory taxa are the lamprologines, which may constitute more than 50% of the species in a local littoral fish community, in which closely related species coexist and are specialized to feed upon different components: plankton, zoobenthos, molluscs, crustaceans and insects. Those species which feed upon identical components of the resource employ a variety of species-specific techniques to do so. These include the collection of similar food items from different sites by hunting either nocturnally or diurnally, by turning over stones and by adopting pursuit or ambush tactics (Hori, 1983).

A consequence of the greater representation of molluscs in Lake Tanganyika is the presence of extensive beds of gastropod shells from 10 to 35 m depth on gently sloping shores. The literature shows that there is a community of shell-dwellers comprising more than ten cichlid species, principally lamprologines which utilize these shells as refuges in which they hold territories, breed and raise their offspring.

The surge zone of Lake Malawi does not appear to have any species that are clearly adapted to live there, but in Lake Tanganyika there are goby-like cichlids of the genera *Eretmodus*, *Spathodus* and *Tanganicodus*. These fishes are negatively buoyant, can use their fins as anchors and have subterminal mouths, all of which appear to be adaptations to life in the wave-washed area.

The most striking difference between the cichlid communities of the two lakes is that those of Lake Malawi are all maternal mouthbrooders whereas those of Lake Tanganyika include maternal mouthbrooders, biparental mouthbrooders and numerous substrate spawners. Behavioural differences associated with these alternative methods of pair formation, spawning and parental care contribute an additional dimension to the utilization of space and other resources (Brichard, 1978; Kawanabe, 1983; Nagoshi, 1983; Yanagisawa and Nshombo, 1983; Taborsky, 1984). Space utilization by substratum spawners appears more intensive than that of mouthbrooders. Whereas a single, territorial mouthbrooder guards a large area, frequently 1 m or more in diameter, the territories of substrate spawners are closely packed and may have diameters of 20 cm or less. The latter territories are occupied by both parents or even a harem, the eggs are laid there and the offspring are reared there. In addition, members of successive broods may be found together within a territory, since 'helpers' from an older brood often remain behind to assist with the new brood (Taborsky, 1984). In maternal mouthbrooders the female carries the eggs away from the territory. The contrast, therefore, is between the intensive use of a small area by a large family (substrate spawners), and the use of a large area by a single individual (maternal mouthbrooder). Although the available data preclude generalizations, it appears that those fishes with large territories such as the maternal mouthbrooders are primarily herbivores, whereas those which depend upon small territories for an extended period, during which they raise several families, are primarily carnivores. If this is correct, then the greater abundance of invertebrate prey in Lake Tanganyika may have been instrumental in the development of communities of substratum spawners.

Lithophilous communities in Lake Victoria

The Mbuna-like lithophilous fish community of Lake Victoria was discovered fairly recently (van Oijen *et al.*, 1981). Only 16 species are known and none has been observed underwater, owing to the poor visibility in the lake. Nevertheless, it has been established that they are lithophilic, trophic specialists (eight trophic groups are recognized) and that each has a specific depth distribution. Morphologically, these fishes parallel the Mbuna very closely (van Oijen *et al.*, 1981).

Sand and mud communities

The demersal cichlids which live over soft substrata are speciose, numerous and of considerable commercial value. Despite their importance to the countries that surround the lakes, these fishes are not well known taxonomically, ecologically or behaviourally. There is no doubt that many species are being exploited which are undescribed and for which there are virtually no ecological data nor any understanding of the role each performs in the complex communities to which they belong. In the last two decades significant scientific advances have been made, but an enormous amount of urgent research is required because of the sensitivity of these fishes to exploitation (Witte, 1981; Witte and Goudswaard, 1984; Coulter *et al.*, 1986).

The general ecological trends demonstrated by the rock-dwelling fishes are paralleled by fishes living in other habitats, but they are not as clearly demarcated. Nevertheless, studies of the fishes over soft substrates show that species composition is strongly influenced by water depth and the nature of the substrate (e.g. Lowe, 1952; Poll, 1952, 1956; Fryer, 1956, 1959; Poll and Matthes, 1962; Coulter, 1967a; Fryer and Iles, 1972; Greenwood, 1974a; van Oijen *et al.*, 1981; Witte, 1981, 1984a; Witte and Goudswaard, 1984; Hoogerhoud *et al.*, 1983; Eccles and Lewis, 1977, 1978, 1979; Turner, 1977a, b; Lewis, 1981; Lowe-McConnell, 1987). Trophic specialization is indicated by anatomical studies, from analyses of gut contents and from limited SCUBA observations of the behaviour of shallow-water species. Furthermore, behavioural inferences drawn from 24 h experimental fishing programmes indicate diel periodicity in feeding and migration patterns of certain species (see also McKaye, 1983 for direct observation of behaviour).

It is becoming increasingly apparent that the remarkable adaptive radiation demonstrated by these fishes is not restricted to their anatomy, but includes numerous behavioural and ecological traits that ultimately lead to a fine partitioning of resources. However, trends in each lake are not identical. The demersal deep-water faunas over soft substrates in Lakes Victoria and Malawi are speciose and stenotopic. In contrast, that of Lake Tanganyika is much less speciose, but its members are more highly differentiated at the generic level and are eurytopic (Eccles, 1986), particularly with regard to their ability to traverse a wide range of depths (Fig. 2.3). Ecological segregation of deep-water demersal cichlids in Lake Tanganyika is more likely to be based on differences in diet than on habitat preferences (Coulter, 1967a; Eccles, 1986). The differences between the demersal fauna of Lake Tanganyika and those of Lakes Malawi and Victoria were attributed to different physical regimes of the lakes

(Eccles, 1986). Lakes Tanganyika and Malawi are both permanently anoxic below 250 m. Lake Victoria, however, is a relatively shallow basin in which the water mass usually mixes completely, though anoxic conditions do develop for short periods in the deepest parts. In the two Rift Valley lakes there are wind-induced upwellings of anoxic water in the cool season. However, basin morphology of the lakes differs so that anoxic water in Lake Tanganyika penetrates into much shallower water than in Lake Malawi. Eccles (1986) suggests that the evolution of stenotopic species which are adapted to deep-water habitats (which in Lake Tanganyika are most extensive at the south end in the region where upwelling occurs) has been inhibited by the need to migrate to avoid anoxic advections. The relative stability of the other two lakes permitted the evolution of stenotopic species in deep, oxygenated bottom habitats. Eccles (1986) considers that the demersal environments of Lakes Victoria and Malawi are essentially similar, but the greater turbidity of Lake Victoria makes the gradients of light intensity and oxygen concentration more marked so that the depth effect is compressed into a narrower range.

Open-water communities

The fishes discussed in this section belong to species flocks or to closely related polyphyletic groups, whose members are collectively, but not exclusively, more closely associated with open waters as adults than with any demersal habitat. Although the life history of most species is unknown, it is probable that they carry their offspring to the shelter afforded by the substrata for release, and therefore each would begin life as a free-swimming demersal fish. Some of these species also come inshore seasonally to hold territories and spawn on a substrate. The fishes of these groups are mainly planktivores or piscivores.

Planktivorous cichlids. Virtually all the fishes of Lake Malawi for which gut contents are known eat zooplankton as juveniles, and as adults they may take it facultatively as an important secondary food resource. However, a species flock referred to by their vernacular name 'Utaka' comprises specialized zooplanktivores (Iles, 1960; Fryer and Iles, 1972). These are small, silvery fish (usually less than 150 mm SL as adults), often with several prominent black patches along the mid-lateral line. They are characterized by a small, narrow mouth and protractile upper jaw which enables the mouth to be converted into a sucking tube during the feeding process. When not engaged in reproductive activities these fishes form large multispecific schools and feed principally on zooplankton and secondarily on phytoplankton. Schools of Utaka commonly congregate near underwater pinnacles of rock or close to islets, where they feed on the plankton carried by currents which are deflected by these steep rocky obstructions. Fryer and Iles (1972) reported up to 11 species of Utaka feeding in such shoals and recorded that the zooplankton diet of all species was essentially identical. They concluded that this observation "emphatically refutes'' Gause's exclusion hypothesis and suggested that, like the Mbuna, the Utaka probably regulate their own population densities.

Utaka species may school together for feeding as adults, but each has its own specific breeding requirements (e.g. *Cyrtocara eucinostoma* in huge leks on shallow sandy substrates, McKaye, 1983; *C. chrysonota* in the water column over sandy and vegetated substrata in shallow water, Eccles and Lewis, 1981; *C. borleyi* on rocky shores, Ribbink *et al.*, 1983a). The ontogeny of each species, from the time fry are released by the parent to the adoption of adult schooling and feeding behaviour, takes place in a different physical and biotic environment. Thus for much of their lives, fish are not part of a mixed-species feeding school. As Gause's hypothesis does not restrict the timing of competition to the adult stage, but embodies the entire life, it is clear from the segregation of these species during their growth periods that the principle is not refuted. It is possible that detailed examinations of the feeding behaviour may show a spatial and temporal segregation of species.

The eight species of zooplanktivore of Lake Victoria feed in the water column, but occur over different substrates (Witte, 1984a). However, four species in the Mwanza Gulf area appear to coexist within a feeding population, rather like that of the Utaka multispecies feeding schools of Lake Malawi. A detailed study of these fishes revealed that, despite overlap, their vertical distribution in the water column during the day differs, and the extent to which vertical migration occurs also varies, so that their relative positions in the water column change at night. Analysis of gut contents of fishes caught during the day and night showed overlap, but there were also clear species-specific differences (Witte, 1984a; Goldschmidt, 1986). Data on reproduction of the zooplanktivores of Lake Victoria indicate that both spatial and temporal segregation of species occur during the breeding season.

All cichlids of Lake Tanganyika appear to feed upon zooplankton for at least the juvenile part of their life cycle, and several species feed on plankton as adults, including the algae-feeders (Mbomba in Kawanabe, 1983). The pelagic fish community is dominated by large populations of endemic clupeids (*Stolothrissa tanganicae* and *Limnothrissa miodon*) and of four centropomid species. The two clupeids are shoaling zooplanktivores which are preyed upon by the centropomids and a number of cichlid species (review, Coulter, 1991). Although there are no mixed-species schools of planktivorous cichlid fishes similar to those of the other two lakes, Lake Tanganyika does have a bathypelagic community of zooplanktivorous cichlids, including the *Trematocara* species flock, *Cyprichromis* spp. and others. These fishes are specialized open-, deep-water planktivores (Poll, 1956; Coulter, 1991).

Piscivorous. The number of piscivorous cichlid species in each lake is remarkably high, particularly in Lake Victoria in which piscivores constitute the most speciose trophic group (van Oijen, 1982) and include *c.* 40% of all species (Greenwood, 1959c, 1965b; van Oijen, 1982). In commenting on the high proportion of piscivores in the African Great Lakes, Fryer and Iles (1972) noted that it was higher than the proportion of raptorial species that one would expect to find in a community of birds or mammals. For such a diversity of piscivorous species to be supported, it is expected that the production of prey species would be substantially greater than that of their predators and that a high degree of resource partitioning would be likely, although overlap would be inevitable. As expected, piscivores do have low population densities, contributing only 1% by number to the trawl catches in the Mwanza Gulf of Lake Victoria (van Oijen *et al.*, 1981). Similarly, a comparable representation of piscivores was found in the shallow (15–40 m) trawls of the south-eastern arm of Lake Malawi, but there was a slightly better representation in the 55 m trawl (nearly 10%), where a substantial contribution was made by the piscivorous *Diplotaxodon* and *Rhamphochromis* species (Lewis, 1981). In Lake Tanganyika there is a similar high percentage of piscivorous species, but in contrast to the other two lakes, total predatory fish biomass (cichlid plus non-cichlid piscivores) equals or exceeds prey fish biomass in the pelagic and in the benthic communities (Coulter, 1991).

Piscivores occupy every habitat within the lakes, but they do not appear to range as widely over various habitats as the patrolling behaviour of pursuit predators might suggest. Indeed, the broad eurytopy previously attributed to many piscivores (e.g. Greenwood, 1967, 1974a; Fryer and Iles, 1972) is not generally found in Lake Victoria (van Oijen, 1982). These fishes are habitat restricted, have species-specific depth distributions (Fig. 2.4), harvest particular food components and use different hunting and food-handling techniques (van Oijen, 1982).

Many piscivores are specialized to feed upon particular prey, so the stenotopy of prey species is imposed on what might otherwise be a more widely ranging predator. The most frequently cited specializations are the morphological and behavioural adaptations of the littoral-zone ambush

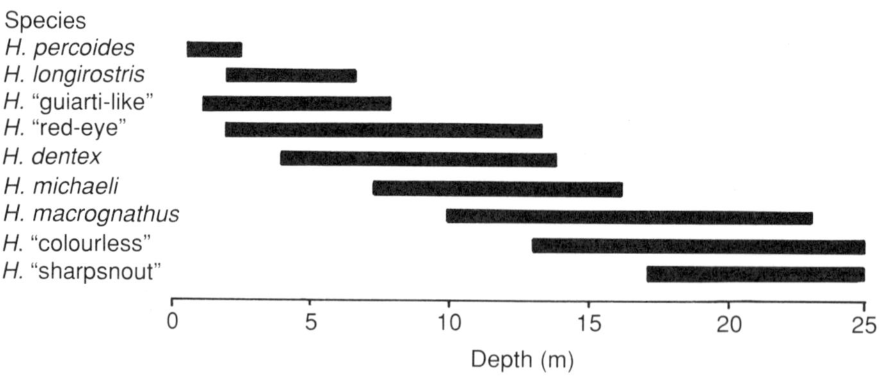

Fig. 2.4. Depth distribution of several adult piscivorous species of the genus *Haplochromis* in the Mwanza Gulf of Lake Victoria (after van Oijen, 1982).

strategists of Lake Malawi (*Cyrtocara linni*, *C. livingstonii*, *C. polystigma* and *C. compressiceps* (Fryer, 1956; Fryer and Iles, 1972; McKaye, 1981a; Ribbink *et al.*, 1983a). All of these fishes feed upon cichlid fry which occupy littoral rocky and vegetated habitats. Others live over sand substrates in deeper water, and the *Rhamphochromis* and *Diplotaxodon* species flocks occur in more open water, *Rhamphochromis* living in shallower water than *Diplotaxodon* (Fryer and Iles, 1972).

2.4 SYNTHESIS

The above description of the distribution of species between and within the principal habitats of each lake, as well as the endeavours to demonstrate how members of communities might coexist, may have given the impression of a greater segregation of intralacustrine communities then actually exists. Within each community there are narrowly stenotopic species which are strictly confined to their specific habitat, but there are also eurytopic members which routinely cross the boundaries between habitats. Thus, certain lithophilous species may make excursions onto sand or may occur among macrophytes, over mud or in the water column. Typical sand-dwellers may appear in other habitats also, as do mud-dwellers, fishes of macrophyte beds, shell-dwellers and open-water species. Thus, in addition to permanent residents, there is a daily movement of visitors across the boundaries between communities. These transition zones between different habitats (the intermediate zones of Fryer, 1959) have fishes which are specifically adapted to live there, having ecological affinities with two or more habitats. Essentially, therefore, every community has its own distinguishing characteristics and is inextricably interlinked with those of other habitats, so that collectively they form a single community.

Each subcommunity is changing constantly through the flux of casual visitors and because of regular daily and seasonal changes in activity and membership. The species which dominate diurnally often differ from those which are active at night. Other cyclical changes also occur, e.g. reproductive activities are prevalent in the morning and feeding takes precedence in the afternoon and early evening (Sharp, 1981; McKaye, 1983). In addition, vertical migrations of planktivores (Witte, 1984a; Goldschmidt, 1986) and piscivores (Coulter, in press) also influence the composition of fish assemblages. All species, even those which breed throughout the year, have distinct peaks of breeding activity (Witte, 1981; Marsh *et al.*, 1986). However, this seasonality is not the same for all species, so the community within each habitat is changing constantly, often losing members from open-sand substrata and open-water habitats to zones which afford more protection for the offspring. Similarly, timing of parental care and the space requirements of fishes which form territories in which to guard their offspring vary seasonally.

Thus, superimposed upon the diurnal and seasonal activities of permanent residents of a subcommunity are those of visitors from other habitats in the lakes, giving rise to dynamic cichlid communities.

The geographic distribution of cichlids within the African Great Lakes is affected by philopatry imposed by the narrow stenotopy of numerous species. Furthermore, parental care in habitat-restricted cichlids tends to keep conspecifics together. Indeed, in most instances such dispersal would be maladaptive. Thus, the behaviour, ecology and relatively small numbers of individuals within each population predispose the cichlids to rapid speciation in the event of population subdivision consequent upon vicariant events. Once they have become subdivided, gene flow between geographically isolated populations is likely to be extremely low or absent.

A tenet of current concepts of intralacustrine speciation is the philopatry of species (e.g. Poll, 1952; Fryer, 1956, 1959; Fryer and Iles, 1972; Ribbink *et al.*, 1983a; Hoogerhoud *et al.*, 1983; Ribbink, 1986). This concept leads to the expectation that species flocks whose members exhibit the most stenotopy would be the most speciose, and in general this appears to be true (e.g. about 200 species of Mbuna). Eccles (1986) has argued that the relative stability of the demersal zones of Lake Malawi and Lake Victoria has enabled fishes inhabiting those environments to become stenotopic and speciose. In contrast, the upwelling of anoxic water into similar habitats of Lake Tanganyika has prevented all but the deep-dwelling eurytopic species from occupying the deep demersal zone. This community is not speciose (Eccles, 1986), though it is highly differentiated. These cichlids are remarkable in their ability to penetrate such depths and to migrate rapidly from them during anoxic conditions. Clearly, their swimbladders can cope with large pressure changes, their metabolism adjusts to extremely low oxygen concentrations (Coulter, 1967b) and they can adapt to very low light intensities.

The very characteristics which make the African cichlids interesting and which accounted for their explosive speciation are also responsible for their extreme sensitivity to Man's activities. The narrow endemicity of insular, small populations of fishes which invest heavily in producing a few offspring makes them particularly vulnerable to exploitation and to predation (Coulter *et al.*, 1986; Ribbink, 1987). The communities are in a state of rapid transition. For example, mechanized fishing has had dramatic effects on the demersal fish community of the south-eastern arm of Lake Malawi, from which 20% of the 140 trawled species disappeared from the catches within 6 y (Turner, 1977a, b). The piscivores were the first to disappear and the community responded with an increase in the number of prey species (Turner, 1977a, b). A similar but more serious situation exists in Lake Victoria, where the combined effects of piscivory by the introduced Nile perch and trawling have had irreparable consequences for community structure and ecology (see Barel *et al.*, Chapter 13, this volume). Even the ornamental fish trade can alter communities, divers select the brightest-coloured, large males and the most attractive females. Such selection may have intraspecific evolutionary consequences for small insular populations. Similarly, the persistent removal of members of only some species from a community while leaving others behind may have long-term ecological implications. The ornamental fish trade is also responsible for the translocation of species from their native areas of a lake to another where they do not occur naturally and where they have become established (Ribbink *et al.*, 1983a; Trendall, 1988). Some of these introduced populations have grown considerably (Trendall, 1988) and must be occupying space and using food resources that would have been available for the indigenous species.

There is every reason to encourage studies that will enhance understanding of the ecology, distribution and behaviour of lacustrine cichlids and lead to an appropriate focusing of the conservation initiatives of these fragile communities.

ACKNOWLEDGEMENTS

I am very grateful to Mike Bruton, Linda Coetzee, George Coulter, O'Thomas Hecht, Vanessa Jones, Miles Keenleyside, Anthea Ribbink, Huibre Tomlinson and Alan Whitfield for their generous help in preparing the script for this chapter. I thank Peter Reinthal for allowing me to refer to information in his PhD thesis.

Chapter three

Ecology of cichlids in South American and African waters, excluding the African Great Lakes

R. H. Lowe-McConnell

3.1 INTRODUCTION

Present-day South America, unlike Africa, lacks large lakes. It is essentially a land of rivers, but because much of it is low-lying the rivers flood extensively into the forest and over the vast savannahs seasonally, in many places filling lateral lakes. So the distinctions between riverine and lacustrine conditions are much less clear in South America than in Africa. The aims of this chapter are to compare the ecological characteristics of cichlid faunas in the rivers of South America and Africa, to examine how cichlids live in rivers as opposed to lakes, and to try to determine what selection pressures have affected their lives in rivers.

Riverine cichlids, a very diverse assemblage in overall morphology, behaviour and ecology, are – apart from a few very specialized rheophilic species – characteristically found in lentic habitats such as on floodplains and in slow-flowing forest brooks. They occur in a great range of habitats, from warm pools open to the sun to shaded forested waters, from clear nutrient-poor rivers to brackish estuaries. Many are very hardy, able to withstand considerable ranges of temperature, poorly oxygenated waters and varied chemical conditions. As cichlids form only a small part of the riverine faunas, their ecology has been little studied in rivers.

The lives of riverine fishes are governed by changes in water level, which affect the physicochemical and biotic conditions. Most riverine fishes are very mobile; the faunas are dominated by otophysans of many kinds (characoids and nematognaths, and in Africa cyprinids). The larger species are migratory, moving long distances up or down river seasonally. Many have well-defined breeding seasons associated with these movements. Cichlid movements tend to be more local in scope – on and off the floodplain, in and out of lateral lakes – and cichlid breeding is less seasonal, but the fish communities in which they live are continually changing as other members of the fauna come and go.

At night, when the nocturnal catfishes and most other predators are active, the sight-orientated, diurnally active cichlids hide away motionless alongside banks, amongst forest litter, or in shallows. South American faunas contain numerous piscivorous fishes, such as the characoid *Hoplias malabaricus*, serrasalmines and small characins of many kinds which prey on larval and juvenile cichlids. Numerous aquatic birds, reptiles and mammals, including otters and men, are amongst other predators.

South American river systems are vast; many are still unexplored. Cichlid information is scattered, often in aquarium literature, and in reports from collections made over a very wide area. Taxonomic revisions have led, and are leading, to changes in nomenclature and re-identification of some of the species on which field observations had been made, so early reports have to be used with caution. The emphasis here is therefore on the places where most information is available on habitats and ecology for cichlids that have been reliably identified. Many of these studies have been made in the Central Amazon basin from the Instituto Nacional de Pesquisas da Amazonica (INPA) research station in Manaus, where the Rio Negro joins the Solimões to form the main Amazon. Other data come from expeditions to tributaries of the Upper Amazon. Personal observations were made in Guyana, where Eigenmann (1912) laid a good foundation for ecological work by providing identification keys to the fishes, and in Mato Grosso streams draining to the Amazon, where, as in much of South America, ecological studies were hampered by the need for much taxonomic work on the material collected (Lowe-McConnell, 1969b, 1987). Field studies on the Paraná–Paraguay and many other South American rivers have concentrated on the dominant otophysan fishes, yielding little ecological information on the cichlids (but see Kullander, 1982), and for reasons of space these reports have been omitted.

3.2 NEOTROPICAL CICHLID FAUNAS

South America probably has *c.* 300 cichlid species (equivalent to 6–10% of the freshwater fish fauna), about half of them found in the Amazon basin. Basic divisions within the family appear to be long established, and perhaps evolved in the more lacustrine conditions which occupied much of the Amazon basin before the present river broke through to the Atlantic. From his revision of South American cichlasomines, Kullander (1983a) concluded that various lines date back to the Tertiary (some possibly even to the Eocene), and that speciation has come about by changes in allopatric populations in different river systems, without much trophic specialization. The Quaternary climatic oscillations probably had little effect on their evolution as riverine cichlids are able to live in many types of habitat and to withstand changes in temperature and other conditions.

The main cichlid groups are well represented in four of the eight faunal regions (Fig. 3.1), namely in the Amazonian/Guianean region which has two centres of evolution, in the Orinoco region to the north and in the Paranean

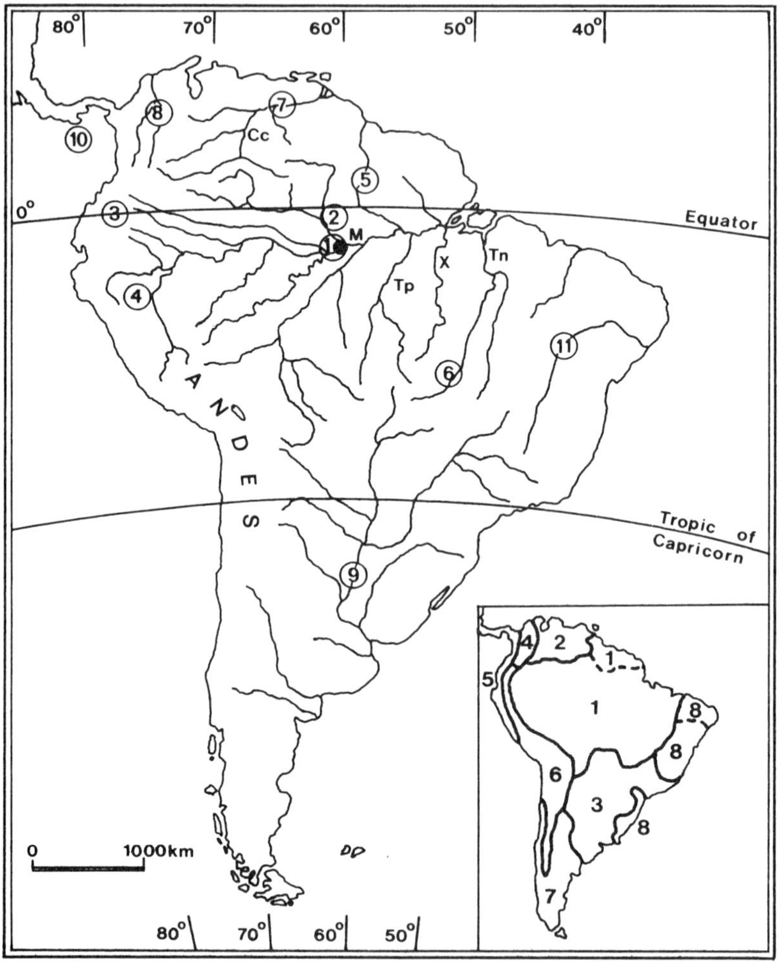

Fig. 3.1 South American sites: (1) Central Amazon; (2) Rio Negro; (3) Ecuador; (4) Peruvian Amazon; (5) Rupununi; (6) Mato Grosso; (7) Orinoco; (8) Magdalena; (9) Parana; (10) Panama; (11) São Francisco. M, Manaus; Cc, Casiquaire Canal; Tp, Tapajos; X, Xingu; Tn, Tocantins. Inset map of ichthyofaunal regions: (1) Amazonian/Guianean; (2) Orinocan; (3) Paranean; (4) Magdalenean; (5) Transandean; (6) Andean; (7) Patagonian; (8) East Brazilian (modified after Lowe-McConnell, 1987).

region to the south, also in East Brazilian rivers. The Magdalenean region in the north-west has four or five cichlid species, the Transandean Pacific coast three or four species (Kullander, 1986); a few *Crenicichla* and *Bujurquina* venture up Andean slopes; *Crenicichla lacustris* occurs in Patagonia far south of the tropics (Kullander, 1981). The wide distributions over the continent of genera such as *Crenicichla* and *Cichlasoma* reflect the former connections of river systems, followed by allopatric speciation in separated river basins.

South America has many cichlid groups with marked specializations for particular trophic strategies or for living under special environmental conditions. These include piscivores (such as *Cichla*), bottom feeders (large geophagines), rheophilic forms (*Retroculus*, *Teleocichla*), leaf litter inhabitants (*Apistogramma* and other dwarf cichlids well known to aquarists), deep-bodied forms living amongst aquatic plants (*Pterophyllum*, *Uaru*, *Astronotus*, *Symphysodon*), and numerous omnivores among the cichlasomines. *Chaetobranchus*, living in floodplain lakes, is thought from its long gill rakers to be a zooplanktivore, but hard evidence is lacking.

In his recent review of Peruvian cichlid genera, Kullander (1986) recognized three groups:

1. Hypothesized, phylogenetically primitive genera the relationships of which are not yet clarified: mainly predatory fishes of more than 70 species, including *Crenicichla* (with over 50 species), *Cichla*, *Teleocichla*, *Acaronia*, *Astronotus*, and derived *Chaetobranchus*, *Chaetobranchopsis*;
2. ***Geophagines***, which possess soft pharyngeal pads hanging from above the gill rakers, thought to be associated with sifting organisms from mud as most are bottom feeders, 12 genera with over 100 species: *Geophagus*, *Gymnogeophagus*, *Satanoperca*, *Retroculus*, *Acarichthys*, *Biotodoma*, 'dwarf cichlid' *Apistogramma* (more than 60 species), *Apistogrammoides*, *Papiliochromis*, *Biotoecus*, *Crenicara*, *Taenicara*.
3. ***Cichlasomines***, more than 17 genera with over 75 species of rather generalized omnivorous fishes: *Cichlasoma*, *Aequidens*, *Bujurquina*, *Caquetaia*, *Heros*, *Mesonauta*, with the deeper-bodied *Symphysodon*, *Uaru*, *Pterophyllum*, dwarf *Nannacara*, also *Hoplarchus*, *Hypselecara*, *Laetacara* and *Tahuantinsuyoa*. Many of these genera were formerly included in *Cichlasoma*, a huge group which also included *c.* 100 Central American species awaiting revision, but the revised genus *Cichlasoma* includes only a dozen species living allopatrically (Kullander, 1983a, Fig. 89): *C. bimaculatum* in the Guianas, *C. amazonarum* in the main Amazon with *C. boliviense* in Bolivian tributaries, and other species restricted to the Paraguay, Paraná, São Francisco, and rivers of north-east and southern Brazil. The morphologically similar genus *Aequidens* has also been subdivided – the '*A. sypsilus* group' of species living in Peruvian forest streams is now ascribed to the new genus *Bujurquina* (Kullander, 1986), and aberrant Surinam species *Cleithracara maronii*, *Krobia guianesis* and *Guianacara geayi* have been

accorded generic status (Kullander and Nijssen, 1989). Habitat data are included in most of Kullander's papers.

Ecological conditions in the Amazon Basin

The Central Amazon is a huge complex of very varied types of water, draining an area exceeding 2.5×10^6 km², lying less than 200 m above sea level and mostly covered by dense rain forest. Here tributary rivers, many of them over 2500 km long, join the main river. This basin forms a very heterogeneous environment, a huge system of anastomosing channels, creeks and streams, oxbows and mouthlakes, lateral lakes and lagoons on the floodplain (**varzea** lakes) or on slightly higher ground filled largely with rain water (**terra firma** lakes). The islands in the main river, with their lagoons and huge areas of forest, are flooded seasonally, both along the nutrient-rich 'whitewater' rivers (**varzea** forest) and the nutrient-poor, acid 'blackwater' rivers (**igapo** zone forest). There are swampy valleys and, in places, flooded grassland (**campo**) and savannahs open to the sun, in which lagoons, pools and swamps remain as water levels fall.

Data on physicochemical conditions in some of the principal aquatic habitats in the Central Amazon have been collated by Kramer *et al.* (1978). In the large rivers the surface temperature changed only *c.* 1–2 °C over the daily cycle and oxygen levels varied little. In the varzea lakes, surface temperatures ranged from *c.* 29 °C at night to *c.* 31 °C in the afternoon; bottom temperatures remained close to 30 °C at low water, and 27 °C at high water. During high water, surface oxygen values ranged from 36 to 71% air saturation at dawn, but declined sharply with depth, reaching less than 5% (often 0) at 4–5 m. The lakes stratify during the high-water season; oxygen concentrations tend to be greater at low water. Floating meadows form a habitat rich in food and cover, but temperature cycles in and under the mat are pronounced (ranging from 28 to 38 °C at the surface). Conditions in the terra firma streams are very stable, both diurnally and annually, temperatures remaining close to 24.5 °C, oxygen at less than saturation, pH *c.* 4.5. Most cichlids are rather eurythermic (able to withstand changes in temperature).

Physicochemical conditions in Upper Amazon tributaries were described by Lüling (1971) for the Rio Huallaga, and Lüling (1975) for the Yarina Cocha, an oxbow flooded with white water from the Rio Ucayali north of Pucallpa (8°40′ S). This oxbow, supporting drifting meadows of *Eichhornia*, was an old river loop 17 km long, 800 m wide, with pH 5.8–7.1 and surface temperatures 25–34 °C. Its fish community of 42 species included a cichlid fauna of 12 species comparable to those in the whitewater varzea lakes in the Central Amazon (Table 3.1).

The open waters of rivers are little frequented by cichlids, but some species may be found in almost any of the other habitats, both in the main basin and in the peripheral mountain streams which carry fish faunas different from those in the central basin. However, cichlids make up only a small proportion of any of the fish communities. The 47 cichlid species recorded from the Rio Negro (Goulding *et al.*, 1988) shared biotopes with over 400 non-cichlid species, most of them otophysan catfishes and characoids, with gymnotoids and 30 species of 11 other families. Many of these Rio Negro cichlids were recorded from several habitats (Table 3.1), distinguished as beach waters, woody shores, flooded forest, island lakes, confluence swamps and rocky pools. The use of different habitats in response to seasonally fluctuating river levels, and according to the size and physiological state of the fish, results in changing associations of species.

Amazon tributaries fall into four groups:

1. Those with Andean headstreams (such as Ucayali and Huallaga tributaries of the Solimões, Upper Amazon) draining the Andes between 3° N and 20° S. These '**whitewater**' rivers, relatively rich in nutrients and with pH about neutral, carry vast loads of sediments which are deposited along the main river, building up the varzea floodplain and islands in which lakes form. Primary production is limited by the turbidity, but calm areas open to the sun support large floating meadows of aquatic plants and grasses;
2. Left-bank '**blackwater**' tributaries (such as the Rio Negro). The clear water, which is stained brown with humic acids leached from flooded vegetation on the Guiana shield which limit microbial activity, is very acid (pH *c.* 4–5), contains almost no electrolytes, and flows over river beds of white sand – the floodplains with seasonally inundated forest into which the fishes move to feed seasonally;
3. '**Clearwater**' rivers, running mainly from the Brazilian highlands (Xingu, Tapajos, Tocantins), have water of variable pH (4.5–7.8) and poor in nutrients; the river beds are often sandy;
4. '**Rainforest**' rivers (e.g. Purus, Jaruá), rising on the floor of the basin and meandering in vast loops.

Throughout most of the central basin the rainfall (*c.* 3000 mm) is well distributed through the year, but tributary rivers north and south of the Equator flood at different times: left-bank tributaries have high water in June, right-bank tributaries in March–April. The huge rainfall causes large variations in water flow, leading to extensive flooding of forest and savannah; the seasonal flood rises to *c.* 6 m at Iquitos, 15 m at Manaus. Shorelines can change by hundreds of metres. Some habitats become seasonally unavailable as pools dry out. Productivity varies, with interacting cycles of nutrient-rich white water and nutrient-poor black water. Daily temperature changes may be larger than annual ones, but changes in size of a water body may affect temperature and oxygen cycles. Many whitewater lakes become covered with floating meadows.

Table 3.1 Habitats of common Amazonian cichlids and their main foods. CA, Central Amazon whitewater varzea lakes*; Rio Negro blackwater habitats†: B, beaches; WS, woody shores; FF, flooded forest; IL, island lakes; CS, confluence swamps, RP, rocky pools. Main foods (†Central Amazon, *Rio negro,—no data): 1, detritus; 2, aquatic plants; 3, terrestrial plants; 4, aquatic invertebrates; 5, terrestrial invertebrates; 6, fish

	Habitats (Rio Negro)							Foods					
Species	*CA*	*B*	*WS*	*FF*	*IL*	*CS*	*RP*	*1*	*2*	*3*	*4*	*5*	*6*
Acarichthys heckelii	†								†	†			
Acaronia nassa	†		*					†			†		*
Aequidens pallidus			*	*		*		*					*
Apistogramma agassizii	†							†	†	†	†		
Apistogramma spp.	†			*	*	*	*	*			*		
Astronotus ocellatus	†	*	*	*						*	*		*
Biotodoma cupido	†												
Biotoecus opercularis			*			*					*		
Chaetobranchus flavescens	†												
C. orbicularis	†												
Cichla ocellaris	†	*											*
C. temensis	†	*											*
Cichlasoma amazonarum	†												
Crenicara filamentosum			*			*		*		*			*
Crenicichla Johanna		*	*	*		*			†		*	*	*
C. lenticulata		*	*	*							*		*
C. lugubris	†	*	*	*							*		*
C. notophthalmus	†		*								†		*
Geophagus altifrons	†	*	*	*	*	*	*	*	*		*		*

Heros severus	†	*	*	*	*			*		*			*
Hoplarchus psittacus		*	*	*	*	*					*		*
Hypselecara coryphaenoides		*	*										*
H. temporale	†												
Mesonauta insignis	†	*	*		*	*	*	*		*			*
Pterophyllum scalare	†												
Satanoperca daemon			*	*							*		
S. jurupari	†	*		*		*		†	†	*	*		*
Uaru amphicanthoides	†	*	*	*		*		*	*	*			

[†]Data from Knöppel (1970), Junk *et al.* (1983).
*Data from Goulding *et al.* (1988).

Reactions of cichlids to ecological conditions

Some cichlids (e.g. *Cichlasoma* species) live mainly in turbid white waters, *C. amazonarum* in floating meadows, others amongst the vast amount of tree litter and debris in the more acid forest streams; the more brightly coloured *Aequidens* live in more acid black or clear waters, *Bujurquina* in forest streams. *Apistogramma* species characteristic of white, black or clear waters, with pictures of their various habitats, are cited by Kullander (1985). Many species, however, can live in several biotopes as riverine cichlids are on the whole eurytopic.

Hypoxia

Oxygen deficiency is widespread in large areas of the Amazon basin. Fish kills during cold spells on the Rio Negro have been attributed to falls in dissolved oxygen (Geisler, 1969). Many tropical waters become deoxygenated seasonally when the vast bulk of plant material in flooded forest and vegetated lagoons decays at the prevailing high temperatures. Many tropical fishes have accessory respiratory organs to cope with hypoxia; others utilize the better-oxygenated surface water, or move away to better-oxygenated water (references in Kramer *et al.*, 1978; Junk *et al.*, 1983; Sioli, 1984). Junk *et al.*, who studied movements away from Lago Camaleão, a macrophyte-dominated floodplain lake near Manaus, suggested that fish species able to withstand low oxygen concentrations would gain sanctuary from piscivorous fishes, as few piscivores remained when oxygen levels fell. Cichlids lack accessory respiratory organs, but some utilize aquatic surface respiration; this has been demonstrated in the Central American *Cichlasoma panamense*, *Geophagus crassilabrus* and *Aequidens caeruleopunctatus* (Kramer and McClure, 1982). The highly vascular stomach has been suspected of aiding the long survival out of water of *Crenicichla saxatilis* and Mato Grosso *Aequidens* (pers. obs.).

In Guyana, diurnal tidal cycling in the coastal stretches of the rivers leads to lowered oxygen and high free carbon dioxide at low water (Cichocki, 1977). Nest-guarding *Biotodoma cupido* become somnolent at low water; such inactivity probably helps them to survive the hypoxic conditions. 'Wriggler-hanging' of larval cichlids is also thought to help combat hypoxia (Courtenay and Keenleyside, 1983).

Trophic relationships

The main food sources of riverine cichlids are fish, aquatic invertebrates (especially juvenile stages of aquatic insects such as Ephemeroptera, Trichoptera, Odonata), together with Crustacea, sometimes molluscs (though these are rare in acid waters), terrestrial invertebrates falling onto the water surface from the forest (ants, other insects, and arachnids), plant material – both from aquatic macrophytes and from terrestrial sources (leaves, fruits, and flowers from riverine vegetation), algae and detritus. Most species have wide diet spectra, and gut contents indicate that individuals may consume many different food items (pers. obs.). *Aequidens tetramerus* stomachs, for example, from Central Amazon streams contained ants, plant material, insect larvae, algae and fish (Knöppel, 1970). Most of these riverine cichlids do, however, have preferred foods, to which they are best adapted. Riverine species have to be adaptable and able to change their diets to make use of whatever is available seasonally – though some savannah pond species, those growing large enough to lay down fat stores, can fast during the low-water season when food is scarce (Lowe-McConnell, 1964).

The main foods found in cichlid stomachs from central Amazon waters (Knöppel, 1970) and Rio Negro habitats (Goulding *et al.*, 1988) are summarized in Table 3.1. Like many of the associated non-cichlids, most of the cichlid species utilized several categories of food, probably opportunistically. Stomach contents of most fish also contained small amounts of other items, demonstrating a lack of trophic specialization. Most large cichlids ate some fish, but only large *Cichla* restricted its diet to fish (juveniles eat other foods). In Rio Negro samples, fish (together with other items) were common in stomachs of cichlids caught off beaches and woody shores, allochthonous items were more frequent in those caught in flooded forests; those from a swamp contained detritus, aquatic plants and aquatic invertebrates, those from a rocky pool mostly detritus and aquatic invertebrates.

Most data are for adult fishes; juveniles use smaller items, such as microcrustaceans and plankton, and few studies have yet been made of ontogenetic changes in dietary composition. The larvae of the large piscivorous *Cichla ocellaris* feed on minute crustaceans and rotifers in shallow water for their first ten weeks, then when they reach 7 cm SL they move into littoral vegetation and feed on insects and palaemonid shrimps, moving to open water when 18 cm long to feed on small fish (Zaret, 1980). In Guyana, *Cichla* stomach contents included many other cichlids in addition to small characoids and *Corydoras*, indicating that they take whatever small fish they can catch (Lowe-McConnell, 1969b). Few South American cichlids make use of aquatic macrophytes as food except as detritus, possibly because macrophyte growth is here very seasonal (Soares *et al.*, 1986).

In streams the terrestrial component of allochthonous foods, invertebrates and plant material falling onto the water from the forest is particularly high (Knöppel, 1970). Comparable conclusions came from another study of stomach contents from Upper Amazon tributaries (Saul, 1975). Saul examined fish from the Rio Aguarico, Rio Napo and associated waters at an altitude of 340 m in Ecuador (00° 06′ N, 76° 51′ W). He also found a lack of specialization, particularly in the small streams, and stressed the importance of the terrestrial component, especially insects (ants). Most of the cichlids here appeared to be "rather omnivorous secondary consumers". A small lake had the highest cichlid populations.

The limnological interpretation of these results was that in these riverine communities, flexibility in diets allows large numbers of a great variety of fish species to find sufficient food. Foods are available in all zones – items from surface and bottom waters are often found in the same fish. Knöppel concluded that these fish do not seem to be in competition for food, i.e. that the food factor appears to have little selective influence on fish living under these riverine conditions. However, personal observations on the behaviour of stream fishes of Guyana and the Mato Grosso, where many species rush to grab any object that falls onto the water, suggest that there may be a lot of competition to get whatever food is available, and that the lack of specialization reflects the need to grab at any food item before another fish does so.

Effects of **Cichla ocellaris** *on food webs*

The most complete autecological information for a South American cichlid is that for the predatory *Cichla ocellaris*, the largest Neotropical cichlid (growing to 70 cm TL, 7 kg), studied in Guyana (Lowe-McConnell, 1969b) and in Panama (Zaret, 1980). In 1965, 100 fingerlings were introduced for a sport fishery near the Chagres River, Panama; they multiplied and spread down the river to Gatún Lake in the Panama Canal. There this predator led to the disappearance of many indigenous fishes and simplified the food web through second-order effects on the zooplankton, insects and vertebrates, including the many fish-eating birds (Zaret and Paine, 1973). Most of the small indigenous fishes (the atherinid *Melaniris chagresi*, poeciliids and small characins) were wiped out from the lake, though they survived in the river where there was more cover. The indigenous *Cichlasoma maculicauda*, however, increased; this effect was attributed to the elimination by *Cichla* of species that formerly fed on *Cichlasoma* fry. *Cichla* spread across Gatun lake in waves, with a leading edge composed of subadults. Adult *Cichla* are very territorial during the breeding season; both parents remain with the schools of fry for 60 d or more; each pair breeds for a considerable part of the year, remaining in the same vicinity during breeding activities, so immigrants to previously unexploited habitats are generally immature fish. These *Cichla* avoid cannibalism even when other small fishes are scarce (Zaret, 1977; Schroder and Zaret, 1979).

Rheophilic cichlids

Kullander (1988) has described a group of small (76 mm SL) *Crenicichla*-like cichlids with specializations that closely parallel those in African rheophilic cichlids (page 82). These goby-like *Teleocichla* (six known species) live in the rapids of the clearwater rivers draining the Brazilian highlands (Tapajos, Xingu, Tocantins). The downwardly-directed snout and deep preorbital suggest that *Teleocichla* obtain food from the upper layers of the substrate, though no field observations have yet been made. Anterior body scales and those on the head are deeply embedded or absent, as in other bottom-grubbing cichlids, and the elongated pelvic fin can support the fish on the bottom as in the African rheophilic genera. The sharing of these features with rheophilic cichlids in Africa is probably a case of parallel evolution rather than indicating any relationships.

The geophagine *Retroculus* (growing to 19 cm SL) are probably the most specialized rheophilic cichlids in South America, with thick lips, broad pelvic fins and small embedded scales. Some *Crenicichla* have similar features. South American rapids have as yet been little investigated.

3.3 THE CICHLID FAUNAS OF AFRICAN RIVERS

African riverine cichlids fall into three major groups, plus a fourth group of genera of as yet undetermined relationships:

1. The **Pelmatochromine** cichlids of western Africa; mostly substrate spawners;
2. The **Haplochromine** cichlids of East and Central Africa, the group which has undergone such spectacular adaptive radiations in the Great Lakes of eastern Africa; maternal mouthbrooders;
3. **Tilapiine** cichlids which fall into three taxa: the substrate-spawning *Tilapia*, the biparental mouthbrooding *Sarotherodon*, and the maternal mouthbrooding *Oreochromis* (Trewavas, 1983). Amongst riverine cichlids the tilapias are best known ecologically as they are of great economic importance; all are specialized for vegetarian diets. There is no cichlid group comparable to the tilapias in South American waters.

The pelmatochromines comprise a monophyletic 'chromidotilapiine' lineage of six genera (*Chromidotilapia, Thysochromis, Pelvicachromis, Limbochromis, Nanochromis, Parananochromis*) of species which have a visor-like hanging pharyngeal pad (compare the pharyngeal pads in South American geophagines), plus two monophyletic taxa (*Pelmatochromis* and *Pterochromis*) which lack such hanging pads (Greenwood, 1987a). The role of the hanging pharyngeal pad, rich in taste buds, has been discussed by Trewavas (1974). *Chromidotilapia* are mouthbrooders; all the other pelmatochromines are substrate spawners. Little is known about pelmatochromine feeding habits, but their dentition and associated structures suggest there is little intragroup trophic specialization; evolution within the group seems to be associated with an increasingly rheophilic way of life. The pelmatochromines are not closely related to the fluviatile haplochromines and there is little geographical overlap between the two groups. From morphological evidence Greenwood (1987a) suggested that many of the riverine haplochromines in East and Central Africa are the ecological equivalents of the pelmatochromines in western Africa, but very little is yet known about the ecology of pelmatochromine species; none of them has enlarged molariform pharyngeal teeth as in the molluscivorous haplochromines.

The fourth group, cichlids of 'doubtful relationships', includes *Tylochromis lateralis*, a maternal mouthbrooder which feeds on insects from bottom deposits, *Heterochromis multidens*, which is also an insect feeder (Matthes, 1964), and the substrate-spawning carnivorous *Hemichromis* species. Initial faunal surveys had indicated that two *Hemichromis* 'species' ('*H. fasciatus*' and '*H. bimaculatus*') were the most widely distributed of any riverine cichlids, from West Africa to the Nile and in parts of the Zaire and Zambezi drainages. Further taxonomic analyses and consideration of distributions and ecology have shown the situation to be much more complex. Loiselle (1979) recognized ten *Hemichromis* species. The most extensively distributed are *H. letourneauxi*, which occurs in savannah rivers across the soudanian region, and *H. elongatus*, restricted to forested waters from Sierra Leone to Northern Angola and also in the Zaire and Upper Zambezi. Other species live allopatrically in forest biotopes in West African coastal basins. In several cases, pairs of species exist sympatrically but seldom syntopically, e.g. from Sierra Leone southwards, *H. cristatus* inhabits forested stretches of river, *H. paynei* the lowest reaches and coastal lagoons.

Ecological conditions in African rivers

The land mass of the African continent is on the whole higher than that of non-andean South America, much of it drained by four great river systems – the Niger, the Nile, the Zaire (Congo) and the Zambezi – together with smaller rivers flowing to east and west coasts (Fig. 3.2). The Niger and Nile share savannah-river species, the central Zaire has mainly forest-river species, and the Zambezi system, south of the equatorial forest, has another fauna of savannah-river species. Large inland drainage areas include the Chad basin between Niger and Nile, and the Okavango swamps which carry a predominantly Zambezi fish fauna. Endemic cichlids have evolved in some smaller river systems, such as the Malagarazi which drains westwards into Lake Tanganyika. Riverine species have given rise to endemics in most lakes, including the small Cameroon crater lakes. Distances are vast and ecological information for many species is sparse as priority has been given to studying lacustrine faunas. The creation of large reservoirs for hydroelectric power has stimulated ecological work in the Volta, Niger, Nile, Bandama (Ivory Coast) and Zambezi systems.

Forest rivers: the Zaire

The Zaire basin straddles the Equator where rain falls throughout the year and much of the river basin is cloaked in equatorial rain forest. Tributary rivers stretching north and south of the Equator flood at different times, bringing their waters into the lower reaches in a bimodal peak, so annual seasons are not as marked here as for fishes living in the savannah rivers. Allochthonous foods from the forest are important in fish diets, as in Amazon tributaries. Although the Zaire is superficially similar to the Amazon, there are many differences. The Zaire has a highly variable slope and drains the watersheds of a succession of plateaux separated by rapids, some steep enough to separate elements of the biota. The Zaire is much shallower than the Amazon, with only

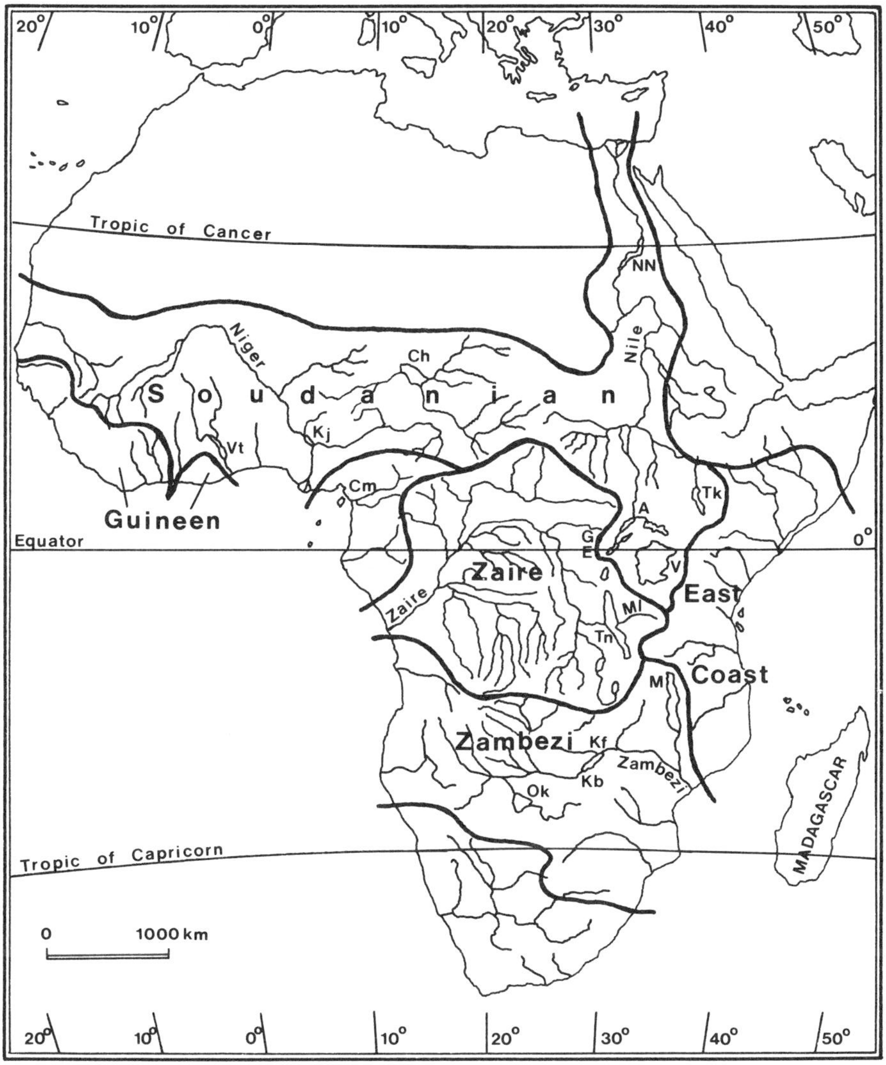

Fig. 3.2 Ichthyofaunal regions of Africa and sites. Ml, Malagarazi River; OK, Okavango swamps; Lakes: A, Albert; Ch, Chad; Cm, Cameroon; E, Edward; G, George, M, Malawi; Tk, Turkaua; Tu, Tanganyika; V, Victoria; Man-made lakes; Kb, Kariba; Kf, Kafue; Kj, Kainji; NN, Nasser/Nubia; Vt, Volta (modified after Lowe-McConnell, 1987).

small fluctuations in level (*c.* 3 m); lakes and swamps in the upper reaches absorb the heavy rain. Geological differences affect the water chemistry – the Zaire lacks the Amazon's continually shifting sediments and turbidity from Andean runoff, but the forest waters are mostly acid, as in the Amazon, and deoxygenation is again a problem in swampy areas.

Analyses of ecological conditions and fish faunas have been made at various points along the huge Zaire river system: in the long stretch of rapids in the Lower Zaire, home of highly specialized endemic rheophilic cichlids (considered later, page 82), from the lake-like Pool Malebo (Stanley Pool) above these rapids (Poll, 1959), from 1500 km up to the main Zaire (Gosse, 1963), and from intermediate stations on the tributary Tshuapa and lateral Lake Tumba (Matthes, 1964). Though the Zaire basin has an estimated 1000 fish species, it has a poor cichlid fauna, only *c.*32 species, and these include members of endemic genera (*Teleogramma*, *Steatocranus*, *Heterochromis*) from the rapids of the Lower Zaire. In the central basin the 13 species of cichlid make up only 6% of the fauna and have to share resources with over 180 otophysans and 42 species of 13 other families. These cichlids include representatives of pelmatochromine genera *Pterochromis*, *Pelmatochromis*, *Parananochromis* and *Nanochromis*, together with species of *Hemichromis*, *Heterochromis*, *Tylochromis*, *Lamprologus* and tilapias. Some of the tilapias could be species introduced into the area for pond culture, but the open waters of Pool Malebo support an endemic subspecies, *Sarotherodon galilaeus boulengeri*, living sympatrically with the herbivorous *Tilapia tholloni*, and an endemic subspecies of the latter, *T. t. congicus*, inhabits Lake Tumba. The *Pelmatochromis* species were caught in swampy places.

Matthes (1964) included data on physicochemical conditions (seasonal variations in water level, flow, temperatures, transparency, pH, oxygen, conductivity and chemical composition) and biotopes (pelagic and littoral zones over rocky, sandy, or vegetable-debris bottoms, in channels and creeks, floating meadows, inundated forest, permanent and temporary swamps). For each species he reported the available information on habitat, feeding and breeding regimes. Many of the cichlid species showed a preference for the more sheltered waters, in bays or near banks with floating meadows, where aquatic insect larvae were a main food source together with smaller invertebrates (crustaceans and hydracharines), small fish and vegetable debris. *Lamprologus congolensis tumbanus* and *Nanochromis nudiceps* frequented lateritic rock zones, where they also fed mainly on aquatic insects and small invertebrates. Another *Nanochromis*, *N. dimidiatus*, an omnivore which ate many insect larvae, inhabited small channels and the swampy inundation zone. Matthes' general conclusion, based on cichlids and non-cichlids, was that the continually changing conditions associated with seasonally changing water levels give a resilience important for the coexistence of so many fish species; although a great diversity of niches exist, the fluctuating riverine conditions can explain the absence of highly specialized species. Most species living here showed a marked preference for one particular biotope, but could survive when conditions change; they have to be eurythermic, able to withstand sudden changes in water temperature of 5–10 °C. Most of these cichlids (in contrast with the otophysans) did not have very restricted breeding seasons. Buccal incubation of eggs was recorded in a *Tylochromis lateralis microdon* female, but the other cichlids here all appeared to be substrate spawners.

Savannah floodplain rivers

The soudanian region

Either side of the equatorial forest, in the soudanian savannah region south of the Sahara, and in the Zambezi system south of the forest, the rivers are open to the sun and have a very much more seasonal flow regime, inundating huge areas of floodplain annually. Over these savannahs, rains occur in the summer months, after daylength and temperatures have been increasing. Peak floods follow, the delay depending on the origin of the main floodwater and how long it takes to travel down river. Rising water runs up creeks onto the floodplain, connecting the ponds and swampy areas which are isolated in the dry season, and releasing their imprisoned fishes. As the floods spread, the water is greatly enriched by nutrients from the breakdown of decaying vegetation and animal droppings. This leads to an explosive growth of bacteria, algae and zooplankton which in turn supports a rich fauna of aquatic insects and other invertebrates. The aquatic vegetation grows very rapidly, fruits while the water is high, and dies down as the water recedes. There is an extremely rapid increase in production and biomass of all kinds of fish food and the fishes grow fast; there is also cover for juvenile fishes. In the dry season, fishes are confined to the river bed which may be broken into pools, or are trapped in small water bodies on the floodplain which become very deoxygenated. Life cycles of the otophysan fishes are closely tied to the flood cycle. The cichlids, which can produce several batches of young through the year, are not so flood-dependent, but nevertheless cichlid growth is arrested in the low-water season (as tilapia studies on the Kafue plain have shown, Kapetsky, 1974), and the fish are all subject to differences in food availability and predation pressure through the year.

The soudanian region has many cichlid species distributed right across Africa south of the Sahara, from the west coast through the Chad basin to the Nile. South of this region, in West Africa, the forest rivers carry other less-widely-distributed 'guineen' species (Daget and Iltis, 1965), often ecological replacement species, such as *Hemichromis elongatus* in forested streams replacing *H. fasciatus* in savannah rivers. Guineen species are generally smaller and more restricted in distribution than the soudanian fishes; the latter undergo extensive longitudinal and lateral migrations, so it is not surprising that they are widely dispersed. River captures and long-term climatic variations have in some cases resulted in soudanian and guineen species living in the same river system, but the soudanian species live in open reaches and the guineen species in forested stretches, indicating that ecological conditions (particularly presence or absence of forest) account for their continued coexistence. Seawards of the forest belt, the coastal lagoons and estuaries of West Africa are inhabited by brackish-water-tolerant species.

In the Niger, cichlids make up only 7–9% of the fish faunas of the Upper Niger and Central Niger deltas (Darget, 1954, 1957); they share biotopes with over 95 otophysans and 17 species of 11 other families. The Chad basin shares this fauna (Blache, 1964); both Daget and Blache give some information on feeding and breeding regimes; ecological studies of some of the same species have been made on the Nile.

Investigations in the Sudan sudd area of the Nile (Bailey *et al.*, 1983; Hickley and Bailey, 1987) showed that *Hemichromis letourneauxi* (to 90 mm SL) lives in open water and submerged and floating marginal vegetation (*Eichhornia*, which has now replaced *Pistia stratiotes*) where it browses on *Caridina* shrimps; this species moves into seasonally inundated grassland to spawn and is tolerant of poorly oxygenated water. The larger *Hemichromis fasciatus* (to 250 mm) was rarely caught among submerged plants and not found in seasonal habitats; it consumed *Caridina*, *Macrobrachium* and live fish. The omnivorous *Thoracochromis loati* (100 mm) was caught in open water, submerged and fringe vegetation, moving into macrophyte beds at high water. *Pseudocrenilabrus multicolor* (70 mm) was abundant in marginal vegetation, weed beds and seasonal habitats. Of the three tilapias, *Oreochromis niloticus*, the largest (50 cm) and commonest, was widely distributed in open water, plant beds in lakes and in the *Eichhornia* fringe; juveniles were abundant in temporary pools on the flooded grassland. Adult *O. niloticus* contained algae; juveniles ate small crustaceans and rotifers; ovaries suggested an extended spawning period through the flood season. *Sarotherodon galilaeus* (to 40 cm) was not as abundant and was more closely associated with submerged vegetation; specimens contained algae with more higher-plant tissue. The smaller *T. zillii* (to 30 cm) was clearly associated with submerged vegetation and the *Eichhornia* fringe, where they fed on the periphytic algae, higher-plant tissue, and some invertebrates.

The Zambezi system

The Zambezi system has a higher proportion of cichlid species (22 of a total of 127) than the Niger or Zaire, and these share habitats with more cyprinid species (37) but fewer other otophysans. The Zambezi has cichlid species different from those in the Niger, Nile and Zaire. Pelmatochromines are absent. Among the haplochromines the 'northern' genera are replaced with 'southern' taxa, serranochromines and *Pharyngochromis*, distinguished by non-ocellate and numerous anal fin spots – a group essentially of the Zambezi, Limpopo and Angolan rivers (though *Serranochromis* is represented in the Upper Zaire). All appear to be maternal mouthbrooders. Environmental conditions differ markedly in the Upper Zambezi above the Victoria Falls (a floodplain river with wide lateral swamps), the Middle Zambezi below the Falls (a 'sandbank' river with little floodplain and few refuges for the fishes at low water), and the Lower Zambezi below Cabora Bassa, where the river grades into brackish estuarine waters. The Kafue tributary of the Middle Zambezi is also a floodplain river and shares with the Upper Zambezi some species not found in the Middle Zambezi.

The serranochromines include two lineages: the larger-growing, more piscivorous *Serranochromis* (*Serranochromis*) and mostly smaller *S.* (*Sargochromis*) species, many of which have the enlarged pharyngeal teeth of molluscivores (Greenwood, 1979). Among the smaller cichlids, *Pseudocrenilabrus philander* (80 mm SL) is very abundant in the Upper and Middle Zambezi, feeding on small fish, crustaceans and insect larvae; the young feed on zooplankton. *Pharyngochromis darlingi* (120 mm SL), a voracious predator (Jubb, 1967) abundant in the Upper Zambezi, has enlarged pharyngeal teeth suitable for mollusc-feeding.

The occurrence and abundance of *S.* (*Serranochromis*) species, which are all basically piscivores,are determined by their preference for shallow reedy lakes or swamps (Bell-Cross and Minshull, 1988). All are maternal mouthbrooders which spawn several times through the rains (October to February); at this latitude and altitude the climate is seasonal. The various species grow to different sizes and choose different habitats. The largest, *S. robustus* (3 kg), is widely distributed and not restricted to a particular habitat; although it is the most piscivorous, its robust pharyngeal teeth enable it to utilize molluscs. *S. angusticeps* (2.5 kg) lives mainly where there are large stands of aquatic plants, preying on small fish and invertebrates. *S. macrocephalus* (1.5 kg) prefers deep pools and feeds on invertebrates with some fish. *S. thumbergi*, piscivore with a preference for open-water habitats, is abundant in the Kafue, but this niche is not available in the Upper Zambezi, where the predatory characid *Hydrocynus* dominates the open water. *S.* (*Sargochromis*) *codringtoni* (1.1 kg) is common in all habitats except small tributaries; primarily a gastropod feeder, it also eats insects, plant seeds and detritus. The smaller *S.* (*Sa.*) *carlottae* (0.6 kg) has similar feeding and breeding habits to *S.* (*S.*) *codringtoni*, and a larger species, *S.* (*Sa.*) *guiardi* (3 kg), widely distributed in the main rivers, feeds on molluscs, including large bivalves.

Among the Zambezi tilapias, *T. rendalli*, which replaces the soudanian *T. zillii*, is a herbivore sharing its range with the smaller, more omnivorous *T. sparrmanii*. *Sarotherodon* species are absent. *Oreochromis* is well represented by *O. andersonii* and *O. macrochir* in the Upper Zambezi (and Kafue), *O. mortimeri* in the Middle Zambezi and its tributaries, and by *O. placidus* and *O. mossambicus* in the Lower Zambezi (Trewavas, 1983). In the Upper Zambezi the omnivorous *T. sparrmanii* (0.25 kg) is the commonest cichlid and primarily a periphyton feeder. *T. rendalli* (1.4 kg), another substrate spawner nesting in shallow water during the rains, is widely distributed and abundant in water with dense macrophytes, where it is primarily a herbivore though it also eats detritus and insects. Of the *Oreochromis*, *O. macrochir* (growing to 1.8 kg) feeds mainly on periphyton and detritus; a maternal mouthbrooder, it breeds at least once before and once during the floods; once juveniles leave the floodplain they are restricted to lagoons and backwaters until they grow large enough to coexist with the predatory *Hydrocynus*. *O. andersonii*, the largest tilapia (males grow to 3.2 kg), occurs in all main habitats except small streams; a less specialized feeder than *O. macrochir*, it eats insects and small crustaceans, and is reported to spawn only once a year between November and January. On the Kafue floodplains, tilapia growth rates have been shown to vary from year to year with flood conditions (Kapetsky, 1974; Dudley, 1979).

The presence of floodplain rivers holding much water throughout the year appears to have been crucial for the evolution of the serranochromines and tilapias within the Zambezi system, possibly aided by the absence of pelmatochromines and the more restricted non-cichlid fauna than occurs in West Africa.

3.4 RIVERINE CICHLIDS ASSOCIATED WITH AFRICAN LAKES

Riverine cichlids associated with East African lakes

The ancestor of some of the species 'flocks' in lakes Victoria and Edward–George (Nile drainage) may have been a generalized riverine form now represented by the *Astatotilapia bloyeti* complex found mainly in the rivers of eastern Africa (Greenwood, 1979). The modern representative of the basal haplochromine stock in Lake Malawi (Zambezi drainage) was once thought to be the structurally similar *Astatotilapia calliptera*, found in Malawi and the Lower Zambezi, but it is now suspected that other lineages (including the Nilotic–Zairean *Thoracochromis*) may also have been involved (and see Eccles and Trewavas, 1989). Lake Tanganyika's cichlids must stem from several sources, including substratum-spawning riverine species such as *Lamprologus*, but some appear to have come from a form similar to *Astatotilapia burtoni*. The inflowing Malagarazi river and the Zaire, to which Lake Tanganyika drains, also carry the derived Zairean genus *Orthochromis*. The basal haplochromines are maternal mouthbrooders, males characterized by ocelli on the anal fin. Most appear to be small, rather omnivorous feeders. Haplochromines of the *A. bloyeti* group in the Pangani and Ruvu Rivers (Tanzania) examined by Bailey *et al.* (1978) were primarily carnivorous but ate some plant material.

Small haplochromines of four species were studied in the Malagarazi River flowing westwards into Lake Tanganyika (Greenwood, 1954). Here a species of the '*bloyeti*-complex' was sympatric with the endemic *Astatotilapia paludinosus*; *A. paludinosus* was caught in swamps, marginal vegetation along the open river, and further offshore among dense stands of water lilies (the young from shallow pools), while *A. 'bloyeti'* kept near the banks; both contained semidecomposed bottom material. *Orthochromis malagaraziensis* was essentially fluviatile, trapped in fast-flowing water along reeds fringing the river, where it fed on aquatic insects, epilithic algae and plant debris. *Astatorheochromis vanderhorsti* lived in shallow inshore swamps and in the river, feeding on snails, insects, ostracods and water lily seeds. *O. malagaraziensis* appeared to be derived from a form resembling the Zaire species *Astatorheochromis straeleni*, and related to the snail-eating *A. alluaudi* found in Lake Victoria and associated rivers, a species relatively tolerant of conditions in papyrus swamps which connect some of the drainages. *Astatorheochromis* all have enlarged pharyngeal bones, as in the Zambezian sargochromines, but this appears to be an adaptational character associated with molluscivory, which may have arisen independently in several lineages.

The relationships of the Malagarazi River fishes with those of the Zaire indicate that the Malagarazi must have been directly connected with the Zaire before faunal continuity was interrupted by the development of lacustrine conditions in Lake Tanganyika (Fig. 3.2). The Upper Malagarazi has seven species of endemic cichlids: the four haplochromines, *Serranochromis janus*, and the tilapias *Oreochromis karomo* and *O. malagarasi* (Trewavas, 1983). *O. malagarasi* is so similar to *O. upembae* (a species of swamps and shallow lakes on Upper Zaire tributaries) that it is probable they had a common ancestor when the Malagarazi was part of the Zaire system; it forms a link between *O. upembae* and *O. variabilis* endemic to Lake Victoria. *O. karomo* is also probably most closely related to *O. upembae* and *O. malagarasi*, but *O. karomo* has a distinctive dentition, well adapted for rasping epiphytic algae off macrophytes in the extensive swamps, whereas *O. malagarasi* feeds by scooping up the soft flocculent bottom deposits consisting of finely divided plant material with many protozoa and bacteria – material passed through the guts of characoid *Alestes* and *Distichodus* which feed on the water lily leaves. The arena breeding behaviour of *O. karomo* in these Malagarazi swamps was described by Lowe-McConnell (1956).

Riverine cichlids associated with Cameroon crater lakes

During ecological studies of West Cameroon crater lakes (Fig. 3.2), Lake Barombi Mbo (Trewavas *et al.*, 1972) and Lakes Kotto and Mboandong (Corbet *et al.*, 1973; Trewavas, 1974), the cichlids in tributaries of the Mungo–Meme river system into which the lakes overflow were also examined. The 17 fish species in Lake Barombi Mbo included 11 endemic species of cichlid (two of new genera), which appeared to have evolved within the lake from ancestral populations which gained access to the lake at different times. None of these endemic cichlids was found in the feeder stream. The only cichlids found in Mungo tributary streams were *Hemichromis elongatus*, *Chromidotilapia guntheri* and *Tilapia camerunensis*, cichlids which all belong to the fauna of the coastal forest belt from South Cameroon to Lagos, none of them even congeneric with the cichlids in Lake Barombi Mbo. Another *Chromidotilapia* (the new species *C. finleyi* apparently confined to flowing water) and *Tilapia mariae* were later found in Meme tributaries.

In Lakes Kotto and Mboandong the fish faunas were less distinct from those of the inflowing Meme tributary streams, and the rim of the shallow Lake Mboandong with its carpet of fallen leaves and its constant rain of terrestrial insects resembled a forest stream (Trewavas, 1974). The six cichlid species within Kotto's crater included *Hemichromis elongatus*, widespread in forest zone rivers, *Tilapia mariae*, abundant in the lower forested reaches of river systems from South Cameroun to the Ivory Coast, the closely related *T. camerunensis*, the widespread *Sarotherodon galilaeus*, together with two endemics, *T. kottae* and a subspecies of *Chromidotilapia guntheri* (*C. g. loennbergi*). *Hemichromis elongatus* was common in the very shallow water, where it fed on chironomid larvae and ephemeropteran nymphs, with some terrestrial insects, fishes and frogs. *Tilapia mariae* also ate chironomid larvae, but contained phytoplankton, a lacustrine food source.

3.5 ECOLOGY OF TILAPIAS

Tilapia species (*s.s.*), with their low number of gill rakers and coarse pharyngeal teeth, are typically macrophyte feeders and substratum brooders; about 29 species are known, mostly of rather local distribution in the rivers of western Africa, though the euryhaline *T. guineensis* inhabits coastal lagoons and estuaries from Senegal to Angola. Three species are widely distributed; the rather similar *T. zillii* and *T. rendalli* are allopatric (page 77); the name '*T. melanopleura*' was at one time used erroneously for *T. rendalli* and for *T. dageti* in West Africa. *Sarotherodon* are microphagous mouthbrooders with minimal sexual dichromatism, in which both sexes or the male brood eggs and young; ten species are known in West Africa from Senegal to the Zaire, with one species (*S. galilaeus*) across the soudanian region to the Nile and Jordan valley (Trewavas, 1983); *Oreochromis* have more numerous gill rakers and are microphagous, polygamous, arena-spawning maternal mouthbrooders, usually with marked sexual dichromatism at breeding time. Trewavas (1983) recognized two subgenera: (1) *O.* (*Oreochromis*), with 15 species mainly in the eastward-flowing rivers (but including *O. niloticus* in the soudanian region and many lakes); (2) *O.* (*Nyasalapia*), in which the breeding male has a genital tassle, with 15 species in Central Africa in westward-draining rivers from the Cunene in Angola to the Lower Zaire, and with endemic species in the African Great Lakes.

Tilapias on the whole exhibit allopatric speciation, with only one or two species of a genus or subgenus in any one river system. They do not show the explosive speciation of the haplochromines under lacustrine conditions; tilapias are already highly specialized for a vegetarian diet, having notched teeth and long intestines. In many parts of Africa, tilapias are riverine fishes but live mainly in backwaters, lagoons, on floodplains and in pools and swamps in the dry season. The more specialized planktivores prefer large expanses of open water and quickly become abundant when rivers are dammed to form new lakes (as did *S. galilaeus* in the Volta lake Ghana, Lake Kainji in Nigeria, and Lake Nasser on the Nile). The macrophyte feeders are able to penetrate further upriver than those restricted to a plankton diet. Tilapias support the main fisheries in most of the African lakes; they are also cultured in ponds so have been moved by Man, often escaping into different river systems and in some cases replacing or hybridizing with, the indigenous species. Tilapias in fish communities, their life histories, growth and production are considered in Pullin and Lowe-McConnell (1982).

Philippart and Ruwet (1982) have collated the immense amount of information available on ecology and distribution of tilapias. They analyse the physical and chemical factors which may influence their geographical distributions: current velocity, depth, temperature, salinity, alkalinity, pH, oxygen and other dissolved gases. They also consider tilapia behavioural ecology – feeding and reproductive behaviour, selection of microhabitats, schooling and movements. Within their original areas of distribution, tilapias have colonized a great range of habitats: permanent and temporary rivers, flowing water, large equatorial and subtropical rivers, swamps, lakes, lagoons and brackish estuaries. These varied habitats represent, both in terms of absolute magnitude and in speed at which fluctuations occur, a very wide range of physical parameters (depth, current velocity, turbidity), and of temperature and chemical composition (salinity, pH, dissolved oxygen). Tilapias appear to be very resistant to low dissolved oxygen; values as low as 0.1 p.p.m. are tolerated by *O. mossambicus* and *O. niloticus*. This enables some tilapias to live and reproduce in swamps and shallow lakes which become deoxygenated, though massive fish kills occasionally occur. Tilapias can also withstand very high levels of carbon dioxide (to 72.6 p.p.m. in *O. macrochir*).

Temperature tolerance experiments have shown some species to be able to tolerate a wide range of temperature, e.g. *T. zillii* (6.5–42.5 °C), *O. mossambicus* and *O. niloticus* (8–42 °C). Other species are less tolerant of high temperatures Temperature preferences depend on size, young tilapias being more tolerant of higher and often of lower, temperatures than the adults. There are also Temperature preferences depend on size, young tilapias being more tolerant of higher, and often of lower, temperatures than the adults. There are also interactions with other factors such as salinity: *O. mossambicus* and *O. aureus* tolerate low temperatures better in saline water (5% seawater) than in fresh water, which helps to explain how these species live at relatively high latitudes. Several tilapias are euryhaline, able to live and reproduce at salinities higher than 30‰; coastal lagoon species *T. guineensis* and *S. melanotheron* in

West Africa; *O. mossambicus*, *O. hornorum* and *O. placidus* in East African coastal rivers. These euryhaline species are widely distributed as they can move from one estuary to another during floods. Feeding regimes within a tilapia species vary with type of food present, size and age of the fish, time of year and presence or absence of competing species. *T. zillii*, for example, feeds on macrophytes when competition is intense but eats plankton and benthos wehre competition is less severe.

3.6 RHEOPHILIC CICHLIDS IN AFRICA

Among African cichlids a number of taxa have evolved forms adapted for life in the rapids of large lowland rivers such as those of the Lower Zaire between Pool Malebo and the estuary – geographically isolated, but often extensive, habitats with perenially swift current over rocky substrata, and turbulent, highly oxygenated water with poor visibility. This rheophilic fauna includes genera known only from rapids, such as *Steatocranus* (seven species) and *Teleogramma* (four species), most of them from the Lower Zaire rapids, which also have four endemic species of *Nanochromis* and an endemic *Lamprologus* (Roberts and Stewart, 1976). Another rheophilic genus, *Gobiocichla*, is represented by *G. wonderi* in the Upper Niger and *G. ethelwynnae* in the Cross River of West Cameroon (Roberts, 1982). Rheophilic haplochromines, *Orthochromis torrenticola*, live in torrential streams of eastern Zaire, and *O. polyacanthus* in the Lualaba system (Thys, 1963). Several tilapias have become adapted for life in fast-flowing water: *Tilapia rheophila* is endemic to the Konkoure River of Liberia, and *T. busumana* lives in streams associated with Lake Bosumtwi in Ghana (Philippart and Ruwet, 1982).

Morphological adaptations in cichlids living in fast-flowing water include an elongate cylindrical body, a single lateral line, small embedded cycloid scales, and reduced squamation on the anterior half of the body (compare adaptations in South American rheophilic cichlids, page 70). In *Teleogramma* and *Steatocranus* the pelvic fins have a long middle ray, thought to aid in clinging to stones. Some rheophilic fishes have very reduced eyes; a cryptophthalmic, depigmented species of *Lamprologus* (*L. lethrops*) lives in the Zaire rapids (Roberts and Stewart, 1976). The genus *Lamprologus*, endemic to the Zaire basin, has several riverine species (*L. congolensis*, *L. moquardi*, *L. werneri*) in addition to the many species endemic to Lake Tanganyika; all are substratum spawners.

Steatocranus casuarius and *S. gibbiceps* live sympatrically in the Lower Zaire rapids. *S. casuarius* has an extremely long, coiled intestine and apparently feeds predominantly on algae; *S. gibbiceps*, which has a relatively short, straight intestine, feeds largely on gastropods, which it swallows whole (Roberts and Stewart, 1976). Readily recognizable by the possession of a large nuchal hump, *Steatocranus* are substrate spawners and fix their eggs to rocks. Little is yet known of the breeding habits of most of these rheophilic fishes. *Gobiocichla ethelwynnae* in the Cross River matures at a small size (50 mm SL) and produces 35–112 eggs, but its reproductive behaviour is unknown.

3.7 SELECTION PRESSURES ON RIVERINE CICHLIDS

To sum up: the key factors affecting the lives of riverine cichlids are: (1) the dynamically fluctuating water levels, which lead to changing physicochemical and biotic conditions and associated changes in the fish communities and population structures, and (2) competition for living space, especially territories for feeding and guarding fry. Water-level fluctuations affect the seasonality of both feeding and reproduction; they also have direct effects on types of breeding behaviour.

Seasonality of feeding

Riverine fishes are subject to more seasonal variations in both numbers and types of food than are lacustrine fishes, so have to be more facultative feeders. There may also be great variations in seasonal abundance of food. These differences are, however, more marked in floodplain rivers than in forest streams. Stomach contents have shown that a wide variety of foods are used by riverine cichlids, often by one individual: from this Knöppel (1970) concluded that Amazon stream and varzea lake fishes do not seem to be in competition for food, that the food factor appears to have little selective influence on fishes living under these conditions. However, it seems probable that a species will thrive on a main diet of the food for which it is best adapted (fish, invertebrates or a vegetarian regime). Better food leads to faster growth; this helps the fish to avoid predation and to obtain the best sites for rearing young, as well as to lay down fat stores which enable the female to produce more or larger eggs with denser yolk.

The young of many cichlids feed on microcrustaceans, whether the adults become piscivores as in *Cichla* or algivores as in *Tilapia*, and competition for larval food may be severe. Zaret (1984) showed that the time available for feeding on zooplankton in Amazonian varzea lakes is restricted by diurnal fluctuations in oxygen levels in the water column, and that many species (particularly small characins) compete for this food source. This may be a 'crunch time' for young cichlids too.

Seasonality of breeding

In many cichlid species, fish ready to spawn may be found at any time of year, producing batches of young at frequent intervals. Peak spawning times depend on the prevailing conditions, and for species found in both rivers and lakes (such as tilapias), breeding appears to be more seasonal in rivers than in lakes at comparable latitudes. In Guyana, *Acarichthys heckelii*, *Satanoperca jurupari*,

Geophagus surinamensis, *Cichla ocellaris* and *Aequidens geayi* bred throughout the year in floodplain lakes where feeding conditions remained good (Cichocki, 1976), though spawning was more seasonal in the eastern floodplain pools in which conditions were poor in the dry season (Lowe-McConnell, 1969b). Both in South America and in Africa, batches of cichlid young are in many cases produced both at low water and then again during the rains (in contrast with most of the cohabiting otophysans which have well-defined breeding seasons geared to the floods). Riverine cichlids need to find places sheltered from the current in which to nest and in close proximity to conditions suitable for the free-swimming young. Sudden changes in water level affect conditions most strongly at the start of the rains or floods. Breeding at low water means that free-swimming young are not likely to be swept away, and eggs and young will not suffer from the turbidity and turbulence as the water rises. But there may be opposed selection pressures to breed at high water, as this is the time when food for the young is most plentiful and abundant plant growth provides cover from the numerous predators.

Effects of fluctuations in water level on cichlid reproductive behaviour

The few detailed field studies of cichlid reproductive behaviour in rivers include those in Guyana of *Cichla ocellaris* and *Mesonauta insignis* ('*C. festivum*' – Lowe-McConnell, 1969b), a comparative study of *Acarichthys heckelii* and *Satanoperca* ('*Geophagus*') *jurupari* in Rupununi floodplain lakes with *Biotodoma cupido* in Guyana's tidally affected coastal rivers (Cichocki, 1976, 1977), and of '*Aequidens vittatus*' in Surinam (Keenleyside and Bietz, 1981). In *M. insignis* the parents attached their wriggler young to a sugar cane stem and moved them up and down the stem in response to water-level changes and threatened danger; when the young were free swimming, itinerant family parties of parents with a cloud of 100–200 young swimming above them wandered in and out of macrophyte beds. '*A. vittatus*', *Bujurquina* species and *S. jurupari* spawn on a movable platform, such as a dead leaf, on which they drag the eggs into more favourable water in response to predator threat or deoxygenation.

The direct effects of water-level fluctuations on many aspects of behaviour were shown by Cichocki's observations. *A. heckelii*, which inhabited floodplain lakes where water levels were fairly stable, made an elaborate, long-lasting nest in water 1 m deep; the brightly coloured aggressive parents guarded eggs (*c.* 2000) and young in this nest for 2.5 weeks; as soon as the nest was vacated, other females competed for it and one was then joined there by a male. In *B. cupido*, living in coastal rivers with daily tidally-caused fluctuations in level, the nest was used for one week; the *c.* 100 fry were moved orally at intervals and when free swimming were moved in itinerant family parties. In *S. jurupari* the oral cavity of the inconspicuously coloured parents became the 'nest' after 1 day, thereby reducing the critical period in which eggs and young were exposed on the substrate. The numerous brood predators included many small characins. The period for which the brooding of young was tied to the substrate (subject to deoxygenation and numerous predators) was of crucial importance; in *B. cupido* the fry had to be moved repeatedly, a problem solved in *S. jurupari* by oral incubation. From his observations Cichocki concluded that the water-level fluctuations exact very stringent nesting requirements in these cichlids, and that these requirements in turn impose restraints on other aspects of reproduction, such as clutch size and the form of parental care.

Importance of cover

Rivers offer cover – in crevices in rocks, tree debris, along banks, in weed beds, etc. – which is lacking in open waters of lakes. Riverine hazards are the current, the continually fluctuating conditions induced by water-level changes (including deoxygenation), and high predation, especially of larval and young stages. The open waters of lakes offer increased living space, opportunities to breed throughout much of the year (reduced seasonality) and new foods, but they lack conditions for establishing feeding territories and cover for the free-swimming young; 'territories' are reduced to small mating plaques in arena-spawning cichlids. It is significant that the only substratum-spawning cichlids in the African Great Lakes are those in Lake Tanganyika, which live mainly over rocky bottoms where cichlids can establish territories. Here Nagoshi (1985) showed the effects of crevice size on growth and survival in larval *Lamprologus*, a genus also represented in the Zaire rapids. Conditions in rapids may have preadapted cichlids for life in lacustrine habitats with rocky substrates (Roberts, 1982).

The lack of suitable cover for free-swimming young in the open water of a lake has been proposed as a strong selection pressure for mouthbrooding in lacustrine cichlids. In African cichlids oral incubation has evolved in *Chromidotilapia*, haplochromines, *Sarotherodon* and *Oreochromis*. Water-level fluctuations appear to have selected for oral incubation in at least two different groups of South American cichlids (geophagines and cichlasomines). Thus within the family Cichlidae, oral incubation has evidently evolved a number of times independently in response to different environmental pressures.

Chapter four

Speciation

P. H. Greenwood

4.1 INTRODUCTION

Since over 70% of the known cichlid species occur in Africa, and since most are members of taxonomically and ecologically complex species flocks, it is not surprising that more attention has been paid to speciation in African than in Neotropical and Asian species of the family (references in Fryer and Iles, 1972; Greenwood, 1984; and papers in Echelle and Kornfield, 1984). Some intriguing problems and pertinent data have stemmed from certain American cichlids (Liem and Kaufman, 1984; McKaye, 1980), but any review of cichlid speciation and its evolutionary consequences must, for the moment, draw heavily on information, hypotheses and speculations generated by studies on their African counterparts.

4.2 SPECIATION: THE PRODUCTS

Species diversity and endemism in African lakes and in rivers

Three of the African Great Lakes, Tanganyika, Malawi and Victoria, have fascinated evolutionary biologists for many years, the centre of their attention being the lakes' adaptively multiradiate and highly speciose cichlid faunas. The ecological and morphological diversity and high levels of endemism amongst these fishes contrast sharply not only with those of species belonging to other families, but especially with the cichlids of neighbouring river systems (Table 4.1). On a more muted scale, but in similar contrast to the riverine fish faunas, are the endemic cichlid faunas of the smaller African lakes, such as Edward and George (Fryer and Iles, 1972; Greenwood, 1981a, b, 1984).

In only two lakes, Turkana and Albert, which must be classed as 'Great' lakes on their size and volume, is the situation reversed (Table 4.1). Here there are not only many fewer cichlid species (albeit with a high level of endemism)

Table 4.1 The number of cichlid species and their percentage endemism, compared with those of all other fish families, in the major African lakes and some river systems*. All figures are approximations, and are based on the number of described species together with newly discovered species awaiting description

Water body	*Number of cichlid species*	*Endemism (%)*	*Number of non-cichlid families*	*Number of non-cichlid species*	*Endemism (%)*
Lake:					
Malawi	> 450	99	8	45	63
Tanganyika	165	99	13	75	70
Victoria + Kyoga	> 250	99	11	38	42
Albert	12	98	13	36	9
Edward + George	> 60	99	7	17	9
Turkana	12	98	14	35	16
River:					
Niger	10	20	25	124	4
Nile	10	20	16	105	20
Zaire	40	65	23	*c.* 650	80

*Modified after Coulter *et al.* (1986).

relative both to the lakes' non-cichlids and to the species flocks of the other Great Lakes, but also a noticeably lower level of adaptive radiation and morphological diversity amongst the cichlids (Greenwood, 1981a). Those contrasts may, in part, be due to these water bodies having recently had, or still having, direct connections with a major river system (the Nile); see Lowe-McConnell, 1969a. The other major lakes have been closed drainage basins for the greater part of their histories, their present outflows being relatively youthful and of a physical nature that still provides a large measure of isolation for the lacustrine faunas (Fryer and Iles, 1972; Beadle, 1981).

It is believed that Lake Tanganyika is the oldest African Great Lake (*c.* 5 m years old, but possibly as much as 20 m y), Lake Malawi somewhat younger, *c.* 2 m y (or as little as 100 000 y; Cromie, 1982), and Lake Victoria the youngest at *c.* 750 000 y (or perhaps only 14 000–25 000 y; Livingstone, 1980; Stager *et al.*, 1986).

Overall, the situation, especially in Lakes Victoria, Malawi and Tanganyika, was well summarized by dominey (1984; 239);

> "The uniqueness of (these) cichlids may not rest so much on the rate of adaptation or even the rate of speciation. What is truely unique is that the extensive speciation and adaptive radiation occurred in relatively small geographic areas (a single lake basin), and resulted in the coexistence of extremely diverse and speciose monophyletic assemblages. That one family could so dominate the process on three separate occasions makes the situation even more extraordinary."

Why the riverine cichlids are less speciose than their lacustrine relatives is a question that has received relatively little attention (Lowe-McConnell, 1969a). Current estimates indicate that there are *c.* 250–300, mostly fluviatile, species in America, and *c.* 90–100 in the rivers of Africa. Even when combined, those figures compare poorly with the estimates of somewhat over 1000 species in the six major African lakes the majority of which are strictly confined to lacustrine habitats, and are endemic to their particular lakes (Table 4.1; Victoria and Kyoga are treated together because they are so close geographically, as are Edward and George).

Assuming that, genetically speaking, the riverine cichlids are like their lacustrine relatives, and that the first cichlid colonizers of the lakes came from the rivers, it may seem, *a priori*, that it is the nature of the riverine environment that has inhibited cichlid speciation there. Since, however, rivers provide a diversity of habitats, ecological and trophic niches, and a variety of physical, chemical and biotic features which could serve to isolate populations at both micro- and macro-allopatric levels (Lowe-McConnell, 1969a), it is difficult to identify a particular inhibitory factor. The issue is further complicated by the fact that a few non-cichlid families, whose species are predominantly riverine, have speciated extensively in that biotope. For example, there are over 461 species of Cyprinidae in the rivers of Africa (G.

Howes, pers. comm.). Clearly, here is a speciation problem which demands more attention than it has received.

Morphological diversity and adaptive radiation in African lakes

Much has been written about the morphological diversity of the African lake cichlids, and about the associated adaptive radiation they have undergone. The word 'adaptive' used in this sense relates to the existence of various species having certain character suites which enable them to utilize particular and seemingly specialized trophic niches that are unavailable to species not having those characters (Fryer and Iles, 1972; Liem, 1973, 1978; Greenwood, 1974a, 1981b, 1984). Paradoxically, most 'specialized' feeders are capable of utilizing a wide range of trophic niches, wider indeed than those open to an anatomically more generalized species (Greenwood, 1974a; Liem, 1980a). Far less is known about habitudinal specialization, but there are several indications that such adaptations exist (Greenwood, 1974a; Ribbink *et al.*, 1983a; Witte, 1984a; Ribbink, 1986; Lowe-McConnell, 1987).

In each of the three major African lakes there is virtually no feeding habit or food organism which is not utilized by one, or more usually several, cichlid species. Neither are there any habitats, except the deoxygenated regions of papyrus swamps and the perpetually deoxygenated hypolimnia of Lakes Malawi and Tanganyika, which are not occupied either permanently or temporarily by several species.

The riverine cichlids of both Africa and America also show a wide range of trophic adaptations, although some of those found amongst the African lake species are apparently absent in both riverine groups (Lowe-McConnell, 1987) and, at least in African rivers, the total number of species in any particular trophic group is much lower than that in the species flocks of Lakes Tanganyika, Malawi or Victoria.

Lakes Tanganyika, Victoria and Malawi

Since it is Lakes Tanganyika, Victoria and Malawi that have attracted most attention from evolutionary biologists, especially those interested in problems of speciation, the discussion which follows will deal mainly with the cichlids of those lakes.

Cichlid diversity in these lakes involves, principally, features of cranial morphology and anatomy, dental morphology, and differences in overall body form which are probably correlated, in a functional way, with the cranial and dental morphology (see discussions and illustrations in Fryer and Iles, 1972; Greenwood, 1974a, 1981a, 1983a, 1984; and Barel *et al.*, 1977, for the species of Lakes Malawi and Victoria, and Liem, 1979; Liem and Stewart, 1976; Yamaoka, 1982; Greenwood, 1983b; and Poll, 1986 for those of Lake Tanganyika).

In all three lakes, the principal trophic niches are exploited by species adapted, in the sense explained above, for feeding on benthic or boring insect larvae and pupae, on other benthic arthropods, for scraping or biting off epilithic and epiphytic algae, grazing or browsing on the 'aufwuchs' cover of rocks, taking scales from other fishes (by biting or scraping), feeding on detritus, eating molluscs (which are crushed either within the pharynx or by the jaws and oral dentition), preying on other fishes, including the embryos and buccal young of cichlids obtained in various ways from the mouths of brooding females, taking newly shed eggs from spawning pairs of cichlids, acting as 'cleaners' by removing ectoparasites from fishes, preying on elements of the zooplankton, and for feeding on phytoplankton, either in suspension or from the substrate (Fryer and Iles, 1972; Greenwood, 1974a; Witte and Witte-Maas, 1981; Witte, 1984a; Poll, 1986; Lowe-McConnell, 1987; and further references in these publications).

The anatomical basis for cichlid morphological diversity and adaptive radiation

The morphological features associated with the different trophic adaptations involve, mainly, differences in skull form and jaw shape, the morphology and arrangement of the teeth in the oral and pharyngeal jaws, and subtle differences in the correlated actions of the functional components employed in prey capture and handling. The total range of intrafamilial diversity shown by these different anatomical elements is outstanding, as is the diversity in body form (Fryer and Iles, 1972; Poll, 1957, 1986; Greenwood, 1974a, 1981a, b, 1983a, b, 1985a, 1987b; Barel *et al.*, 1977; Stiassny, 1981b, 1982; Kullander, 1983a, 1986, 1988; Liem, 1973, 1978, 1980a; Liem and Osse, 1975; Liem and Kaufman, 1984; Liem and Stewart, 1976; and chapters by Liem and Yamaoka in this volume).

It has been argued by Liem (1973) and Greenwood (1973a, 1974a, 1981a, b) that the basic cichlid 'bauplan', the fundamental body plan and organization, especially that of the neuro- and splanchnocrania, and of the associated musculature, combined with the nature of the oral and pharyngeal dentition, provide a uniquely plastic substrate for evolutionary change. That is to say, relatively slight alterations to the shape or size of component parts in one or more of these systems can produce far-reaching effects with regard to the potential feeding capabilities of the individual so modified. In our view, this feature is an important factor in explaining the range of cichlid diversity (both morphological and ecological), and the speed with which this diversity has apparently evolved in the species flocks of the African Great Lakes. We do not, however, consider that it is a factor directly affecting rates of speciation *per se* (see Dominey, 1984; and page 96).

A detailed examination of the Lake Victoria cichlid species suggests that even the most extreme morphotypes seen in any of the African lakes could be achieved through relatively gradual modifications involving slight heterochronic shifts in the ontogeny of different anatomical features (Greenwood, 1974a, 1984; Strauss, 1984; Meyer, 1987a). In Lake Victoria, and also in several of the lineages in Lake Malawi, there are species which, at least morphologically speaking, fill the 'gaps' between what are considered to be anatomically generalized taxa and those which are taken to be highly derived, anatomically and trophically specialized forms.

Stasigenic and anagenic products of speciation

If the seemingly gradualistic pattern of evolution through minor changes cumulatively leading to major ones ('cladistic gradualism' of Greenwood, 1981b) does account for the adaptive radiation in the lacustrine flocks, then these anagenic speciation events have occurred together with essentially stasigenic ones. That conclusion is reached because in all three major 'cichlid lakes' there is, at any one level of morphological and anatomical derivation, a cluster of several species which barely differ from one another except in details of their male breeding coloration. Besides the numerous groups of taxonomically 'difficult' species in Lake Victoria (Greenwood, 1981a, b), the phenomenon of stasigenic speciation is well illustrated in the several sibling species recently recognized by Ribbink and his co-workers in Lake Malawi (Ribbink *et al.*, 1983a; Ribbink, 1986).

A predominance of stasigenic over anagenic speciation is apparently a characteristic feature of the tilapiine cichlids in Africa (i.e. the genera *Tilapia*, with 10 species; *Oreochromis*, with 32 species and 17 subspecies; and *Sarotherodon*, with 10 species and 3 subspecies; data from Trewavas, 1983). As a lineage, in all probability a truly phyletic one, the tilapiines are speciose, but within each genus there is little morphological diversity, particularly in anatomical and dental features.

The tilapiine lineage, as restricted here, has apparently given rise to only two species flocks, that in Lake Malawi, a group of just four or five species (Trewavas, 1983) and the morphologically more diverse assemblage of nine species in Lake Barombi Mbo (Trewavas *et al.*, 1972). As Trewavas points out, the tilapiine 'bauplan', especially that associated with its feeding mechanisms, is a specialized one. It is not surprising, therefore, that it is not suited for producing trophically multiradiate species flocks.

It is surprising, however, that a lineage with 52 species (and several subspecies) has not undergone some form of morphologically stasigenic speciation in more than one lake. For example, the geomorphological history of Lake Victoria would seem to provide an ideal background for classical allopatric speciation (see page 94). However, there are only two endemic tilapiine species in that lake, and the species do not seem to be closely related (Trewavas, 1983).

A start can only be made on the whole question of tilapiine versus haplochromine (*sensu* Greenwood, 1979 and 1980) speciation when more genetical information is available, and when more is known about intrafamilial phylogenesis. Certainly the old idea that several endemic lineages in Lake Tanganyika and elsewhere (Regan, 1920; Fryer and Iles, 1972) are related to the tilapiines (as here restricted) must be viewed with suspicion (Greenwood, 1979, 1987a, b).

Recent work by Sage *et al.* (1984) is of particular interest with regard to morphological diversity in cichlid fishes. These authors, working on samples from a small group of 11 Lake Victoria haplochromine species (and virtually all the taxa constituting the North American percoid family Centrarchidae) showed that, in relation to intragroup protein divergence, the morphological divergence of the cichlids was far higher than that of the centrarchids. Sage *et al.* (1984: 192) concluded that possibly "... morphological evolution within the cichlid species flocks [of Lake Victoria] is not concentrated in speciation events", and went on to propose other mechanisms to explain the greater extent of morphological than protein divergence in the Lake Victoria cichlids, whose status as species the authors do not doubt (see page 99). Incidentally, Sage *et al.*'s index of morphological divergence in the cichlids would have been even more disparate, relative both to that for the centrarchids and to the index for protein divergence in the cichlids, had the raw data on which it was based included anatomical features as well as the relatively few morphometric characters actually used.

Be that as it may, there are few who would doubt that high levels of speciation are a feature of the Cichlidae, particularly those in the Great Lakes of Africa. Whether the equally high level of morpho-anatomical differentiation occurred during or after speciation events is a topic for future research.

The repeated occurrence of numerous and varied anatomical features, often replicated very closely, in the endemic species of more than one lake, raises questions of convergent or parallel evolution and thus of the interrelationships of the different flocks, and also of a flock's mono- or polyphyletic origin (Fryer and Iles, 1972; Greenwood, 1983a). The critical phylogenetic analyses necessary to resolve those problems are still in their early and exploratory phases (Stiassny, 1981b, 1982, and this volume; Greenwood, 1978, 1979, 1980, 1983a, b, 1985a,b, 1987a,b).

4.3 SPECIATION: THE PROCESS

As a preface to any discussion of this topic, one cannot improve on Ribbink *et al.*'s (1983a: 298) statement, "A great deal has been written about speciation in African cichlid fishes and the nature of the subject is such that much is speculative, some aspects are controversial while others remain enigmatic." Those views were based, essentially, on summaries given by Fryer and Iles (1972) and Greenwood (1974a), but even had Ribbink and his co-workers been able to read the several more recent papers dealing with, or touching on the subject in Echelle and Kornfield's (1984) book *Evolution of Fish Species Flocks*, their sentiments would still hold. Scenarios, some more suited to Hollywood than *cinéma vérité*, abound, but hard facts (especially genetical ones) and testable hypotheses are in short supply. Ribbink has recently (1986) written a short, well-balanced review and critique of several controversial points, including the reawakened controversy over the species concept (Masters *et al.*, 1987; also page 94.

Probably the earliest comment on speciation in cichlids stems from Plate (1913) who suggested that the Lake Tanganyika cichlids might demonstrate a mode of speciation differing from that of the usual kind, an idea voiced more recently by Trewavas *et al.* (1972) and by Sage *et al.* (1984) for other groups of cichlids. The example which prompted the comments of Trewavas *et al.*, namely the species flock in Lake Barombi Mbo (Cameroon), is, however, probably explicable on the basis of micro-allopatric speciation, a process studied in some detail by Ribbink and others (Ribbink, 1986) in Lake Malawi. Sage *et al.*'s ideas are discussed below (page 99).

External influences

As more information became available on the cichlids of the African lakes, especially following Regan's (1920, 1921, 1922a) fundamental studies, and through the influence of ideas developed in *The New Systematics* (Huxley, 1940), attention was focused largely on the question of how, in those lakes, the spatial isolation necessary for allopatric speciation was achieved, and whether or not the species flocks are examples of sympatric speciation (Brooks, 1950, Mayr, 1942, 1976). Associated with this debate was the suggestion that lacustrine cichlid flocks could be the results of multiple invasions from neighbouring river systems as much as the consequence of intralacustrine evolution.

The role of non-cichlid predators, especially the Nile perch (*Lates*) and tigerfish (*Hydrocynus*), in supposedly suppressing cichlid speciation (particularly in those major lakes, like Albert and Turkana, with depauperate species flocks) was hotly and at times emotionally debated during this period. Various lines of evidence, including the fossil record fail to support the idea that predators have on overall inhibiting effect on speciation (see for a full discussion of the debate Fryer and Iles, 1972) and that, indeed, they may promote speciation (Greenwood, 1959b; Lowe-McConnell, 1969a). Few conclusions can be drawn from the introduction of *Lates* into Lake Victoria in *c.* 1960 (Coulter *et al.*, 1986), and the species' subsequent serious inroads into the fishes, both cichlid and non-cichlid, of the lake, provide an ironic footnote to that debate. But, since the introduction was made into an already highly evolved cichlid species flock. At least for the present that unfortunate experiment throws little light on the evolutionary role such a predator could have played in the context discussed by Worthington (1937) and others (see Fryer and Iles, 1972), that is, an element suppressing speciation.

To return to the other contentious points already mentioned. Little support has been found for the idea that, over time, multiple invasions from affluent rivers could, as it were, provide the necessary degree of allopatry to account for the development of speciose cichlid species flocks within a single lake basin. The debate on allopatric v. sympatric speciation continues, however, as does that on the related, and perhaps more semantic argument about extra- *v.* intralacustrine speciation, especially with regard to the species flocks of Lakes Victoria, Malawi and Tanganyika. It is argued that, because the history of Lake Victoria has involved repeated fragmentation and reunion of its waters consequent upon local climatic changes and tectonic activity, the species flock of the present lake is for the most part of extralacustrine origin. Since, in contrast, the basins of Lakes Malawi and Tanganyika have, for the greater part of their histories, been single water bodies, their species flocks are considered to be of intralacustrine origin (discussions in Witte, 1984a, and other papers in Echelle and Kornfield, 1984).

Consensus opinion on the allopatric *v.* sympatric speciation issue is that, even within the lakes, allopatric speciation predominates. However, in Lake Malawi (and, from its overall ecological and geomorphological similarity with that lake, I would suspect in Lake Tanganyika as well) speciation is thought to be of a micro- rather than a macro-allopatric kind, the latter supposedly characterizing much of the speciation in Lake Victoria. The pronounced stenotopy and philopatry of many cichlid species in Lake Malawi (Ribbink *et al.*, 1983a; Ribbink, 1986) would seem to provide a sound biological basis for the concept of micro-allopatric speciation, but its wider applicability, particularly to the history of Lake Victoria's flock, awaits the results of more field research. Recent studies on that lake, however, indicate that micro-allopatric speciation could have played a significant role (Witte, 1984a), especially in the origin of the deep-water species. To date there has been no evidence of truly sympatric, that is stasipatric, speciation in any cichlid species. But as there has been very little karyological investigation of the family, stasipatric speciation cannot be overruled.

Behavioural and anatomical influences

Although, on the whole, there is general agreement about what might be called the external factors involved in speciation, there is less agreement, or at least more varied views are held, on the other factors and processes involved.

Species recognition and sexual selection

Almost without exception, authors who have discussed speciation in cichlid fishes have done so with Mayr's biological species concept in mind (Mayr, 1963, 1982; Wiley, 1981), and have sought, consciously or unconsciously, to explain the origin of interspecific isolating mechanisms. That is an important issue when species seemingly originated in allopatry, but is more so when sympatric or intralacustrine speciation is implicated. Recently (Ribbink *et al.*, 1983a; Ribbink, 1986, 1988) problems of cichlid speciation have been considered from the viewpoint of Paterson's 'recognition concept' (Paterson, 1985; Masters *et al.*, 1987). That thesis places its emphasis on the origin, not of isolating mechanisms *per se*, but of those mechanisms which ensure syngamy amongst members of the nascent species itself. These conceptual differences are best illustrated by two quotations. First, from Mayr (1963: 89) on the biological species: "The mechanisms that isolate one species reproductively from others are perhaps the most important set of attributes a species has because they are, by definition, the species criteria ... there is a whole set of special devices by which the gap between species is maintained." And then, from Masters *et al.* (1987: 535), on the recognition concept: a species on that basis is "...a group of organisms that share a common system of fertilization mechanisms. A subset of these mechanisms includes the means by which conspecific mating partners recognize each other – Paterson's Specific-mate Recognition System (SMRS). In his [Paterson's] view, it is not barriers to hybridization, but positive recognition between conspecific mates that represents the crucial species criterion."

Masters *et al.* provide a detailed comparison of the two concepts (and their underlying philosophies), concepts which, at first sight, might seem to be merely different sides of the same coin. But I would accept that there is a fundamental difference, and that the recognition concept reflects more accurately the real nature and process of speciation, and is well illustrated by the cichlids, especially the members of a species flock.

It has for many years been accepted that changes in mating systems have played a fundamental part in cichlid speciation (Baerends and Baerends-van Roon, 1950). Particular emphasis in that respect has been placed on sexual dichromatism in mouthbrooding species since, in all the taxa whose live coloration is known, sexually active males have species-specific colours (Fryer and Iles, 1972; Greenwood, 1981a; Ribbink *et al.*, 1983a; Trewavas, 1983; Staeck, 1974, 1977, 1983). Species-specific coloration is also a feature of many non-mouthbrooding but pair-bonding species as well, although it is not necessarily, or usually, expressed through sex-limited dichromatism, (Dominey, 1984).

Dominey (1984) has argued at length for the importance of sexual selection (*sensu* Darwin) not only in bringing about speciation, but also in doing so rapidly. A major element in his hypothesis involves female choice between particular variants in male coloration (or other features). This situation, he believes, could lead to a rapid divergence of those characteristics at both intra- and interpopulation levels, and to a consequent divergence of genotypes within the assortatively mating individuals. In that way, too, species-specific mate recognition signals could evolve. Such a sequence of events might will explain the evolutionary history of the several sibling species complexes among the cichlids of Lake Malawi (Ribbink, *et al.*, 1983a; Ribbink, 1986), especially since there is some, but by no means enough, evidence of the rapidity with which male coloration can change (Fernald and Hirata, 1977; Greenwood, 1965a). However, neither from personal experience nor from the literature (Fryer and Iles, 1972; Schröder, 1986) can I find many unequivocal examples of intrapopulational polymorphism in male coloration amongst mouthbrooding cichlids, the one intrafamilial reproductive guild in which male-limited sexual dimorphism is of prime importance (discussion and references in Dominey, 1984).

There is no doubt that some form of intrapopulational assortative mating and competition among males occurs in cichlids. But could not this result in a stabilizing selection for a particular mate recognition system rather than the evolution of new features? Certainly there is no record amongst cichlids of the 'rare male effect' (i.e. females preferring atypial males; Spiers, 1982) reported in other animals.

According to Dominey (1984: 233): "Sexual selection has special significance with respect to speciation because sexually selected traits are likely to diverge rapidly..." and thus "should be particularly effective at facilitating reproductive isolation." In reaching that conclusion, Dominey drew heavily on the views of West-Eberhard (1983), but no precise data for rapidity of change in cichlids are cited in her paper. Such data on the speed with which male coloration can change in cichlids, as noted above, are rather scant at present.

Paterson's (1980: 330) ideas on the origin of modifications to specific-mate recognition systems stand in complete contradistinction to those of Dominey. Paterson is of the opinion that such changes may occur indirectly and be mediated by pleiotropy, and that effectively "...speciation occurs as an incidental consequence of the adaptation of small isolated populations to a new habitat".

Until we know more about cichlid genetics, particularly at the Mendelian level (Schröder, 1986), it is not possible to discuss the different genetic mechanisms implicit in Dominey's and Paterson's interpretations of how specific-mate recognition systems evolve.

Morphological plasticity

Dominey (1984) considers that the major competing explanation for rapid speciation not consistent with his sexual selection hypothesis is an idea developed by Liem (1973) and Greenwood (1973a, 1981b), namely, that the cichlid 'bauplan' provides a unique substrate for rapid anatomical change, especially in the evolution of different feeding mechanisms (see page 90). In his rejection of the 'bauplan' idea, however, Dominey is confusing two issues, that of speciation *per se*, and that relating to the speed with which anatomical and functional changes can be effected. The existence, in all the major cichlid flocks, of species groups (at every level of trophic specialization) showing virtually no interspecific differences in their anatomical and other morphological features (except male coloration) would seem to emphasize the difference between, indeed the unrelatedness of, the two concepts. Granted, some sort of genetic or epigenetic reorganization is required to produce the anatomical changes, and a speciation event seems a probable time at which that could happen. But such changes are, unless Paterson's (1980) ideas on pleiotropy can be substantiated, more likely to be the results rather than the primary cause of speciation. Until the Patersonian sequence can be established as invariable or common, Dominey's model is as valid as any other put forward to explain the rapidity of speciation in lacustrine species flocks. It is not sufficient, however, to explain the speed with which non-sexually related morphological changes could occur, nor does it impair Paterson's species concept.

Colour polymorphism. The part played by coloration in the life history of cichlid fishes requires some comment on colour polymorphism, especially since such polymorphism would be a feature, albeit a transient one, if the sexual selection model were operative in speciation.

Sex-limited polychromatism is not uncommon amongst African cichlids (Fryer and Iles, 1972; Greenwood, 1981a), although some examples of male polychromatism are now known to be based on errors of taxonomy, the different morphs actually being distinct species (Marsh *et al.*, 1981; Ribbink, 1986) and the phenomenon is usually confined to females. The only fully documented case of supposed polychromatism in an American cichlid, *Cichlasoma citrinellum*, certainly suggests that the two colour morphs provide an example of sympatric speciation in progress (McKaye, 1980), since one form breeds in deeper water than the other, and there are differences in the nature of their pharyngeal jaws (Meyer, 1987).

Examples of supposed polychromatism in which the 'morphs' proved to be distinct species (Marsh *et al.*, 1981; Ribbink, 1986) provide further support for the idea that male coloration is an important part of the specific-mate recognition system in non-pair-bonding cichlid species. (It is apparently of reduced or different significance in pair-bonding species; Dominey, 1984). These examples would also seem to support the view that colour polymorphism, either in itself or when linked with other and ecologically important differences, could be a likely starting point for the kind of micro-allopatric speciation thought to have taken place in some of the African Great Lakes (see page 94).

Interestingly, and for the moment inexplicably, no examples of intra-populational polychromatism are known from the fluviatile cichlids of Africa, and the phenomenon is seemingly very rare in the American species as well. However, there are distinctive interpopulational differences in the male coloration of the southern African species *Pseudocrenilabrus philander* (Ribbink, pers. comm.). Whether or not this is another example of incipient, or of actual, speciation is currently under review.

Anatomical polymorphism in Cichlasoma minckleyi.. The only instance of genetically based anatomical polymorphism in cichlids involves the newly recognized Mexican species *Cichlasoma minckleyi* (full discussion in Liem and Kaufman, 1984). One of the two morphs is characterized by its greatly enlarged and stout pharyngeal jaws, enlarged and molariform pharyngeal teeth, certain hypertrophied elements in the musculature associated with the pharyngeal jaws, and in osteological features associated with the hypertrophy of those muscles. The second morph has relatively delicate pharyngeal jaws, fine and papilliform teeth in both pharyngeal jaws, no hypertrophy in the pharyngeal musculature (thus it lacks the associated neurocranial modifications), and it has a longer gut.

By the usual taxonomic criteria applied to museum specimens, the two morphs would each be given specific status. But in nature there is no indication of assortative mating between the forms, nor of any consistent ecological differentiation either, except, seasonally, a more durophagous diet in the morph with enlarged pharyngeal jaws. In fact, there are no indications that this might be another case of incipient sympatric speciation as has been suggested for *Cichlasoma citrinellum*, the other polymorphic (i.e. polychromatic) species (McKaye, 1980; page 97).

None would deny, on the anatomical evidence available, and the absence of assortative mating between morphs, that *C. minckleyi* is a truly polymorphic species, but historically there is the interesting suggestion that it could have originated as the result of introgressive hybridization between two species, populations of which are still extant in the region (Trewavas, pers. comm.; Liem and Kaufman, 1984).

Polymorphism and hybridization. The potential contribution of hybridization to speciation in cichlids, either directly or indirectly as the soure of polymorphism, has not been given much attention. In part that may result from the apparent rarity of naturally occurring hybrids, even within species flocks (Fryer and Iles, 1972; Greenwood, 1974a); hybrids, however, have been produced experimentally in laboratory stocks of Lake Victoria species (Crapon de Caprona, 1986). The possibilities raised by *Cichlasoma minckleyi*, and the fact that some of the species used in the hybridization experiments are morphologically disparate and may represent different phyletic lineages, should help to focus future attention on the part hybridization could play in speciation. Crapon de Caprona's work (1982, 1986; Crapon de Caprona and Fritzsch, 1984) is also very pertinent to the question of specific-mate recognition cues, both visual and chemical.

Non-genetic polymorphism. Naturally occurring or artificially induced ecophenotypic polymorphism (that is, polymorphism which is not genetically fixed) has been reported in a few cichlid species (Greenwood, 1965b; Hoogerhoud, 1986b; Witte, 1984b; Meyer, 1987a). The degree to which these ecophenotypes depart from their 'normal' conspecifics is sometimes very marked. This has led Meyer (1987a) to believe that the taxonomic, ecological and evolutionary analyses of species flocks may be in need of revision, especially with regard to the number of species involved; similar sentiments were expressed by Sage and Selander (1975) when the true polymorphism of *Cichlasoma minckleyi* was first discovered.

Meyer appears to confuse ecophenotypic variation (which he studied by its artificial induction in *Cichlasoma managuense*) with true polymorphism, and he seriously underestimates contemporary taxonomists' appreciation of phenotypic variability in cichlids. He also seems unaware of the attention these workers pay to ecological and non-anatomical characters, like coloration, before they formulate their taxonomic conclusions. Indeed, recent work on once supposedly polymorphic species suggests that we may be underestimating rather than overestimating species numbers in flocks like those of Lake Malawi (see Ribbink, 1986 for references).

Be that as it may, ecophenotypically induced plasticity exists, and its role in speciation and evolution cannot be ignored either. That would be particularly true if environmentally induced, and therefore potentially reversible changes could be incorporated in the genome through some form of genetic assimilation (*sensu* Waddington, 1975), and if those genetic changes were, as Paterson (1980) suggested, pleiotropically linked with the specific-mate recognition system.

There is also the possibility, as Meyer (1987a) suggests on the basis of ideas developed by Sultan (1987), that plasticity may be a form of inertia against speciation. That I feel is a highly debatable point. Nevertheless, as a concept apparently opposed to those in which plasticity, in the form of polymorphism, is an integral part, it must be considered. In particular, the inertia concept would run counter to Rosenzweig's (1978) idea of competitive speciation, a model incorporated by Dominey (1984) and others in their theories relating to the evolution of, and differentiation within, species flocks (see various discussions on competitive speciation in Echelle and Kornfield, 1984).

Behaviour and learning

A very different approach to the problems of speciation and morphological diversity in cichlids is that of Sage *et al.* (1984). These authors showed that electrophoretic comparisons of proteins in 11 morphologically diverse members of the Lake Victoria species flock indicated extremely small genetic differences between the species (*viz.* 0.006 electrophoretically detectable substitutions per locus), despite normal levels of intraspecific variability. In marked contrast, the sample's index of interspecific morphological distance was high. This disparity between the genetic and morphological indices of divergence in the cichlids is emphasized when it is compared with those obtained for members of the Centrarchidae (American freshwater sunfishes) by the same authors. Here there is a very much higher index of genetical divergence, and only a slightly higher one for morphological divergence.

Sage *et al.* (1984), impressed by the genetical/morphologial disparity in cichlids, and by the presumed speed at which speciation had occurred in Lake Victoria, believe that morphological differentiation might not have been concentrated in speciation events, but took place after these had occurred. In their view, speciation would be effected, and rapidly so, through the process of sexual selection proposed by Dominey (1984), that process being accelerated by socially transmitted shifts in mate recognition systems (see also Lande, 1982; West-Eberhard, 1983; Dominey, 1984).

Morphological differentiation in the Sage *et al.* model would result from a behavioural drive mechanism (Wyles *et al.*, 1983), a mechanism which would involve quick intrapopulational propagation, by social learning, of any novel and successful feeding behaviour which arose in a member of the population. The spread of that particular behaviour pattern would, Sage *et al.* believe, set up selection pressures favouring the genetic fixation of mutations complementing the new behaviour, especially mutations which affected structures associated with feeding mechanisms. Sage and his co-workers then go on to develop their idea by suggesting that, for example, a new feeding behaviour might, through social learning, be transmitted to other species. They conclude that "... successful new behavioural combinations may be more likely to appear in a species flock that experiences crosstalk than in a lake with large populations of fish that do not engage in interspecific learning."

As Sage *et al.* rightly point out, their speculations underline the need for further studies. At least one of the parameters in their model, large effective population sizes, is unlikely to be satisfied in nature. The estimated total number of cichlid individuals in the Lake Victoria flock, 10^{12} or 10^{13}, given by Fryer and Iles (1972) and quoted by Sage *et al.* (1984), in no way reflects individual population sizes. To judge from some field studies, population sizes can be very small, as low as 100 fishes (Ribbink, pers. comm. *re*: Lake Malawi), and few if any species in Lakes Malawi and Victoria have spatially continuous populations (Fryer and Iles, 1972; Ribbink *et al.* 1983a; Witte, 1984a; pers. obs.).

4.4 CONCLUSIONS AND SUMMARY

It is not yet possible to draw any solid conclusions about the methods and processes involved in the speciation of cichlid fishes. That the Cichlidae is one of the most speciose and morphologically diverse families of perciform fishes is accepted by most ichthyologists. Even the doubters (like Meyer, 1987a) would have difficulty in substantially reducing the number of described species.

That most cichlid speciation, and adaptive radiation, has happened in lacustrine rather than riverine environments is also very obvious. It could be argued that, if the fossil record (van Couvering, 1982) for the family was better, we might discover that the riverine progenitors of the present-day lake flocks were more speciose and anatomically diverse than their modern descendants (as, indeed, if only by a narrow margin, the modern American fluviatile species are more diverse than their African counterparts). Gross diversity might be determinable in fossils, but most of the characters used to identify living fishes would not be preserved, and so the number of species would be impossible to estimate accurately.

The anatomical features underlying the many different types of feeding mechanisms represented amongst the cichlids were very probably brought about through slight changes in dentition and syncranial anatomy, either on their own or in various interrelated combinations. We are, however, almost totally ignorant of the genetical factors and epigenetical processes through which such changes were effected.

Our ignorance of cichlid genetics undermines some of the ideas put forward to explain the mechanism of speciation. Few of those ideas are testable hypotheses, and our understanding of the biotic and abiotic factors which could have provided the physical isolation necessary for speciation is uncertain, especially when it is dependent on historical records. No evidence has yet been adduced for truly sympatric speciation amongst the Cichlidae, but again we are very ill informed about cichlid karyology and are thus unlikely to have detected examples of stasipatric speciation.

On a broader canvas, we are still in the exploratory stages of investigating an important outcome of speciation, intrafamilial phylogenesis, so we can rarely distinguish between convergent and parallel evolution with certainty. A better knowledge of cichlid phylogenetic relationships would also help in the choice of taxa for comparative studies, be they of anatomy, behaviour or hybridization.

Supporters both of the biological and of the recognition species concept generally concur that specific-mate recognition systems are essential to speciation. Species-specific male coloration may be particularly important in speciation, especially in mouthbrooding taxa, but we still do not know how females recognize conspecific males on this basis. Coloration is also probably of importance in the speciation of pair-bonding, usually substrate spawning and guarding taxa. Other components of the specific-mate recognition system, such as pheromones, and courtship behaviour patterns have not been studied in as much detail as coloration, and have rarely been considered in the context of speciation. To the human ege, for example, courtship patterns amongst supposedly related – and even distantly related – species belonging to the same reproductive guild are apparently identical, and thus could be dismissed as important components of a species-specific fertilization system. But need that be the case when these visual signals are received and interpreted by the fishes involved? Would it be stretching an analogy too far (especially one based on a comparison between classes) to suggest, however, that a newcomer to ballet, lacking the knowledge of a balletomane, is unlikely to appreciate the subtle differences of style which, for the latter, would make or mar a performance?

Estimates of the time taken to produce the highly endemic and diversified species flocks of Lakes Victoria, Malawi and Tanganyika vary considerably, but few would cavil at the use of adjectives like 'rapid' or even 'explosive' to describe the speciation rates in these lakes, especially that in Lake Victoria which has been variously estimated as between 14 000 and 750 000 years old.

This speed of evolution has raised still-unanswered questions about the relative efficiency, in that regard, of the different speciation mechanisms that have been postulated. Some 80 y ago it was suggested that perhaps the speciosity of the Cichlidae could be attributed to a unique mode of speciation. Subsequent research has not provided any direct evidence to support this idea, and even some of the more speculative suggestions have been based on the extension of ideas formulated from the study of other animal groups.

The evolutionary success of the cichlids could lie in the family, as we know it today, having arisen as the results of several different modes of speciation, none of which can be considered peculiar to the family. But why cichlids should have had, or may still have this pluripotentiality, and whether it is present in other groups, are unanswered questions whose solution will involve disentangling a complex of genetical and environmental factors. The existence within the Cichlidae of both species-rich and species-poor lineages could, however, provide an entry into this field of research.

All in all, the only firm conclusion that can be drawn about cichlid speciation is the now hackneyed one, that these animals provide biologists with unusual opportunities for studying a wide range of evolutionary phenomena. Even what seems to be the biological disaster resulting from the effects of an introduced predator on the endemic cichlid species flock of Lake Victoria (Coulter *et al.*, 1986) opens up rare, and one hopes not to be repeated, chances for studying evolution in action.

ACKNOWLEDGEMENTS

I would thank all those, past and present, who have provided data, ideas, irritation and encouragement to those working in this field. In particular I must express my gratitude to Ethelwynn Trewavas, Geoffrey Fryer, Tony Ribbink, Karel Liem, Rosemary Lowe-McConnell, Gordon Howes, Kees Barel and the HEST team, all of whom have contributed in good measure to my thoughts on this intriguing subject.

Chapter five

Genetics

I. Kornfield

5.1 INTRODUCTION

The genetics of cichlid fishes can be loosely divided into three broad and somewhat overlapping areas: (1) descriptive studies characterizing attributes of the family or groups of taxa, (2) studies examining genetics as they bear on systematic problems, particularly questions of phylogeny and taxonomy, and (3) studies exploiting genetics to examine the processes of evolution in the family. Here I present genetic data applicable to all these areas, and emphasize the evolutionary perspective. I summarize recently published information and include data from on going research.

My perspectives on the genetics of cichlids are biased in several dimensions. First, I consider that many molecular methodologies provide a means to erect accurate phylogenies; with a few caveats, molecular systematics can be used as the benchmark against which other data sets may be evaluated. Second, because of major differences in the reproductive biology of Old and New World cichlids, I consider it inappropriate to extend generalizations uncritically from one fauna to the other. While basic genetic properties, e.g. linkage groups or DNA base composition, may be shared throughout the family, evolutionary mechanisms controlling processes may differ importantly (McElroy and Kornfield, 1989). Evaluations of hypotheses about differentiation in Old World faunas, particularly those of the African Great Lakes, are appropriately made using Old World taxa. Note that this concern extends to comparisons including tilapias and haplochromines, whose reproductive behaviour differs considerably. Likewise, evolutionary generalizations applicable to all elements of an endemic fauna should be approached cautiously. For example, though all haplochromines in Lake Malawi are closely related, basic differences in mouthbrooding may impose evolutionary constraints. Non-Mbuna fry, e.g. *Cyrtocara* spp., are guarded after prolonged incubation while free swimming, while Mbuna are released among rocks (Trendall, 1988b). Mbuna fry therefore cannot imprint visually on their mother (Siepen and Crapon de Caprona, 1986). Nevertheless, perhaps some findings in one of these systems may be applicable to all. Third, because many fruitful studies have been conducted on elements of the endemic fauna of Lake Malawi, I tend implicitly to evaluate information from other systems with respect to these baselines. Finally, I am critical of *ad hoc* hypotheses. Selection, both natural and sexual, can explain so much that it can easily become the default (Gould and Lewontin, 1979); it is important explicitly to formulate testable predictions and fully to articulate assumptions.

5.2 DESCRIPTIVE GENETICS

Kornfield (1984) summarized descriptive genetics. Since 1983, considerable progress has been made in several genetic areas related to the aquaculture of tilapias (overview, Pullin, 1988); note again that basic genetic insights made for these taxa may extend to other members of the family.

Gross chromosomal characterizations have been extended to additional taxa and support earlier generalizations about the conservative nature of karyotypic evolution in the family. Feldberg and Bertollo (1985) studied the previously uncharacterized New World genera *Batrachops*, *Chaetobranchopsis* and *Gymnogeophagus*, further extending conservation of the basic Neotropical karyotype ($2n = 48$). In the Old World fauna, Foerster and Schartl (1987) and Meyer and Schartl (1984) examined *Aulonocara* spp. and additional *Pseudotropheus* taxa from Lake Malawi; diploid number ($2n = 44$) and gross morphology conform to previous findings on this endemic fauna (Kornfield, 1984: Fig. 7). Chromosomal studies on tilapias (Majumdar and McAndrew, 1986) have extended the number of characterized taxa and agree with earlier findings (Kornfield *et al.*, 1979). In aggregate, these data reinforce the dichotomy in chromosome number between the New and Old Worlds. Within faunas, cytological studies investigating comparative banding patterns still have not been undertaken; given the conservative nature of the cichlid karyotype and the increasing availability of acute molecular techniques to provide less-ambiguous phylogenetic data, such studies may now be of little value. The accuracy of phylogenetic inferences based on karyotypic changes (e.g. fundamental number) remains speculative and requires independent evaluation. Nonetheless, it would be useful to survey chromosomal diversity in taxa that have not been investigated – most obviously, lineages within the endemic faunas of Tanganyika and Madagascar. Characterization and comparative analyses of the Malagasy endemics in particular could help to establish the pleisomorphic condition of the cichlid karyotype.

Understanding of sex determination in the family may be central to successful long-term aquaculture. The malleability of the sex-determining system may already have retarded the culture of tilapias in regions where production needs are greatest. Sustained all-male cultures can significantly increase yield through greater somatic (as opposed to gonadal) coversion and faster growth. Despite the attraction of monosex culture, females have occurred persistently and unpredictably in crosses designed to produce only males.

Females occur in monosex systems for at least two reasons. First, autosomal genes contribute to sex determination (Avtalion, 1982; Tave, 1987, 1988; Lester *et al.*, 1989; Avtalion and Don, 1988). While alleles at supposed sex-determining loci can be manipulated, e.g. by hybridization, simple crosses may not control for these additional sources of allelic variation, because the genetics of autosomal loci have not been established. Autosomal variation need not be discrete, e.g. possessing simple dominance among alleles, but could be expressed incompletely or regulated in some cascading manner (e.g. Baker, 1989). The identification of sex-specific regulatory loci in cichlids may progress rapidly if findings in other vertebrate systems are exploited (Wachtel, 1988) or if the genome is examined with sufficiently resolved molecular probes. Second, spontaneous sex reversal in tilapias, though infrequent, is ubiquitous. For example, the production of 'YY supermales' in *Oreochromis niloticus niloticus* by Scott *et al.* (1989) was made possible by use of a phenotypic female which was a genetic (XY) male; this individual was discovered accidentally. The frequency of spontaneous sex reversal has not been estimated, but seems to be greater than that of occurrence of new colour variants. Anecdotal reports of sex reversal from female to male in tilapias are not common; in *Oreochromis aureus*, we have observed the production of fry in tanks which held only mature females. Such sex reversal does not occur, or at least is at an extremely low frequency, in commercial culture of Malawi haplochromines where sexes are routinely maintained in separate holding facilities; fry have never been reported in such segregated systems. Histological studies should be conducted to evaluate claims of sex reversal (e.g. Naish and Ribbinnk, 1990). As with tilapias, sex reversal in haplochromines can be induced by hormonal treatment early in development (DeMason, 1988). Nonetheless, the mechanism(s) regulating spontaneous sex reversal has not been characterized. It may be possible to eliminate sex-reversed phenotypic females from experimental systems by careful monitoring and selected breeding. However, in applied monosex culture systems, natural selection would reward spontaneously sex-reversed fish with extreme fitness: this is particularly important in less-developed areas where augmentation of brood stock is infrequent.

Much experimental work has focused on the production and performance of polyploids in tilapias. The principal rationale is the expectation that in polyploids, gonadal development may be suppressed so that additional energy may be channelled into increased somatic growth (Wolters *et al.*, 1982). Triploids have been produced in *Oreochromis niloticus*, *O. messambicus* and *O. aureus* by cold and heat shocks (Chorrout and Itskovich, 1983; Don and Avtalion, 1986, 1988) and tetraploids have been induced in *O. niloticus* and *O. massambicus* by thermal and pressure shocks (Myers, 1986). Optimal treatment differs both between and within species; the origin and genetic background of experimental animals may be critical to the results. Nevertheless, despite the potential of the technique, the polyploids have not performed better than diploids.

An alternative strategy for increasing production in tilapias involves gene transfer or transgenic modification. Brem *et al.*, (1988) demonstrated the expression of human growth hormone in *Oreochromis niloticus niloticus* that had been microinjected with DNA early in development. Complete transmission of this genetic material and its expression in F_1s has not been reported. Though this technique is attractive, potential problems associated with the release of such genetically engineered fishes presents numerous problems, including hybridization and competition with native fauna, which need to be considered objectively prior to research investments.

More objective assessments of performance and selection in tilapias have been possible because of concern for the identity of the fishes being used. Careful consideration of the origins of stocks that have been dispersed globally in aquaculture has enabled evaluation of the potential for lineage mixing and the possibility of founder effects (Pullin, 1988). Macaranas *et al.* (1986) provide a particularly clear example of the application of molecular methods to the aquaculture of tilapias; widespread hybridization of *O. mossambicus* with cultured *O. niloticus* was clearly demonstrated by a simple allozyme assay. Stocks of tilapias used in production studies must be identifiable unambiguously if replication elsewhere is desired. It is particularly desirable that past (un) intentional hybridization events be identified, because they may have introduced additional and perhaps novel genetic variation. The International Center for Living Aquatic Resources Management (ICLARM) in Manila has taken the lead in fostering research to support this objective. While the problem has been technically exacerbated because of low levels of intraspecific and intersubspecific allozyme differentiation in many taxa of interest, alternative molecular techniques are being developed (p. 120). An integrated approach to the genetic identification of cultured cichlids should include (1) known and constantly reliable genetic lineages, (2) standardized molecular techniques and nomenclature, and (3) infrastructure necessary to deliver information and methodology to workers in the field as well as to research groups. Given the global scale of cichlid aquaculture, these are necessary but tractable long-term goals.

5.3 GENETICS OF THE PHENOTYPE

Differentiation in cichlid fishes, unlike that in many other teleost groups, has captured the imagination of systematists and evolutionary biologists because of the ubiquity of polymorphism and plasticity as well as rapid evolution. It is now clear that both characterization of, and experimentation with, intraspecific variation are prerequisites to appreciating the significance and origin of interspecific morphological and trophic diversity. Intraspecific variability has at least two discrete foci: genetically based variation, and phenotypic plasticity. Although spectacular advances in molecular biology have provided insights into the fundamental nature of differentiation at the cellular level, a comprehensive understanding of phenotypic expression at the organismic level remains elusive. Confounding this is the interaction of genetics and environment, particularly during ontogeny. Attempts to understand the balance between genetic determinism and environmental influence in cichlids requires rigorous, well-planned and often long-term breeding experiments.

Morphology

Polymorphism in cichlid morphology, first suggested by Sage and Selander (1975) for *Cichlasomaminckleyi* of Cuatro Cienegas, Mexico, has now been demonstrated in other Neotropical taxa. In *C. citrinellum*, the polychromatic species endemic to the lakes of Nicaragua (Barlow, 1976), and in *C. haitensis* endemic to Haiti, a trophic polymorphism involving pharyngeal dentition has been discovered by Meyer (1989, pers. comm.) As in *C. minckleyi*, two alternative dental types are present together in natural populations: individuals with slender jaws and papilliform teeth, and those with robust jaws and molariform teeth. However, unlike *C. minckleyi*, Meyer (pers. comm.) believes that phenotypic plasticity controls the polymorphism; unfortunately, the results of formal inheritance experiments are not yet available for these taxa. (In *C. minckleyi*, formal breeding studies have confirmed a genetic basis for the polymorphism (Kornfield and Taylor, unpubl.). The progeny of single-pair crosses between molariform and papilliform trophic types have displayed both pharyngeal morphs when raised on uniform soft-food diets. Sex linkage in this system, suggested by Kornfield and Taylor (1983), has yet to be confirmed).

In the Old World fauna, a polymorphism in the pharyngeal dentition of *Labidochromis caeruleus* from northern Lake Malawi was suggested by Lewis (1982); however, controlled dietary or breeding experiments have not been conducted. Additional morphological polymorphisms have been suggested for *L. mylodon* and *L. ianthinuj* (P. Reinthal, pers. comm.). In all these systems, an association between phenotype and resource availability, principally the abundance of gastropods, has been suggested.

In contrast to pharyngeal dentition, no discrete polymorphisms involving oral dentition have been confirmed, but the probability of finding them seems high, given (1) the plasticity in these characters and (2) the correlation between variation in the pharyngeal jaw apparatus and the oral jaw apparatus (Barel, 1983; Hoogerhoud, 1986b). Dietary shifts, perhaps mediated by climatic and hydrological changes, may have cascading effects upon all portions of the trophic morphology.

In contrast to discrete polymorphisms, there is now considerable experimental evidence for more-continuous, but equally distinctive, phenotypic plasticity in both New World and Old World cichlids. In the New World cichlid fauna, several important studies on plasticity of tropic morphology were completed in the late 1980s. Meyer (1987) manipulated the diet of *Cichlasoma managuense* from Nicaragua and quantified changes in head morphology. Split broods received diets either of *Artemia* nauplii or of commercial flake food plus nematodes for eight months; the differences in these two treatments involved both diet and mode of food capture. Significant differences were observed in univariate attributes of head morphology reminiscent of 'biters' and 'suckers', two of the basic cichlid functional trophic groups (Barel, 1983; Meyer, 1987a). These differences could be reversed after an additional eight months on a common diet. Since the two treatments involved both diet and mode of food capture, the stimuli explicitly responsible for the observed plasticity have yet to be characterized. Nevertheless the flexibility in the development of trophic morphology in *C. managuense* suggests that phenotypic plasticity should be considered first when examining natural trophic variation.

More recently, P. Wimberger (unpubl.) has conducted split-brood plasticity experiments on the Neotropical species *Geophagus brasiliensis* and *G. steindachneri*. Fish were fed either *Artemia* nauplii or blood-worms; again, the treatment involved both diet and mode of feeding. Using truss analysis (Bookstein *et al.*, 1985), Wimberger demonstrated small but important responses in both these taxa; nauplii treatment resulted in head-shape differences, while the worm treatment produced caudal modifications. Wimberger (pers. comm.) notes that nutrition, particularly the availability of vitamins, may profoundly influence morphology.

The classic observations of Greenwood (1965b) on pharyngeal dimorphism in *Astatoreochromis alluaudi* have been experimentally extended and confirmed by Hoogerhoud (1986b). Molariform pharyngeal dentition in this widely distributed Old World species can be induced by feeding snails to papilliform fish. As with pharyngeal polymorphisms (Liem and Kaufman, 1984), correlated changes include alterations to the vertebral apophysis and to associated muscles and ligaments.

Witte (1984b) has demonstrated considerable plasticity in the oral dentition of *Harpagochromis* (= *Haplochromis*) *squamipinnis*, an endemic of Lakes Edward and George, East Africa. Comparisons of wild-caught specimens, which are nominally piscivorous predators, with fish that had been reared in aquaria on soft-food diets, revealed significant changes in simple shape descriptors of the premaxillae. Witte (1984b) suggested that these changes were epigenetic (induced by diet or feeding), converting suckers into biters. Unfortunately, the number of wild-caught specimens used in the aquarium experiments was very small and the study has not been repeated.

Our work with *Oreochromis aureus* suggests that pharyngeal dentition may be particularly malleable in tilapiine lineages. Broods of *O. aureus* from single-pair matings were split into two groups approximately one month after yolk-sac absorption and were reared in a common recirculated water system. Fish were fed *ad libitum* on 4 mm trout pellets (Rangen FPSS, 38–490, Zeigler Bros., Inc.). However, the form of the food differed: in one treatment (hard food), fish were given unmodified pellets, and in the other (soft food), pellets were crushed in a mortar and the uncompacted food was provided. Pellets could readily be crushed by hand; the average pressure required to shear pellets was less than 5 N. It is important that the mean weight of pellets was < 20 mg and the maximum force applied by fish to crush the food pellets was approximately one-tenth that experienced by some snail-eating cichlids (Fig. 5.1). Treatments were administered for seven months, then the fish ($n = 60$) were killed and their pharyngeal jaws were examined. On average, fish fed intact trout pellets grew slower than did their sibs fed soft food. Although the form of individual teeth was the same, the shape of the dentigerous area of the lower pharyngeal jaw differed significantly between groups ($P < 0.001$; Fig. 5.2). In addition, the relative size of the pharyngeal horns (PHW; Kornfield and Taylor, 1983) was significantly greater in the hard treatment ($P < 0.005$).

Given the general plasticity of vertebrate bone (Currey, 1984), some of the changes observed in natural and experimental systems noted above are not unexpected. Additional critical study in simulated natural or stressed environments would clearly be of value.

Coloration

Their varied coloration has been a principal attraction of cichlids for both biologists and aquarists. Colour has long been considered by most as integral to specific recognition; indeed, many taxa in the African lake systems have been elevated to specific rank principally (and occasionally, solely) on the basis of male breeding coloration (Greenwood, 1965b; Ribbink *et al.*, 1983a; Witte and Witte-Maas, 1987). It has been established from spectral absorption data of visual pigments and retinal morphology that cichlids can indeed see colour (Levine and MacNichol, 1979; Levine *et al.*, 1981; Fernald, 1984), although their spectral acuity may be less than that of higher vertebrates. Despite the common perception that cichlids are a colourful group, critical studies of the magnitude of colour diversity are limited. Investigations of diversity within the Malawi Mbuna by McElroy *et al.* (in prep.) have suggested that this group displays limited innovation in colour diversity; portions of visual spectrum nominally available for the group are not occupied. Diversity in colour pattern may also be restricted in this fauna, possibly because of selective constraints

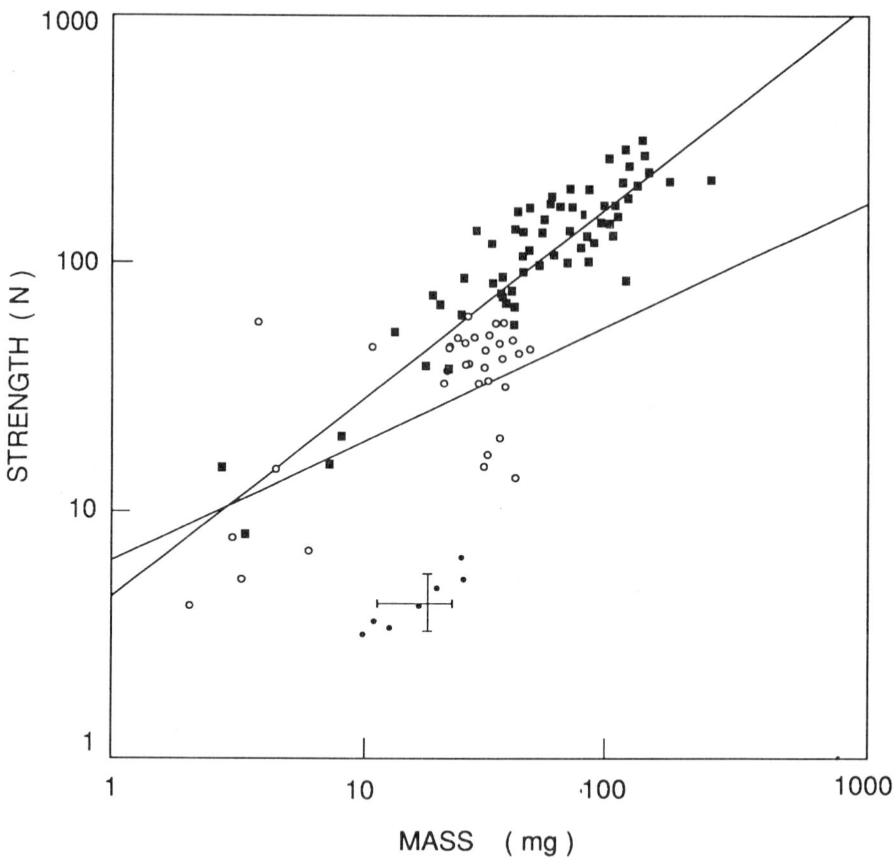

Fig. 5.1 Absolute strength (Newtons) required to fracture gastropod shells and food pellets. Forces were measured on formalin-preserved snails using an Instron Testing machine traversing at 5 mm min^{-1}. Weight is highly correlated with snail size (discussion, Slootweg, 1987); each snail weighted less than 300 mg. ●, strength of food pellets. Cross represents 95% confidence limits to food pellet means. Curves, observed regressions for Cuatro Cienegas, Mexico, endemics; ■, *Nymphophilus minckleyi* and *Mexithalma* sp., log strength = 0.703 + 0.790 log mass; ○, *Mexipyrgus* sp., log strength = 0.826 + 0.490 log mass. Slopes of the two regressions differed significantly ($P < 0.001$). See Vermeij and Currey (1980), Ash (1989), and Meyer (1990) for methods and additional considerations.

imposed by community composition (McElroy and Kornfield, unpubl.). That is, isolated communities may independently contain unrelated endemic taxa with similar colour patterns, e.g. yellow flanks in *Pseudotropheus* and *Petrotilapia*.

While geographic variation in coloration is ubiquitous, true colour polymorphisms (multiple sympatric colour morphs unrelated to sex) are rare. Putative colour polymorphisms in some Malawi Mbuna (Fryer and Iles, 1972) have

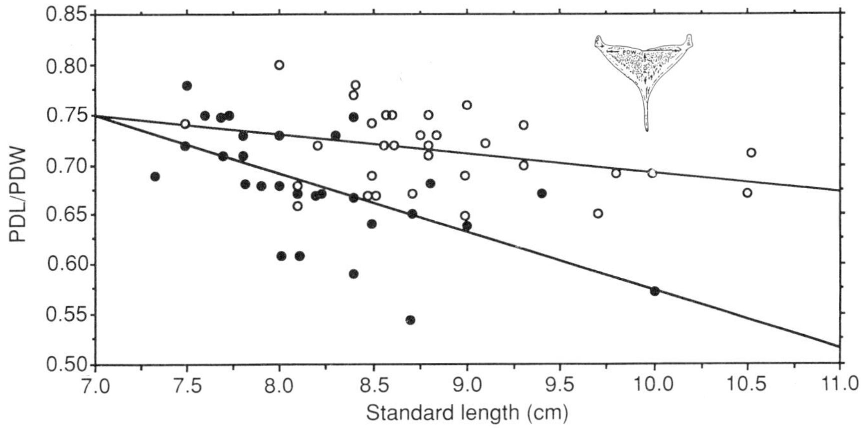

Fig. 5.2 Changes in the pharyngeal dentition of *Oreochromis aureus* induced by feeding intact (hard food ●), or ground (soft food ○) trout pellets. The relative shape of the dentigerous area of the lower pharyngeal plate (PDL/PDW; see Barel *et al.*, 1977) differed significantly between groups (ANCOVA, $P < 0.001$). Slopes of the two regression lines did not differ significantly.

been shown unambiguously to consist of sibling species complexes (Holzberg, 1978; Marsh *et al.*, 1981). One type of discrete colour variation that may be common in cichlids involves the 'sneaky male' phenotype. Instead of the normal adult phenotype, reproductively mature males may possess female or juvenile coloration enabling them to steal fertilizations during the reproduction of others (Keenleyside, 1972); this behaviour may be facilitated by oral fertilization (Mrowka, 1987b) in some mouthbrooders. While the sneaking strategy involves only-transient colour changes in some fishes, maybe some cichlid species possess permanent sneaker phenotypes. Goldschmidt (1989a) has presented preliminary evidence on intraspecific variation in egg spot coloration among Lake Victoria haplochromines. Since egg spot coloration appears to match transmission characteristics of water masses, colour differences among allopatric conspecifics may not be unexpected if the value of egg spots as signals is high. The extent of genetic variation and phenotypic expression have not been characterized.

In general, adult colour can be thought of as a stable attribute of a particular taxon, despite changes associated with behaviour. Transient alterations in body coloration and in colour patterns are the rule in most cichlids, and colour phenotype can be used to assess behavioural or physiological status (Baerends and Baerends-van Roon, 1950; Lanzing and Bower, 1974; Carlstead, 1983). However, male reproductive coloration is generally well defined and is equivalent in many taxa to aggressive coloration so it can be induced relatively easily by stimuli such as mirrors (for territorial species) or hormones (Fernald, 1976). The histological basis of colour variation in brightly coloured and normal conspecific cichlids has been investigated (Dickman *et al.*, 1980, McAndrew *et al.*, 1988; Lanzing and Bower, 1974). These histological studies have established that coloration results from the differential development of chromatophores; the absence of overlying dark melanophores exposes underlying bright xanthophores.

The general extent of environmental modification of coloration in cichlids appears limited. Diets in captivity can affect the intensity of coloration, particularly reds (Schmidt-Focke and Homburg, 1977; Matsuno *et al.*, 1980); aquarium specimens are generally less colourful than their wild counterparts, e.g. *Heros citrinellum* or several species of Mbuna (L. DeMason, pers. comm.). Nonetheless reproductive coloration has remained constant over many generations in virtually all studied New and Old World taxa (Witte and Witte-Maas, 1987; DeMason, pers. comm.; Kornfield, unpubl.). A possible exception is Turner's (1988) study on *Oreochromis niloticus*, in which two thermal/feeding treatments applied during early growth resulted in differences in the distribution of red coloration in adult males. However, since the absolute size of fish in the two treatments differed greatly, it is difficult to assess the systematic significance of this observation. Considerable genetic divergence has been demonstrated between subspecies of *O. niloticus* for similar-sized individuals exhibiting just such colour differences (Seyoum and Kornfield, in prep.). Phenotypic expression of breeding coloration in tilaptas may nevertheless be modulated by individual size during ontogeny. Experimentation involving environmental gradients may be required to learn whether such variation is continuous or discrete. Whereas the expression of colour may be labile in certain environments, changes in colour (hue) have not been documented, except in *Astatotilapia burtoni* of Lake Tanganyika (Crapon de Caprona and Fritzsch, 1984). In aquarium studies, some male *A. burtoni* may change from yellow to blue morphs. The causes of this transition and their phenotypic stability merit additional study.

The genetics of coloration has been investigated in several cichlid taxa, but the variation examined has generally involved changes that alter colour of the entire body. In the convict cichlid, *Cichlasoma nigrofasciatum*, pink body coloration is recessive to normal (= blue) body colour (Itzkovich *et al.*, 1981). In *Etroplus maculatus*, lack of melanophores produces a red body colour which is also inherited as a simple recessive (Ward, 1978). Red body coloration has appeared independently in *Oreochromis* species on numerous occasions. In *O. niloticus niloticus*, red is an autosomal dominant (McAndrew *et al.*, 1988), while in red tilapias produced by hybridization between *O. mossambicus* and *O. niloticus*, red body colour is inherited as a sinple locus with incomplete dominance (Huang *et al.*, 1988). In *O. mossambicus*, gold body coloration is controlled by a single autosomal gene with incomplete dominance (Tave *et al.*, 1989). In general, genetic control over cichlid body coloration involves single loci, a situation similar to that in other teleosts (Kirpichnikov, 1981;

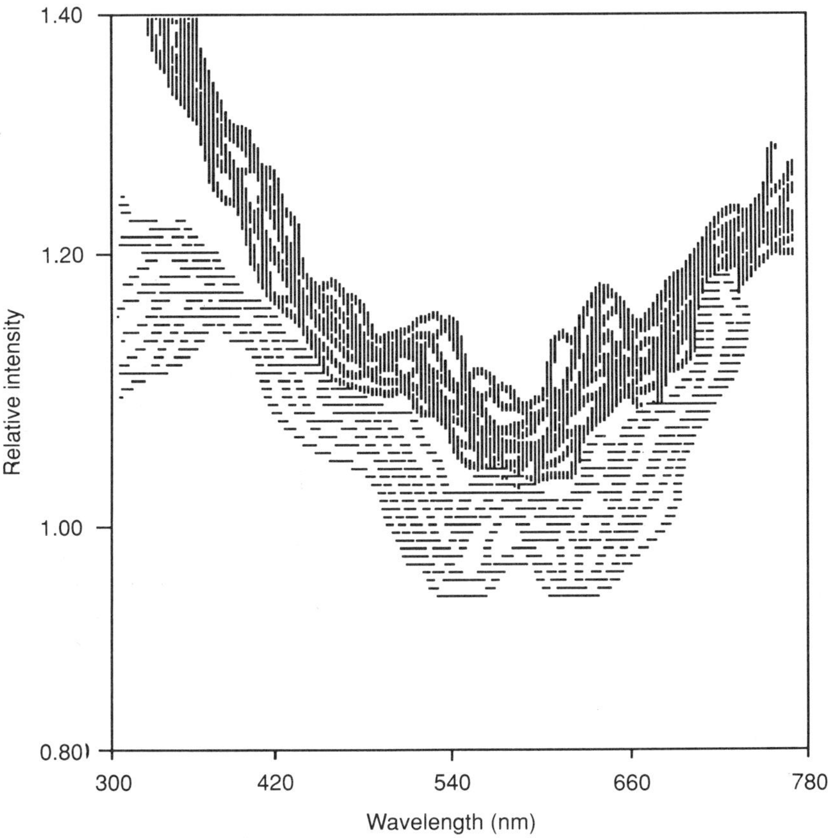

Fig. 5.3. Blue flank coloration in allopatric *Pseudotropheus zebra* from Lake Malawi. Spectral scans of reflected colour from five males from Thumbi Island West (vertical lines) are less intense in colour at all wavelengths than five males from Zimbawe Rocks (horizontal lines) (discussion, Ribbink *et al.*, 1983a). Coloration was quantified on live specimens under solar illumination using a Spectrum CE390 fast spectral scanner with a Unicolor grey card for the standard.

Borowsky, 1984). It is therefore reasonable to hypothesize that differences in the relative intensity of body coloration are also under simple genetic control, though some environmental influence is possible (Fig. 5.3).

We are exploiting a simple system to study the genetics of colour differentiation in Lake Malawi haplochromines. A series of crosses have been made between two taxa in the *Pseudotropheus zebra* species complex; *P. zebra* from Thumbi Island West (Lewis *et al.*, 1986, Fig. 1), and P. *zebras* 'red dorsal' from Nabantenga Island (Lewis *et al.*, 1986, Fig. 4). The latter taxon is a narrow endemic, differing conspicuously from its more broadly distributed

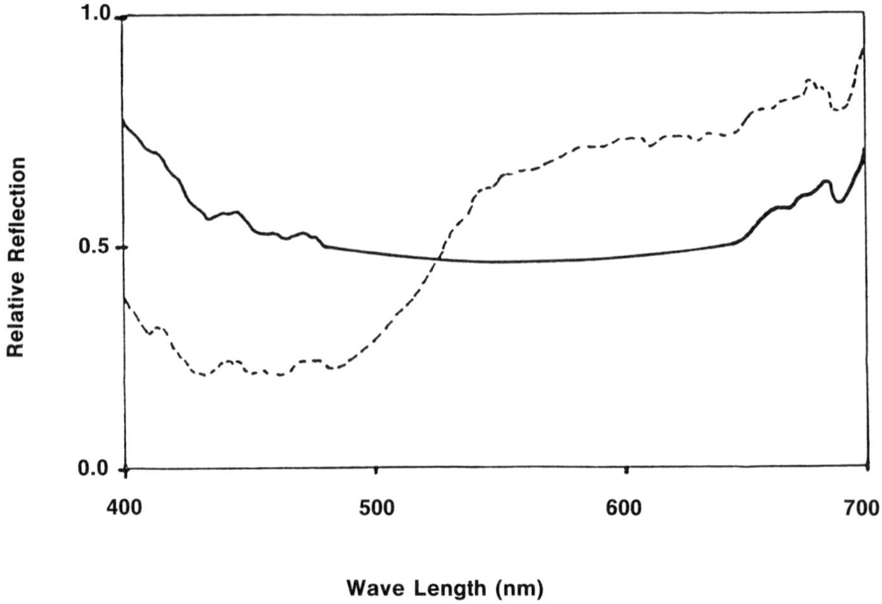

Fig. 5.4 Dorsal fin coloration in territorial male *Pseudotropheus zebra* from Lake Malawi. Spectral scans of reflected colour of 'blue dorsal' *P. zebra* from Thumbi Island West (solid curve) and 'red dorsal' *P. zebra* from Nakantenga Island (broken curve). Methods as in Fig. 5.3.

allopatric congener in the coloration of the dorsal fin; the dorsal of Thumbi Island males is close to 'light sky blue' (Smithe, 1981, colour no. 168D), while Nakantenga males have dorsal fin coloration close to 'spectrum organge' (Smithe, 1975, colour no. 17) (Fig. 5.4). Aside from the dorsal fin, coloration in the two taxa is identical. The colour difference observed between these two taxa is similar to geographic variation common in many Lake Malawi species.

Reciprocal single-pair crosses were made between the two taxa, and progeny were reared to reproductive maturity for expression of fin coloration. In five of six crosses, all F_1s had 'blue' dorsal fins, but in one cross, both 'blue' and 'red' dorsal fin phenotypes were observed among the male progeny (Table 5.1). Segregation of these two forms within one cross and the absence of any progeny with intermediate dorsal coloration suggest a simple two-allele system with complete dominance. Thus, the 'blue' dorsal mother in cross 149 which produced both 'blue' and 'red' dorsal offspring would be a heterozygote. However, there is a serious deficiency with this model; in natural populations of 'blue' dorsal fish, matings between heterozygotes would produce 'red' dorsal individuals. *P. zebra* with 'red' dorsal fins have never been observed at Thumbi

Table 5.1 Inheritance of dorsal fin coloration in six crosses of *Pseudotropheus zebra*

Cross		*Observed phenotypes*		*Genetic model**
135 141 148	P	male 'blue' × female 'red dorsal'	P	AArr × aaRR
	F_1	'blue'	F_1	AaRr
139 147	P	male 'red dorsal' × female 'blue'	P	aaRR × AArr
	F_1	'blue'	F_1	AaRr
149	P	male 'red dorsal' × female 'blue'	P	aaRR × Aarr
	F_1	'red dorsal' and 'blue'	F_1	aaRr and AaRr

*The genetic model assumes a colour locus with two alleles (A, a) and a colour-regulation locus with two alleles (R, r); R and A are dominant. Under this model, the regulatory locus determines whether the colour locus is expressed. RR or Rr individuals express the colour locus; rr individuals do not. Individuals exhibit the 'red' phenotype only if their red genotype (aa) is expressed; that is, if they are RR or Rr. In contrast, a 'blue' phenotype can arise in two distinct ways: (1) individuals carry the blue genotype (AA or Aa) and are RR or Rr; (2) individuals are rr and are thus blue regardless of their genotype at the colour locus.

Island West. Non-genetic factors cannot be responsible for the observed segregation of progeny fin colour because these experiments were conducted in a common environment. Therefore any alternative explanation must involve additional genetic components. One simple alternative model involves two diallelic loci, one providing colour and the other regulating the expression of the colour (Table 5.1). In this system, and simple alternatives to it, there is an epistatic interaction between the two loci. Although this model explains the observations, less parsimonious explanations may in fact be correct. At a minimum, however, at least two gene loci are involved in the expression of dorsal fin coloration.

5.4 MOLECULAR SYSTEMATICS

Rationales for phylogenetic construction are manifest (Sneath and Sokal, 1973; Wiley, 1987), but they are particularly well defined for the Cichlidae. As in all groups, systematic interests range from familial affinities to acute queries, some involving conspecific relationships. However, we need to establish accurate evolutionary histories for cichlids because phylogenies are essential to understanding the processes underlying their speciation, radiation and diversification. For example, if a 'true' phylogeny were known for the Malawi

Mbuna, basic questions concerning ecological convergence, niche diversification, and behavioural radiation could be evaluated objectively. As Futuyma and Moreno (1988) have remarked, "Knowledge of phylogenetic history is crucial for determining the reality of macroevolutionary patterns."

The methods employed for investigating problems of systematics depend on the goals of the study, the materials available and the investigator's biases. Molecular approaches to systematics have many limitations but can be particularly useful for resolving problems uncovered by more classical methods. It is critically important that the limitations of molecular techniques be generally appreciated; the following review of methods may help. For a general introduction to methodologies, see Hillis and Moritz, 1990.

Certain types of molecular data are preferable to morphological data, particularly if statistical approaches to phylogenetic reconstruction are pursued (Felsenstein, 1988). The general appeal of molecular data sets is that they offer very many characters for analysis. In addition, most types of data can be used as 'molecular clocks' to establish relative rates of divergence within and among lineages. Though controversy surrounds absolute calibration of such clocks, their utility is undisputed. The application of molecular techniques to cichlid systematics capitalizes on these cryptic timepieces. Molecular investigations will not, however, preempt studies based on morphology; indeed, morphological study is at the root of cichlid systematics and will continue to provide the avenue through which evolutionary patterns on almost all scales are perceived. In general, it is desirable to integrate molecular and morphological studies best to approach cichlid evolutionary history; it is the comparisons between these independent data sets that provide a basis for our enriched understanding of evolution (Hillis, 1987). Thoughtful treatments of the relative roles of molecules and morphology in systematics can be found in Patterson's (1987) review and in other contributions to that volume.

Three types of techniques used in molecular studies of cichlids will be considered: electrophoretic analysis of isozymes, restriction analysis of mitochondrial DNA (mtDNA), and nucleic acid sequence analysis (principally of mtDNA). It is most important to bear in mind the assumptions and potential problems associated with each technique to appreciate any particular investigation critically.

Isozyme analysis

Electrophoretic analysis of water-soluble proteins, the most ubiquitous of molecular techniques, has had a considerable impact on systematics (Avise, 1974). Conventional surveys generate character states (alleles or electromorphs) for a large suite of independent characters (protein loci) which are homologous among taxa. Electrophoresis is particularly suited to comparisons among taxa at early-to-intermediate levels of differentiation. By permitting rapid characterization of variation within and among populations, isozyme analysis has allowed us to view some of the processes of evolution.

Techniques and limitations

The utility of standard starch gel electrophoresis is constrained by at least four major factors. First, it significantly underestimates the amount of variation within a taxon, perhaps by 30% or more (Ramshaw *et al.*, 1979). This is not a concern for general taxonomic application of the technique, but may be important in the reconstruction of evolutionary histories among conspecific isolates. Second, estimates of genetic divergence are highly dependent upon the number and identity of loci examined (Nei, 1987). This dependence implies both that genetic divergence estimates will have large standard errors, and that independent studies of identical taxa may not produce concordant divergence estimates. Third, small sample sizes may significantly bias estimates of variation and reduce the ability to detect alleles (Hillis, 1987). This may have a great effect on phylogenetic analyses if synapomorphies for fixed alleles are rare. Finally, conventional electrophoresis may underestimate true divergence because co-migrating bands may not be identical, or buffer systems may not resolve slight differences in net charge.

Electrophoresis is particularly easy to conduct, and therefore can easily be abused. Meaningful application demands the ability to interpret gels correctly, to use appropriate techniques of analysis and to understand the limitations of the data. An electrophoretic study on the haplochromines from Lake Victoria provides an illustration of these concerns. In this study (van Rompaey *et al.*, 1988; see also Verheyen and van Rompaey, 1986), the general protein banding patterns of muscle homogenates from ten Victorian taxa were used to infer trophic divergence and to suggest a sympatric mode for speciation in the fauna. However, because the variation observed was incorrectly interpreted, any systematic discussion is not credible. That is, the fundamental assumption of all molecular systematic studies, that variation be genetic in origin, was violated. The 'genotypes' observed at the postulated *Pr-1* locus differed significantly from simple Hardy–Weinberg expectations (*Haplochromis* 'velvet black', $P = 0.003$; *Hoplotilapia retrodens*, $P = 0.025$); in the absence of breeding data for electrophoretic variants, a fit to Hardy–Weinberg is necessary, but not sufficient, to support Mendelian inheritance for electrophoretic variants. Furthermore, the density differences among presumably homologous bands, both within and among the electrophoretic patterns, is not consistent with simple allelic variation. In general, interpretation of gels stained for general proteins, as well as esterases and peptidases, should be approached cautiously because variation may be due to overlapping loci, transient ontogenetic production, differential expression in males and females, environmental induction, or parasitic contamination. It is thus quite unlikely that the bands observed for *Pr-1* are allelic or that the general protein patterns are interpretable in any conventional genetic manner. Regardless, a single locus or banding pattern cannot reasonably be used to infer phylogenetic relatedness or taxonomic placement as was done in that study.

Despite these constraints, electrophoresis has been quite useful for investigations on cichlids, although there have been few studies. Electrophoresis has been used for analyses extending from populations (e.g. Sage and Selander, 1975) to intergeneric comparisons (p. 119). It can be of value in generating preliminary estimates of cladistic relationships for much of the New World cichlid fauna, particularly taxa in the diverse *Cichlasoma* assemblage. Though the technique has supported resolution of problems involving specific status in some Old World faunas from a systematic perspective, it has limited utility for comparisons among both closely related taxa and distantly related lineages.

Applications

An inversion in the relative migration of lactate dehydrogenase isozymes in Neotropical cichlids may have a significant phyletic distribution products of the *Ldh-A* locus which are usually cathodal to (slower than) *Ldh-B* are anodal to *Ldh-B* in some taxa (Scholl and Holzberg, 1972). Oppenheimer *et al.* (1989) have demonstrated heterogeneity in the occurrence of the inversion among taxa within the genus *Geophagus*, and on the basis of this character have recognized a potential clade involving several other genera (the inverted loci also display conventional allozyme variants in some taxa.) Note the interpretation of this molecular variation: the reversal synapomorphy is qualitatively more informative than simple allelic mobility variants, given the relative occurrence of the two types of mutations. The utility of LDH for the systematics of Neotropical cichlids may be further increased by analysis of the distribution of a putative fourth locus (Leibel and Peairs, 1989).

A principal finding from electrophoretic analyses involves the relationships among the cichlid complexes of Lakes Malawi, Tanganyika and Victoria. Several studies have demonstrated extremely low genetic distances ($D < 0.10$;) among morphologically divergent cichlid lineages within Lake Malawi (Kornfield, 1978; T. Kocher, unpubl.) and among lineages within Lake Victoria (Sage *et al.*, 1984; Verheyen *et al.*, 1985). Virtually all these endemics lack diagnostic alleles and can only be distinguished by differences in allele frequencies. The implication here is clear: these radiations occurred extremely recently. For Lake Malawi, these results are consistent with findings on lake-level fluctuations (Scholz and Rosendahl, 1988) which suggest that some narrow endemics did not exist 25 000 y ago. Indeed, a number of endemic Mbuna species from southern Lake Malawi may have evolved less than 1000 years ago (Owen *et al.*, 1990). By contrast, comparisons among species from different genera in Lake Tanganyika (Kornfield *et al.*, 1985; Nishida, 1985) have

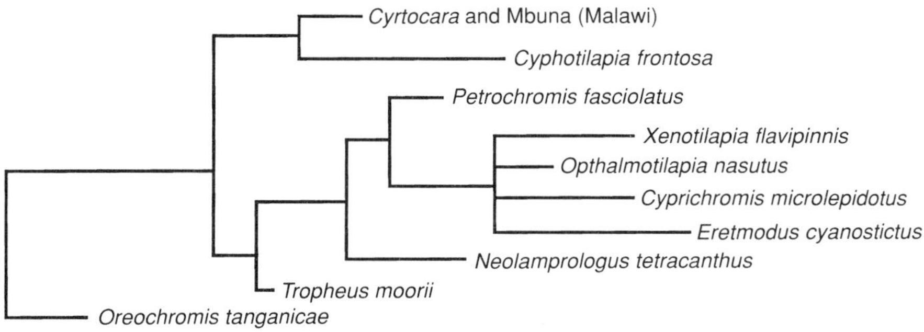

Fig. 5.5 Cladistic relationships among haplochromine cichlids of Lake Tanganyika and Lake Malawi. This analysis establishes that morphological similarity between genera from Malawi and Tanganyika is due to convergence, i.e. *Petrotilapia-Petrochromis*, *Labeotropheus-Tropheus*, and *Pseudotropheus-Cyprichromis*. The association of *Cyphotilapia* with the Malawi lineage confirms the polyphyletic nature of the Tanganyikan fauna. This consensus cladogram summarizes 15 equally parsimonious trees generated by PAUP (Swofford, 1985) from 18 informative allozyme loci; consistency index = 0.818, length = 33.00. The tree was rooted using *Oreochromis*. Nomenclature for Tanganyikan species from Poll (1986).

revealed substantial genetic distances, e.g. $D = 0.847$ for *Cyphotilapia–Cyprichromis* (= *Limnotilapia*) and $D = 0.935$ for *Tropheus–Neolamprologus* (= *Lamprologus*). These levels of differentiation indicate long periods of independent evolution since divergence from common ancestors. A cladistic analysis of electrophoretic data demonstrates the extent of diversification within Lake Tanganyika (Fig. 5.5). The polyphyletic nature of this fauna (Regan, 1920; Liem, 1981a) is confirmed by the presence of a clade containing *Cyphotilapia* and Lake Malawi haplochromines (Kornfield *et al.*, 1985). Although the analysis is preliminary, the relationships among genera are at variance with those suggested by Fryer and Iles (1972; Fig. 337); there is clearly no indication of tilapiine affinities for any of the Lake Tanganyika taxa. Furthermore, contrary to earlier suggestions (Greenwood, 1983b, 1986b; Yamaoka, 1983b), cladistic and phenetic analysis unambiguously establish that the morphological similarity between the Tanganyika–Malawi generic pairs is due to convergence specialized trophic morphologies involving cranial and dental architectures have been invented repeatedly.

Preliminary allozyme comparisons of Lake Victoria and Lake Malawi haplochromines suggest extremely close affinities between these two faunas (Verheyen *et al.*, 1985; Verheyen and van Rompaey, 1986). Estimates of genetic distance between Lake Victoria taxa and *Melanochromis auratus* from Lake Malawi are smaller than those observed among some haplochromine species pairs in Lake Victoria, though the error associated with estimates of this magnitude precludes such resolution. These findings are consistent with the idea that the phylogenies of the African species flocks cut across present lake boundaries (Barbour, 1973; Greenwood, 1983a). The close affinities between

Lakes Victoria and Malawi and in contrast to the distinctiveness of the Lake Tanganyika haplochromines.

Mitochondrial DNA restriction analysis

Restriction enzyme analysis of mitochondrial DNA (mtDNA) has had a substantial positive impact on systematic study (Avise *et al.*, 1987; Moritz *et al.*, 1987); some acute taxonomic problems that had proved intractable by standard electrophoretic examination have been resolved (e.g. Avise *et al.*, 1986; p. 121 herein).

Techniques and limitations

Restriction analysis of mtDNA involves the digestion of purified DNA with a series of enzymes which cleave the DNA when they encounter short, specific nucleotide sequences. The products of digestion are separated by electrophoresis and visualized directly with uv stains or indirectly with radio-isotopes. There are several reasons for the appeal of mtDNA restriction analysis. First, mtDNA as a small molecule which is homologous among widely divergent taxa. Because it is extranuclear and inherited via the maternal cytoplasm, it provides material for assessment of phyletic affinities independent of nuclear genes. In mammals the rate of evolution of mtDNA is considerably higher than that of DNAs coding for nuclear genes; the molecular clock for cichlids may also run fast, but has not yet been formally tested. Thus mtDNA is potentially capable of detecting differentiation that is not revealed by conventional electrophoretic analysis. Finally, it is relatively easy to isolate mtDNA and perform restriction analysis. Theoretical and numerical treatments of mtDNA evolution have produced much information for analysing and understanding variation visualized by restriction analysis.

For informative application of mtDNA restriction analysis, it is essential to recognize the assumptions involved and the technique's limitations. First, since mtDNA is inherited maternally, phylogenies based on inferred site changes might not be congruent with those estimated from nuclear data sets; since mtDNA is transmitted asexually, clonal lineages within populations can have dynamics very different from those of recombining nuclear genes. In fact, since clonal diversification can pre-date speciation, under some conditions mtDNA analysis may incorrectly suggest gene flow among taxa that are known to be reproductively isolated (Nei, 1987). This is probably not very important among relatively divergent taxa, but may be serious at and below the species level. In addition, because of maternal transmission, F_1 hybrids between well-differentiated species cannot be distinguished from either parent; clearly this is a double-edge concern, since it would thus be easy to determine the direction of gene flow in hybridization (e.g. Herke *et al.*, 1990). Second, restriction analysis detects only a portion of the variation that is present in the mtDNA molecule, so individuals that are judged to be identical may well differ at some nucleotide positions in the molecule. This concern is generally not a problem for systematic application of the technique. Third, mtDNA may not evolve in a strictly neutral manner (Moritz *et al.*, 1987; MacRae and Anderson, 1988). Neutrality is a desirable property for a molecular marker used in phylogenetic analysis because selection could inflate homoplasy. Fortunately, most mtDNA clones so far examined behave in a nearly neutral manner. Fourth, the interpretation of restriction-digest banding patterns may require caution. Restriction fragments may behave anomalously under certain electrophoretic conditions (Lalande *et al.*, 1988), partial digestions of purified DNAs are common, and heteroplasmy or length mutation may be present. However, although it requires experience, gel interpretation is objective. Finally, restriction analysis is not appropriate for taxa that have sequence divergence in excess of perhaps 15% (Kessler and Avise, 1985). Beyond that level, fragment similarity can be produced by chance alone, so the signal-to-noise ratio is unacceptable.

Applications

Restriction enzyme analysis of mtDNA has been applied to systematic problems in the Old World cichlid fauna where protein electrophoresis has not provided sufficient sensitivity. In Lake Victoria, Dorit (1986) demonstrated heterogeneity of mtDNA phenotype frequencies for samples of *Neochromis* (= *Haplochromis*) 'velvet-black' around the Mwanza Gulf region (map, Fig. 13.9). Several locality-specific clones were identified, but small sample sizes may have precluded detection of rare, but ubiquitous, variants. mtDNA cleavage patterns were compared among 11 endemics assigned to six haplochromine genera; intraspecific variability was low. While some taxa could not be individually distinguished, the molecular data provided support for the generic revisions of Greenwood (1979). However, alternative tree topologies are quite possible because the number of informative restriction sites was very limited.

mtDNA restriction analysis has been extensively applied to the endemic haplochromines of Lake Malawi by Moran and Kornfield (unpubl.). Though our assessment is incomplete, several generalizations are possible. Restriction analysis does not provide sufficient resolution to discriminate unambiguously among species of Mbuna; good biological species may have identical cleavage profiles. Secondly, some endemic non-Mbuna haplochromines possess distinctive mtDNA composite profiles that are identical to members of the Mbuna (Fig. 5.6). This identity indicates that as currently defined, the Mbuna is a paraphyletic group. Third, the presence of a well-differentiated mtDNA polymorphism confounds cladistic analysis of the Mbuna. Within the group, two divergent ($p = 1\%$) and statistically distinct mtDNA lineages (α and β) are present. Most species examined are either α or β, but in a number of taxa, individuals at one locality are α while allopatric conspecifics are β (Fig. 5.7). This pattern suggests 'lineage sorting' (Avise *et al.*, 1987), the random fixation of ancestral mtDNA clones within isolated populations. Thus, cladistic analysis

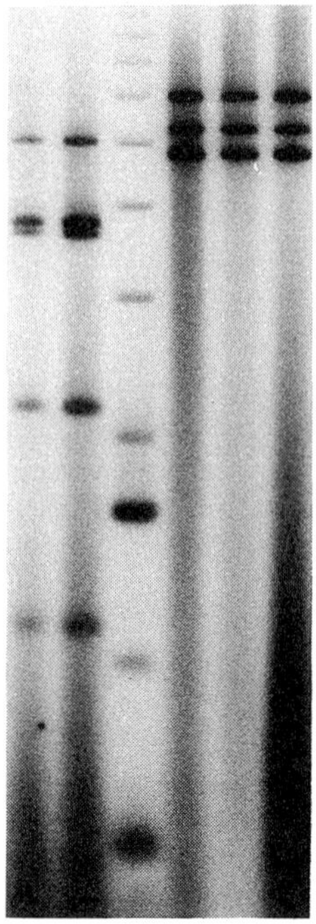

Fig. 5.6 Mitochondrial DNA restriction fragment patterns from Malawi haplochromines digested with *Ava*I. From left to right, taxa are: *Lethrinops gossei*, *Petrotilapia genalutea*, molecular weight standard. *Chilotilapia rhoadessii*, *Taeniochromis holotaenia*, and *Buccochromis lepturus*. The similarities between *L. gossei* and *P. genalutea* suggest paraphyly for the Mbuna. These taxa (and other Mbuna) form a distinct group well separated from the non-Mbuna haplochromines (all other lanes). Data from Moran and Kornfield, unpubl.

of mtDNA within the Mbuna produces 'gene trees' rather than 'species trees' (Nei, 1987); it is the 'species tree' that we wish to reconstruct. Finally, mtDNA comparisons have provided sufficient information to erect phylogenies depicting higher-order relationships within the lake; lineages within non-Mbuna haplochromines are well-differentiated from the Mbuna (Fig. 5.6). Further, some endemic lineages, for example *Rhamphochromis* and *Diplotaxodon* are well separated from the other haplochromines.

Extensive mtDNA characterizations of tilapias from East Africa have been completed (Seyoum, 1989; Seyoum and Kornfield, in prep.). The acuity of mtDNA for resolving difficult questions of systematics is seen with *Oreochromis niloticus*, where conventional protein electrophoresis could not discriminate subspecies (Seyoum, 1990). Composite restriction phenotypes could be used to distinguish all seven described subspecies unambiguously (Table 5.2). These genetic markers have direct applicability for defining wild taxa and cultured

Table 5.2 Dichotomous key to the described subspecies of *Oreochromis niloticus* from East Africa and Egypt, based on mitochondrial DNA restriction enzyme phenotypes*. Restriction-endonucleases are in italics, phenotypes are in bold face

1.	A.	*Dra* I '**H**'	*O. n. eduardianus*
	B.	All other phenotypes	2
2.	A.	*Hinf* I '**D**'	*O. n. sugutae*
	B.	All other phenotypes	3
3.	A.	*Rsa* I '**F**'	*O. n. vulcani*
	B.	All other phenotypes	4
4.	A.	*Bcl* I '**B**'	*O. n. filoa*
	B.	All other phenotypes	5
5.	A.	*Ava* I '**A**' or '**E**'	6
	B.	All other phenotypes	7
6.	A.	*Ava* I '**A**'	*O. n. cancellatus*
	B.	*Ava* I '**E**'	*O. n. cancellatus* (Lake Tana population)
7.	A.	*Nci* I '**X**'	*O. n. niloticus*
	B.	*Apa* I '**I**'	*O. n. baringoensis*

*Data from Seyoum and Kornfield (in prep. a).

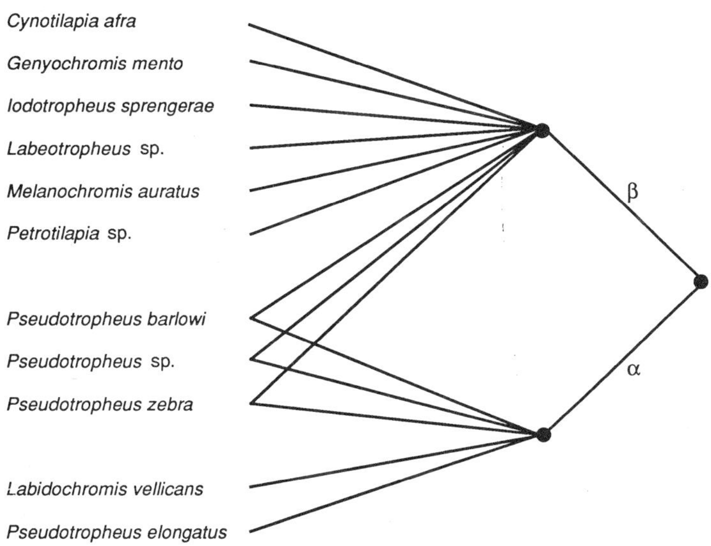

Fig. 5.7 Lineage sorting of mitochondrial DNA restriction phenotypes among 11 species of Mbuna (Moran and Kornfield, unpubl.). Estimated sequence divergence between mtDNA clades is 1.0%. For *Pseudotropheus barloi, P.* sp. and *P. zebra*, allopatric conspecifics were allied with divergent mtDNAs.

stocks. The taxonomy of the *O. niloticus* subspecies complex (Trewavas, 1983) was examined by phenetic and cladistic analyses of the extensive mtDNA character set. All trees recognized a distinctive clade containing *O. n. cancellatus* and *O.n. filoa* that was widely separated from the other subspecies (Fig. 5.7), suggesting that taxonomic modifications are warranted. Additional comparative mtDNA studies among representatives of the dominant tilapiine genera (*Oreochromis*, *Sarotherodon* and *Tilapia*) indicated broad concordance with conventional taxonomy (Trewavas, 1983) of the group. Results of these continuing investigations are under review (Seyoum and Kornfield, unpubl.).

DNA sequence analysis

With the development of methods for rapid amplification, sequencing and data analysis, comparative analysis of DNA sequences has begun to be exploited by evolutionary biologists. This approach will soon dominate molecular systematics as lab protocols become increasingly simplified and systematists gain familiarity with the techniques. Of particular significance is application of the polymerase chain reaction (PCR), an automated procedure for enzymatic amplification of native DNA; PCR obviates the need for conventional cloning (Kocher and White, 1989).

Techniques and limitations

One of the clear advantages of DNA sequencing over other methods in molecular systematics is data standardization. In contrast to isozyme banding patterns or restriction enzyme cleavage patterns, where expression of the phenotype may depend upon experimental electrophoretic conditions and interpretation may require considerable expertise, DNA sequences are universally and independently replicable and understood. Given homologous genes or established landmarks, such as highly conserved regions, DNAs can be compared at all taxonomic levels. The ability of DNA sequencing to resolve phylogenies depends critically upon the choice of the region examined. In the mitochondrial genome, for example, cytochrome b genes evolve slowly and are useful for comparisons involving distantly related taxa, e.g. New and Old World faunas. Resolution necessary for interspecific comparisons involving recently separated taxa may be provided by sequencing in the displacement loop (D-loop) region (but see above). Homologous genes or sequences in portions of the nuclear genome can also be analysed.

A suite of difficulties may exacerbate the analysis of DNA sequences (see Heijne, 1987), for an introduction to sequence analysis). Many of the limitations noted for restriction endonuclease analysis of mtDNA apply to sequencing. The more distantly related the taxa, the more the complications that may arise. Aligning sequences for comparisons can present problems for regions with low similarity or where sequence deletions are present. If the DNA sequences of interest are translated into proteins, comparative systematic analyses may be valid only when there are no important constraints on amino acid substitutions (Lewontin, 1989). That is, analysis would appear to be limited to positions involving synonymous substitutions. There is substantial heterogeneity in the types of mutations that accumulate in DNA lineages. Substitutions involving transitions (purine ⟷ purine or pyrimidine ⟷ pyrimidine) are much more frequent than transversions (purine ⟷ pyrimidine). Furthermore, as molecular divergence proceeds, DNA positions that have already mutated can again acquire new substitutions; such 'multiple hits' cannot be distinguished from simple substitutions. In consequence, at early stages of divergence it may be appropriate to weight substitutions differentially, while at later stages transitions may best be excluded from calculations. Similarly, because of the degeneracy of the genetic code, site changes accumulate rapidly at the third position in codons; data from these positions should often be excluded when divergent genomes are compared. As with other types of molecular analyses, confidence estimates of divergence or relationships among lineages should be provided for the findings. Enough base pairs must therefore be examined for conclusions to be robust. This is particularly critical, both where lineages are very closely related (such that infrequent errors in sequencing (Krawetz, 1989) may contribute to perceived relationships), and to comparisons involving distant lineages where the number of informative sites may be limited. Statistical treatment of such problems is an active field of investigation.

Applications

Only a few results are available for sequencing studies on cichlids, although the floodgates should be opening directly. Several independent groups are using PCR amplification and rapid sequencing to clarify questions of cichlid systematics and to study basic mechanisms of genomic evolution in the family. The creation of universal mtDNA primers for PCR studies (Kocher *et al.*, 1989) has facilitated much of this research.

Comparative studies of cichlid interrelationships at continental scales by A. Meyer and T. Kocher (Kocher *et al.*, 1989; Meyer *et al.*, 1990; Kocher, unpubl., pers. comm.; Meyer, unpubl., pers. comm.) have established first estimates of divergence between New World and Old World faunas. As anticipated, the estimated separation is substantial; using cytochrome b primers, they noted sequence divergence of approximately 7% occurs between *Juliochromis* sp., a Lake Tanganyika endemic, and *Heros* spp. from Middle America. Preliminary characterizations of the endemic cichlids of Madagascar from specimens newly collected by M. Stiassny and P. Reinthal have been made (Fig. 5.8), and analyses are under way (Mayer, pers. comm.). Within faunas, comparisons among a few related New World taxa have been completed (Kocher *et al.*, 1989; Meyer, unpubl.), but no large-scale systematic surveys have yet been initiated. The mtDNA sequencing study by Meyer *et al.* (1990) has elegantly established the relationships between the endemic haplochromine faunas of

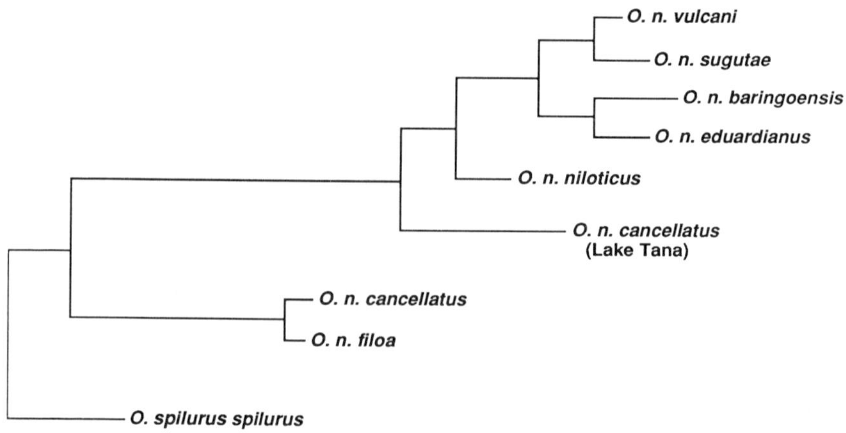

Fig. 5.8 Cladistic relationships among the subspecies of *Oreochromis niloticus* from East Africa. Note the distinctiveness of the clade containing *O. n. cancellatus* and *O. n. filoa*. The cladogram was generated by PAUP (Swofford, 1984) base on 192 informative mtDNA restriction fragments; consistency index = 0.652, length = 293. *O. spilurus* was used as an outgroup to root the tree. Data from Seyoum and Kornfield (in prep. b).

Lakes Victoria and Malawi. Despite very recent origins and close affinities, it is clear that morphological similarities between taxa from these lakes is due to convergence. Using D-loop primers for PCR, specific lineages were discriminated within Lake Victoria. In Lake Malawi, two major groups representing the Mbuna and non-Mbuna were recognized (see Fig. 5.6). The relations between the faunas of these lakes are distinct from those of Tanzania, an observation consistent with the earlier electrophoretic studies. For Lake Malawi, mtDNA sequence comparisons among selected Mbuna (Meyer, pers. comm.) suggest divergence similar to that estimated from restriction analyses (Moran and Kornfield, unpubl.); the α and β lineages noted above are distinguishable.

Sequencing studies on nuclear DNA are scarce. Wright (1989) has characterized size variants of repetitive DNAs from *Oreochromis* of hybrid origin. Comparative examination of the sequences of these repeats among representatives of other tilapiine genera has permitted construction of phylogenetic trees (Wright, pers. comm.); these trees appear broadly concordant with those independently established by restriction analysis of mtDNA (Seyoum, 1989a), validating the use of these satellites for broader phylogenetic studies in the family (Wright, pers. comm.). Sequences for additional regions of nuclear DNA will soon be acquired for broad systematic application in the family.

5.5 CONCLUDING REMARKS

Genetic knowledge of the Cichlidae is limited despite both pure (evolutionary) and applied (aquaculture) interests. Several research directions are clearly indicated. One obvious need is for a rigorous system of taxonomic identification that can be applied to cultures of tilapias world-wide. Given the importance of *Oreochromis* species to global aquaculture, and the vast number of independent selection experiments in progress, efficient identification of genetic lineages could accelerate increases in performance. Similarly, lineage identification may be critical to the success of research on sex determination. Deposition of cell lines of DNAs from known taxa in a central repository, for example the American Type Culture Collection, would promote standardization of many studies. Some of the molecular methods reviewed, particularly characterizations of nuclear DNA, seem appropriate for this purpose. Because of their leadership, perhaps ICLARM and FAO ought to co-ordinate the necessary efforts.

There are numerous acute taxonomic problems in the Cichlidae that can be addressed using molecular methods. Critical examination of relationships among the major New World cichlid lineages, particularly clades allied to *Cichlasoma*, is long overdue. While such investigations would appear straightforward, this is not the case for relationships among endemics within the African lakes. Phylogenies there are of great importance because of their utility for evaluating questions related to speciation and diversification. However, cladistic analysis of restriction profiles or sequences for mtDNA may be of limited value in very recently evolved faunas. While genetic fingerprinting is sufficiently sensitive to discriminate cryptic relationships (Jeffreys *et al.*, 1985), broad application of the method has been questioned (Lynch, 1988). However, judgement may best be reserved until the technique has been tested in these systems. One alternative approach that has worked well with humans (Schumm *et al.*, 1988) is screening random cosmid clones for restriction-fragment length polymorphisms (RFLPs). Because of the potential to identify polymorphic markers useful for tracking gene flow and local differentiation, this approach should be encouraged. However, it is becoming reasonably clear that relationships among some of the endemic taxa may simply not be resolvable.

The questions posed by Meyer (1989) regarding the role of trophic polymorphism in cichlid speciation merits attention. Studies designed to force morphological changes should be balanced against the realities of the speciation process. If sexual selection is indeed a driving force in speciation, the role of trophic polymorphism becomes less clear. In the Lake Malawi Mbuna, where colour differences characterize many allopatric conspecifics (Ribbink *et al.*, 1983a), populations are not trophically diversified. That is, trophic divergence may be secondary to speciation. Nonetheless, it is important to formalize the genetics of morphology and coloration so as to understand the processes mediating diversification. In particular, knowledge of the number of loci controlling these traits and the extent of genetic covariation between the sexes may be crucial to modelling the divergence process (Butin and Richie, 1989; Lofstedt *et al.*, 1989). For example, genetic information is essential for critical evaluation of the role of sexual selection in cichlid evolution; in the absence of supporting genetic information, sexual selection models should not be uncritically embraced despite their unifying appeal. Understanding these characteristics will require detailed, tedious and long-term breeding experiments.

ACKNOWLEDGEMENTS

The co-operation and interest of the Malawi Department of Fisheries and the National Research Council of Malawi have greatly facilitated much of the genetic work reported here. I am especially grateful for the support provided by S. Grant in Salima, Malawi. L. DeMason and J. Taylor of Old World Exotic Fish, Homestead, Florida, have been consistently valuable sources of information and support for investigations into the genetics of cichlids; they have my continuing gratitude. I greatly acknowledge the assistance of T. Kocher, A. Meyer and P. Wimberger for clarifying my thoughts on several problems and providing unpublished data. I acknowledge the intellectual leadership of G. Barlow in understanding the biology of cichlids. D. McElroy, P. Moran, T. Goldschmidt and Peter Reinthal kindly commented on drafts of the manuscript. My research on cichlids has been supported by the National Science Foundation (BSR 84-16131 and 85-17040), the Smithsonian Oceanographic Sorting Center and the University of Maine.

Chapter six

Functional morphology

K. F. Liem

6.1 INTRODUCTION

Space does not permit an in-depth and comprehensive synthesis of the functional morphology of cichlids. Consequently this discussion will be restricted to the functional morphology of the cichlid feeding apparatus. Such an approach seems timely and appropriate because the diversification of cichlids is especially expressed in their trophic biology, and modern analytical tools have been applied to the study of the feeding apparatus (e.g. Liem, 1973, 1978, 1979, 1980a; Aerts *et al.*, 1987). From a functional morphological perspective, the feeding apparatus of cichlids is better known and understood than that of any other teleostean group. Important concepts on adaptive radiation (Fryer and Iles, 1972; Greenwood, 1974a, 1981a), adaptive plasticity (Greenwood, 1956; Meyer, 1987), polymorphism (Sage and Selander, 1975; Kornfield and Koehn, 1975; Liem and Kaufman, 1984), constructional morphology (Barel, 1983) and ecomorphology (Hoogerhoud and Barel, 1978; Hoogerhoud, 1986a) are primarily based on the feeding apparatus. Therefore a review of the state of the art of functional morphology of the feeding apparatus seems timely as a foundation for future research.

In this chapter I will present an up-to-date summary of the functional morphology of the cichlid feeding apparatus and its possible implications for ecology and evolution. On the basis of Lake Tanganyika cichlids, I will make a critical assessment of the historical or genealogical factors and ecological or economic factors underlying trophic adaptive radiations.

6.2 FUNCTIONAL DESIGN OF THE FEEDING APPARATUS

Functionally the feeding apparatus can be subdivided into three major subunits: (1) the prey-capture apparatus; (2) the prey-processing mechanism; and (3) the hydrodynamic tongue (Bemis and Lauder, 1986).

Prey-capture apparatus

Suction feeding, ram feeding and bitting

From a functional perspective, cichlids feed in three basic ways: (1) **suction feeding**, in which the prey is sucked and moved into the mouth cavity by the stationary predator; (2) **ram feeding**, in which the predator opens its mouth and simply overtakes its prey; (3) **biting**, in which the teeth of the upper and lower oral jaws are applied to the prey in various ways. Many cichlids combine suction and ram feeding in various proportions. Likewise cichlids that capture their food primarily by biting can simultaneously create suction. Variations on these three basic prey-capture themes are described by Yamaoka in Chapter 7.

The three basic prey-capture modes are clearly reflected in the functional design of the cichlid head.

The expanding cone and cylinder models

Experimental studies (Lauder, 1980; 1983a, b; Lauder and Clark, 1984; van Leeuwen and Müller, 1984; Müller *et al.*, 1982; Müller and Osse, 1984; Brainerd, 1985) have shown that buccal cavity expansion is the primary force behind suction feeding. Following Brainerd (1985) the cichlid buccal cavity can be modelled as a truncated cone (Fig. 6.1) that can be expanded in various ways. This model, combined with empirical pressure data (Brainerd, 1985), predicts that the magnitudes of pressure changes within the buccal cavity depend on the shape of the truncated cone at the resting stage. Cylindrically shaped cones tend to generate relatively low suction pressures, especially when the gape of the mouth is large. In general, ram feeders have cylindrical buccal cavities with large mouth openings (Fig. 6.1(a)). Cones with very narrow truncated tips and wide bases are capable of generating very large negative pressures. All suction-feeding cichlids have such attenuated cone-shaped buccal cavities with small mouth openings (Fig. 6.1(c)). Cichlids combining ram- and suction-feeding tactics have the shape of their buccal cavity in the intermediate range between the cylinder and attenuated truncated cone (Fig. 6.1(b)).

Most biters have buccal cavities of the attenuated cone type (Barel, 1983), mainly as a result of accommodating the greatly enlarged adductor mandibulae muscles and shortening the jaws to increase the biting force.

Osteological adaptations

Whether a cichlid is a pure suction feeder, a pure ram feeder, a combined suction/ram feeder, or a biter is most clearly reflected in the overall design of its truncated expanding cone (Fig. 6.1). The various shapes of the truncated cone are correlated with the shape of several mechanical units (Gans, 1989) of the skull, especially the neurocranium, suspensory apparatus and jaw apparatus (Barel, 1983).

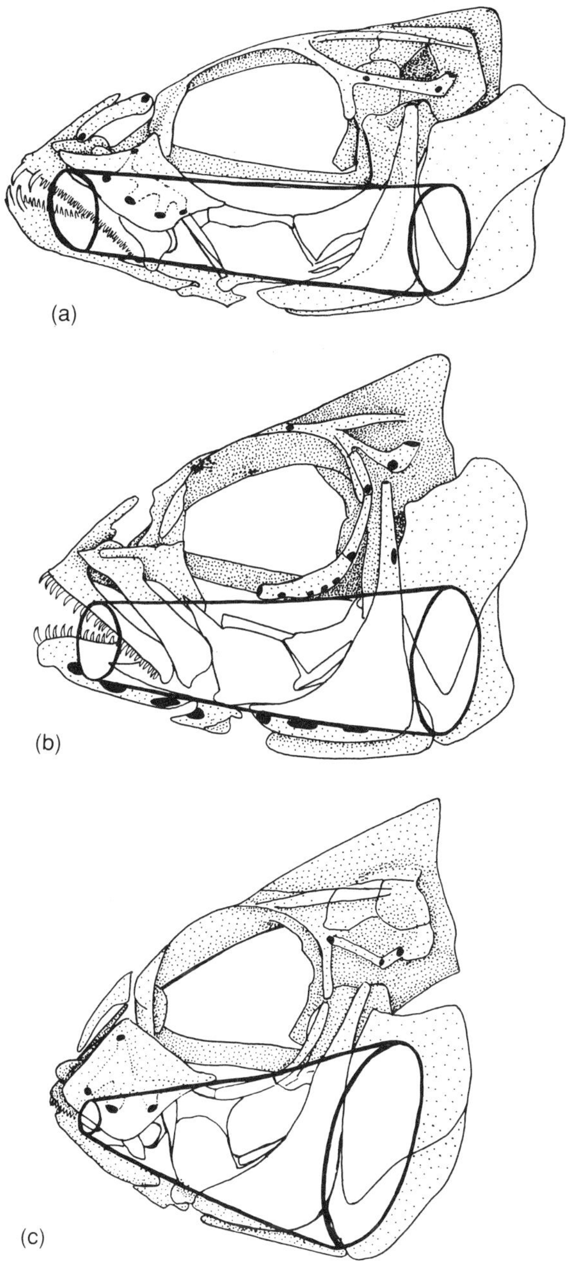

Fig. 6.1 Models of three differently designed buccal cavities in cichlids. Buccal cavities are outlined in heavy lines. (a) Model of a typical ram feeder, e.g. *Lamprologus elongatus*, with a cylindrically shaped buccal cavity. (b) Model of a combined ram and suction feeder, e.g. by *Grammatoria lemairei*. (c) Model of a typical suction feeder, e.g. *Tylochromis microlepis*. All these figures are drawn to the same scale as Figs. 6.1 and 6.2.)

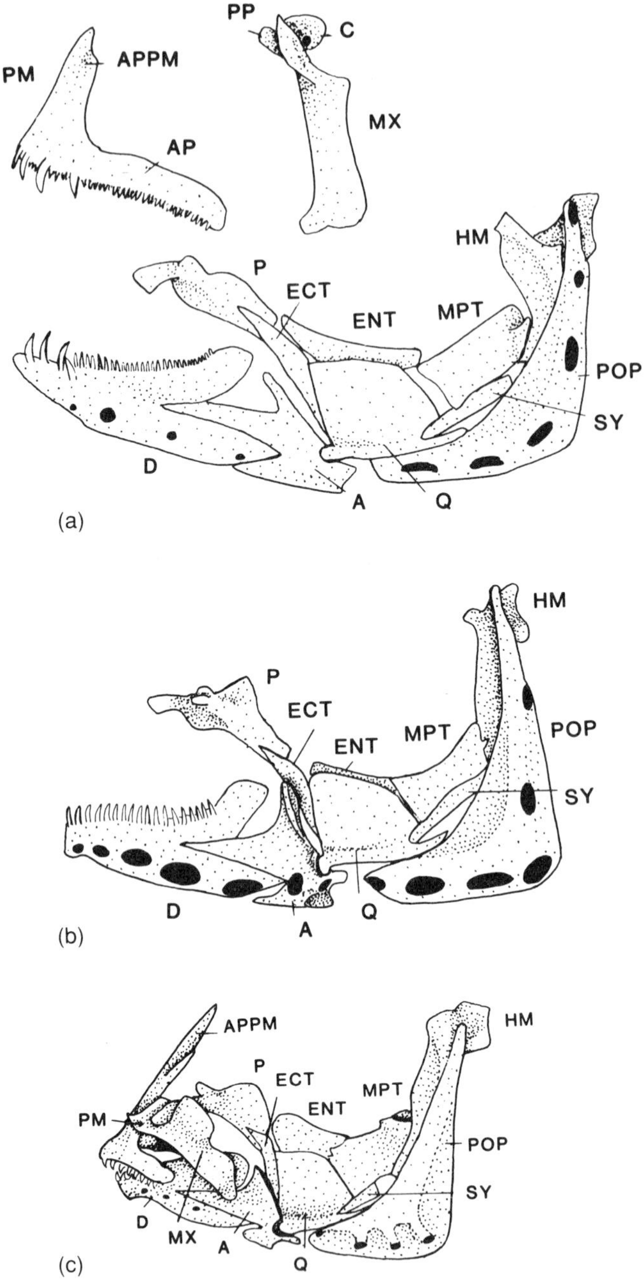

Fig. 6.2 Lateral aspect of suspensory apparatus and jaws of (a) *Lamprologus elongatus*, a ram feeder with a shallow palatopterygoid, relatively short symplectic process of the hyomandibula and elongate slender jaws; (b) *Grammatotria lemairei*, a ram/suction feeder with a deeper palatopterygoid arch, a longer symplectic process of the

Ram-feeder design. Ram feeders are characterized by elongated, relatively shallow neurocrania with attenuated preorbital regions (Fig. 6.1(a)). Often the supraoccipital crest is shallow and the ethmoid is positioned in a more horizontal configuration (Liem, 1978). The lower jaw is typically long and slender with the ascending process of the dentary and angulo-articular widely separated (Fig. 6.2(a)). Likewise the maxilla is a long, relatively slender bone with smaller processes. The premaxilla has a long alveolar ramus and a relatively short ascending process. The suspensory apparatus is elongate and shallow, serving as the origin of the thin layer of the adductor mandibulae muscle complex, which is characterized by long muscle fibres (Liem, 1978). The shallow, elongate suspensory apparatus, together with the long jaws and long preorbital region of the neurocranium, will result in the cylindrical configuration of the buccal cavity so characteristic of ram feeders which are, in general, pursuit hunters (Liem, 1978; Barel, 1983). Convergence in skull design of ram feeders is pervasive among cichlids (Barel, 1983; Brainerd, 1985).

Suction-feeder design. Suction feeders are characterized by relatively deep neurocrania with abbreviated preorbital regions in which the ethmoid is positioned in a more vertical configuration (Fig. 6.1(c)). The lower jaw is typically short and stout with the dentary and angulo-articular closely connected (Fig. 6.2(c)). The maxilla is short and possesses very large processes for articulations with the palatine and premaxilla. The premaxilla has a short alveolar process and often a very long, slender ascending process. The suspensory apparatus is deep and furnishes an expansive surface area for the hypertophied adductor mandibulae muscle complex. The depth of the suspensory apparatus results in the typically cone-shaped buccal cavity (Fig. 6.1(c)) of suction feeders.

Combined suction/ram-feeder design

Cichlids employing dual strategies of prey capture represent the majority. They possess skulls of intermediate design (Figs 6.1(b), 6.2(b)).

Biter design. The biter design can best be characterized as an exaggeration of the suction design. Thus the neurocranium has even a shorter preorbital

---

hyomandibula and relatively long but stouter jaws; (c) *Tylochromis microlepis*, a suction feeder, with a very deep palatopterygoid arch, a very long symplectic process of the hyomandibula, shorter stouter jaws with very prominent processes including a long ascending process of the premaxilla. Abbreviations: A, angular; AP, dentigerous ramus of premaxilla; APPM, ascending process of the premaxilla; C, cranial condyle; D, dentary; ECT, ectopterygoid; ENT, entopterygoid; HM, hyomandibula; MPT, metapterygoid; MX, maxilla; P, palatine; PM, premaxilla; PP, premaxillary process of maxilla; POP, preoperculum; Q, quadrate; SY, symplectic.

region and deeper profile, and the very stout, short mandible has the ascending processes of the dentary and the angulo-articular sutured into one unit (Liem, 1979; Liem and Stewart, 1976; Barel, 1983). The depth of the suspensory apparatus is exceptional, furnishing a very large origin for the massive adductor mandibulae complex (Liem, 1979).

Muscles and their core functions

This review is based on the study of *Hemitilapia oxyrhynchus*, which exhibits a full repertoire of prey-capture mechanisms and can therefore serve as a model for the many specialized cichlid species. For comprehensive anatomical and functional studies, the reader is referred to Aerts (1985), Aerts *et al.* (1987), Anker (1978), Barel *et al.* (1976), Chardon and Vandewalle (1971), Dutta (1987), Goedel (1974a, b), Otten (1983), Liem (1978, 1979, 1980a), Liem and Osse (1975), Vandewalle (1976) and Yamaoka (1982, 1983a).

Functional analysis applying high-speed cinematography, electromyography and pressure transducer recordings (e.g. Lauder, 1983a, b) have revealed the following primary functions of the major head muscles of *Hemitilapia* (Fig. 6.3) during suction feeding of zooplankton. Prey capture by suction occurs in two phases: (1) the **expansive phase** (Fig. 6.3(a)), during which a sudden extensive expansion of the cone-shaped buccopharyngeal cavity produces suction and a water current into the opened and protruded mouth; (2) the **compressive phase** (Fig. 6.3(b)), during which the protruded mouth is closed and the expanded cone-shaped buccal cavity is compressed. The expansive phase is driven by activity of the epaxial muscles (Fig. 6.3, EP) responsible for head lifting; the sternohyoideus muscle (SH), which lowers the floor of the buccal cavity; the levator arcus palatini and dilatator operculi muscle which abduct, respectively, the palatopterygoid or suspensory apparatus and the opercular apparatus. The combined actions of the epaxial, sternohyoideus, and levator arcus palatini muscles are to expand the cone-shaped buccal cavity. The levator operculi, sternohyoideus and epaxial muscles will open the mouth (see also Liem, 1978; and Aerts *et al.*, 1987, for a discussion of the levator operculi muscle as a triggering device). During the expansive phase, the upper jaw is protruded (Fig. 6.3) by various mechanisms (see Liem, 1979 for detailed discussion) caused in cichlids by actions of the levator arcus palatini and epaxial muscles or by jaw depression. The mechanism postulated by Dutta (1987), in which the maxilla is thought to twist to push the premaxilla forward, would require action of part A_1 of the adductor mandibulae muscle (Fig. 6.3, AM_1). However, in all cichlids studied the muscle remains inactive during the expansive phase, falsifying Dutt's hypothesis. Compressive muscles are the adductor mandibulae complex (Fig. 6.3, AM_1, AM_2) responsible for adduction of the jaws (Fig. 6.3(b)); the adductor arcus palatini muscle (Fig. 6.3, AAP) compresses the side walls (suspensory apparatuses) of the buccal cavity, and the geniohyoideus muscle (GH) lifts the floor of the buccal cavity. The result of the actions of these

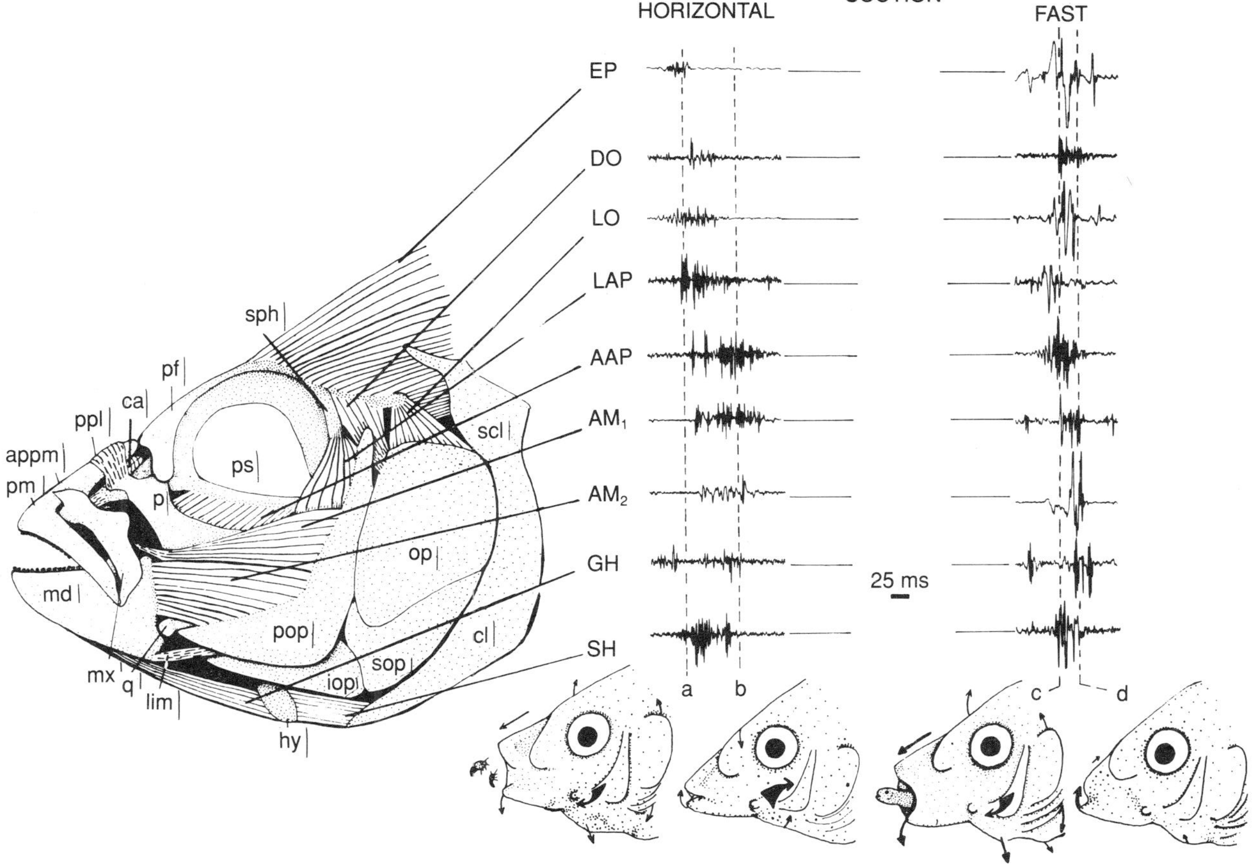

Fig. 6.3 Myology, electromyographic profiles, and kinematic profiles of *Hemitilapia oxyrhynchus* when feeding from midwater on zooplankton (HORIZONTAL) and small live fish (FAST). At left, lateral aspect of the cephalic musculature, Electromyograms are correlated with two major kinematic events ((a) and (c), expansive phases; (b) and (d), compressive phases). Abbreviations: AAP, adductor arcus palatini; AM_1, A_1 portion of adductor mandibulae; AM_2, A_2 portion of adductor mandibulae; APPM, ascending process of premaxilla; CA, rostral cartilage; CL, cleithrum; DO, dilator operculi; EP, epaxial muscle; GH, geniohyoideus; HY, hyoid; IOP, interoperculum; LAP, levator arcus palatini; LIM, mandibulointeropercular ligament; LO, levator operculi; MD, mandible; MX, maxilla; IP, operculum; p, palatine; PF, prefrontal; PM, premaxilla; POP, preoperculum; PPL, palatopalatine ligament; PS, parasphenoid; q, quadrate; SCL, supracleithrum; SH, sternohyoideus; SOP, suboperculum; SPH, sphenotic.

compressive muscles is a drastic reduction in the buccal cavity, forcing the food backwards into the pharynx.

These core functions of the major head muscles are maintained throughout the evolutionary radiations of the Cichlidae (Liem and Osse, 1975; Liem, 1978, 1979, 1980a). However, the pattern of muscle actions can vary intra- and interspecifically. The variations in onset and offset of muscle activity, duration and amplitude of muscle firings are often correlated with specific patterns of feeding behaviour (Liem, 1980a).

Prey-dependent variations in suction and biting patterns

The following is based on experiments with *Hemitilapia oxyrhynchus*, which is characterized as a periphyton collector because it scrapes material from the leaves of *Vallisneria*. "The fish turns on its side, takes the flat, grasslike leaf between its jaws and nibbles" (Fryer and Iles, 1972: p 74). Unfortunately it was not possible to simulate natural conditions (i.e. with *Vallisneria*) in the laboratory, where *H. oxyrhynchus* feeds on zooplankton, small fish, arthropods, worms, and algae on rock surfaces. Apparently this species exhibits a feeding repertoire that includes zooplankton, arthropods, worms and algae in its natural diet in Lake Malawi (K. R. McKaye, pers. comm., unpublished data 1982).

When feeding on zooplankton, *H. oxyrhynchus* uses suction to capture individual prey items from mid-water (Fig. 6.3(a) and (b)), with the electromyographic profile described above.

Elusive prey is captured by a high-velocity suction combined with forward locomotion. The electromyographic profile is characterized by higher amplitudes and shorter and overlapping bursts of activities of the head muscles (Fig. 6.3(c) and (d)).

When feeding from the surface the fish does not protrude its upper jaw, and the gape is limited (Fig. 6.4(a) and (b)), while the accompanying electromyographic profile is characterized by inactive epaxial muscles and early action of the A_1 part of the adductor mandibulae muscle (Fig. 6.4, EP, AM_1).

When sucking from the bottom, the fish minimizes the gape while protruding its upper jaw fully (Fig. 6.4(c) and (d)). Strong suction is generated by extended, high-amplitude actions of the epaxial, dilator operculi and levator arcus palatini muscles (Fig. 6.4(c) and (d), EP, DO, LAP).

When *H. oxyrhynchus* is biting an arthropod limb, it manipulates the prey item for extended periods (Fig. 6.5(a) and (b)). It can protrude and retract the jaws with an astonishing degree of control. The electromyographic profile is irregular in that some muscles show rhythmic activity (e.g. the levator operculi, levator arcus palatini, and adductor arcus palatini muscles, Fig. 6.5, LO, LAP, AAP) while the other muscles show irregularly overlapping bursts.

When biting hard, for example on fins of fishes, the fish produces a large gape and a protruded upper jaw (Fig. 6.5(c) and (d)). The dilator operculi, levator arcus palatini and adductor arcus palatini muscles are not active

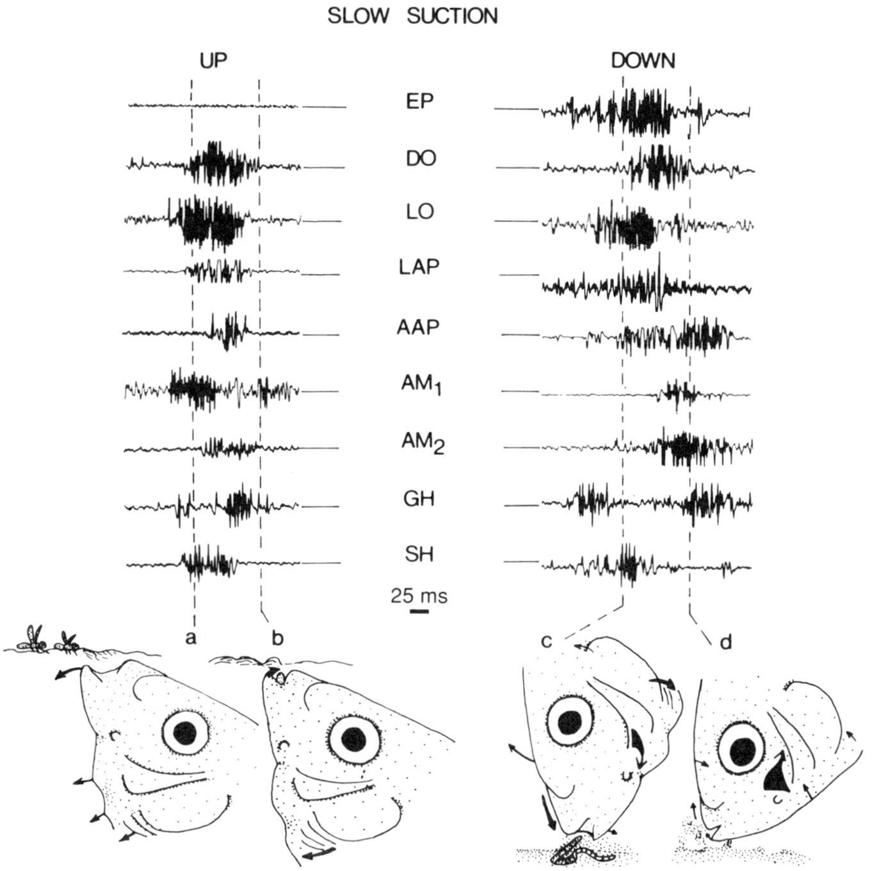

Fig. 6.4 Electromyographic and kinematic profiles of *H. oxyrhynchus* when suction feeding from the surface (UP) and from the bottom (DOWN). Note small gape and great expansion of the posterior part of the buccal cavity. Labels as for Fig. 6.3.

(Fig. 6.5 (c) and (d), DO, LAP, AAP), while the adductor muscles exhibit bursts of high amplitude and frequency (AM_1, AM_2, GH). The large gape seems to be produced by synchronous actions of the epaxial and sternohyoideus muscles (EP, SH).

Scraping of algae is a form of biting. It is effected by a very large gape but with limited upper jaw protrusion (Fig. 6.6). The electromyographic profiles are highly irregular, even though rhythmic jaw movements ('nibbles') can be discerned. Some muscles, e.g. the epaxial, levator operculi, adductor mandibulae A_1 and A_2, do show cyclical actions. However, the other muscles, including the sternohyoideus, are arrhythmic in their actions. It seems that the fish collects the algae from the rock surfaces by rhythmical jaw

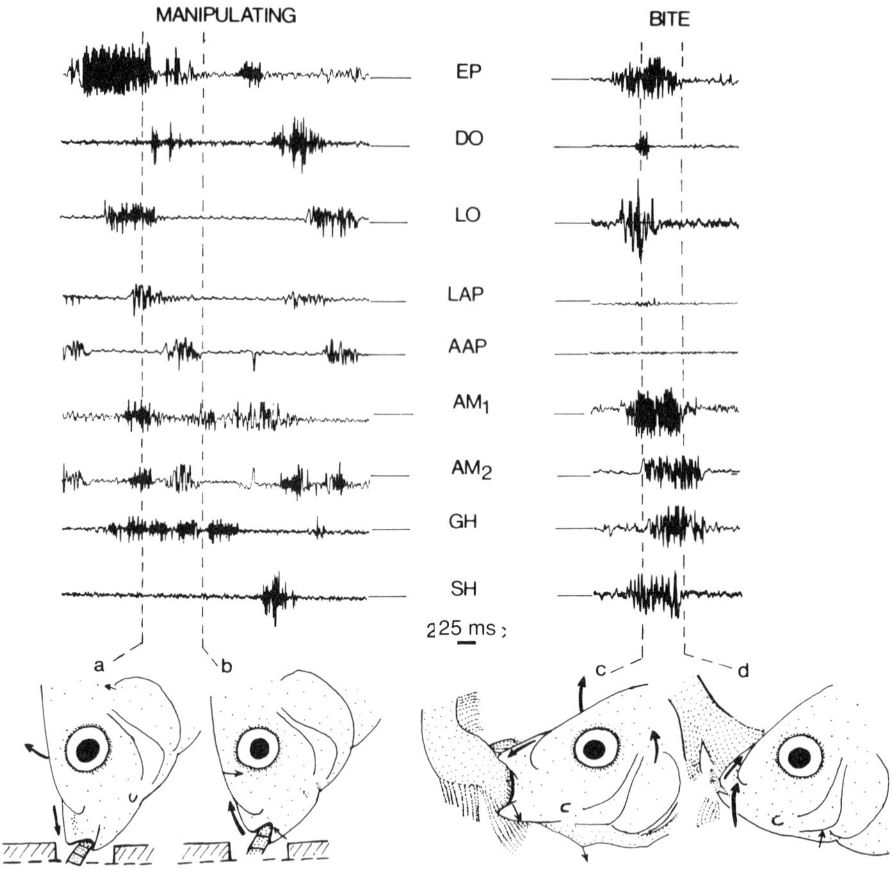

Fig. 6.5 Electromyographic and kinematic profiles of *H. oxyrhynchus* when biting and manipulating. Note the ability of the fish to protrude its upper jaw fully without increasing the gape of the mouth (a). Labels as for Fig. 6.3.

movements, but the collected algae are sucked deeper into the mouth by arrhythmic suction movements at irregular intervals.

Possible significance of the observed variations

The different patterns of muscle actions (Figs. 6.3–6.6) in *Hemitilapia oxyrhynchus* produce different deformations of the cone-shaped buccal cavity, which can be observed on X-ray cine films. During suction feeding, the muscles produce an attenuated cone with a very broad base. When suction feeding from the bottom, the mouth is protruded but the gape kept small (Fig. 6.4(c) and (d)), while the epaxial, levator arcus palatini, adductor arcus palatini and adductor mandibulae muscles play dominant roles. This pattern foreshadows the design of specialized suction feeders with very large epaxial muscles and

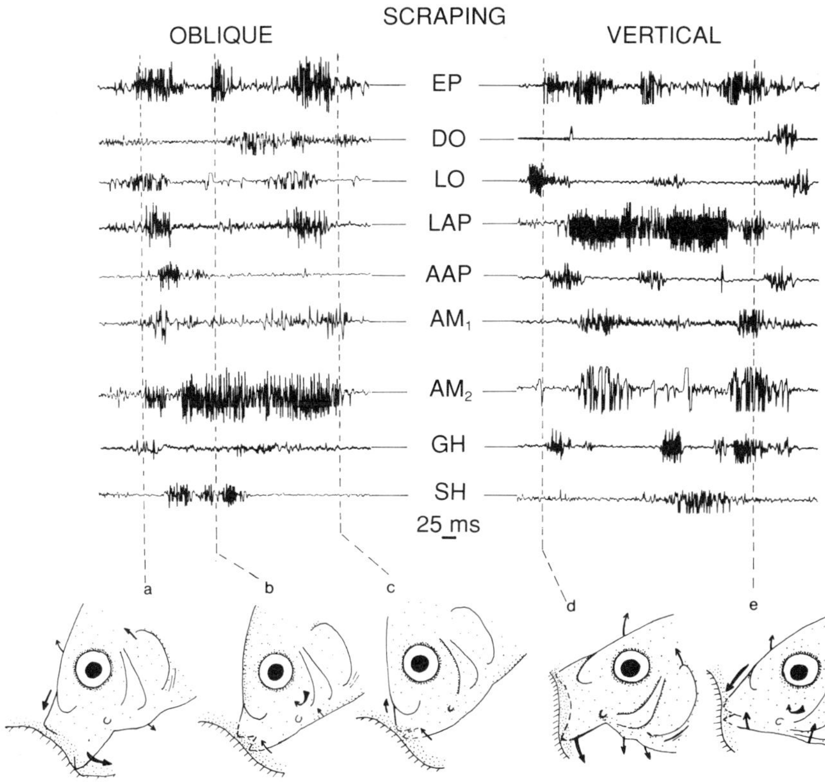

Fig. 6.6 Electromyographic and kinematic profiles of *H. oxyrhynchus* when scraping algae. Note the very large gape of the mouth with limited jaw protrusion and little expansion of the buccal cavity. Labels as for Fig. 6.3.

high supraoccipital crests, deep palatopterygoid arches offering expansive surfaces for the levator and adductor arcus palatini and adductor mandibulae muscles and long ascending processes for jaw protrusion.

When *H. oxyrhynchus* feeds on elusive, rather large prey in mid-water, it uses both ram- and suction-feeding strategies, opening its gape to the maximum. The sternohyoideus plays a dominant role (Fig. 6.3 (c) and (d), SH). The pattern of muscle actions during the capture of large elusive prey produces a more cylindrical configuration of the buccal cavity. Specialized ram-feeding cichlids do exhibit dominant sternohyoideus muscle mechanisms, large gapes, cylindrical buccal cavities and shallow palatopterygoid arches, offering limited space for the adductor mandibulae and levator arcus palatini muscles (Liem, 1978).

Biting, scraping, manipulating and grazing in *H. oxyrhynchus* (Figs 6.5, 6.6) are accompanied by the dominance of the actions of the adductor mandibulae muscle complex (AM_2), the epaxial muscles (EP), and geniohyoideus muscle.

Biting is accomplished either with a small gape (Fig. 6.5(a) and (b)) or with a large gape (Fig. 6.5(c) and (d); Fig. 6.6(c) and (d)). Accordingly, biters, scrapers and grazers are characterized by very large adductor mandibulae muscle complexes accommodated on the deep suspensory apparatus, large epaxial muscles and high occipital crests, and more massive geniohyoideus muscles in conjunction with a greater width of the lower jaw and head (Barel, 1983).

Thus, the different patterns in the feeding repertoire of *H. oxyrhynchus*, which resemble those of *Petrotilapia* (Liem, 1980a), reflect, on a smaller scale, the specialized patterns and morphologies of the highly specialized cichlids such as the grazers (e.g. *Labeotropheus, Petrochromis*), scale scrapers (e.g. *Perissodus*, Liem and Stewart, 1976), invertebrate pickers (*Tanganicodus*, Liem, 1978) and suction feeders (e.g. *Tylochromis*).

Prey-processing apparatus

Cichlid morphological design

Cichlids possess a specialized design of their pharyngeal jaw apparatus (PJA): the upper pharyngeal jaw (composed of pharyngobranchials 2 and 3 and the toothplate of the fourth) articulates with the pharyngeal apophysis of the skull base by means of a synovial joint (Liem, 1973, 1986; Greenwood, 1974a; Ismail *et al.*, 1982). The lower pharyngeal jaw is formed by the variously sutured left and right fifth ceratobranchials, which possess prominent muscular processes (Fig. 6.7). The principal muscles acting upon the pharyngeal jaws are the fourth levator externus, levator posterior, retractor dorsalis, pharyngohyoideus and pharyngocleithralis externus and internus muscles (Fig. 6.7). In all cichlids, the fourth levator externus generally inserts on the muscular process of the lower pharyngeal jaw, while in many species the levator posterior muscle does likewise. As a result, the lower pharyngeal jaw is suspended by a muscle sling. For discussions on the evolutionary origin of the muscle sling the reader is referred to Aerts (1982), Claeys and Aerts (1984) and Liem (1986).

Many interspecific variations are found, affecting not only the size and shape of the pharyngeal apophysis (Greenwood, 1978) and pharyngeal jaws (Poll, 1956; Greenwood, 1974a), but also the size, shape and architecture of the muscle sling (Liem, 1973, 1978; Liem and Kaufman, 1984). Extraordinary plasticity of the pharyngeal jaw apparatus in response to the nature of the prey has been reported for *Astatoreochromis alluaudi* (Greenwood, 1965b) and other cichlids (Hoogerhoud, 1986a). When *Astatoreochromis* is fed hard prey, it develops a massive pharyngeal mill with hypertrophied muscles, while a consistently soft diet leads to a dramatic reduction of the size of the pharyngeal jaws and associated musculature. In *Cichlasoma minckleyi* there are two distinct morphs: (1) the molariform morph, with massively differentiated

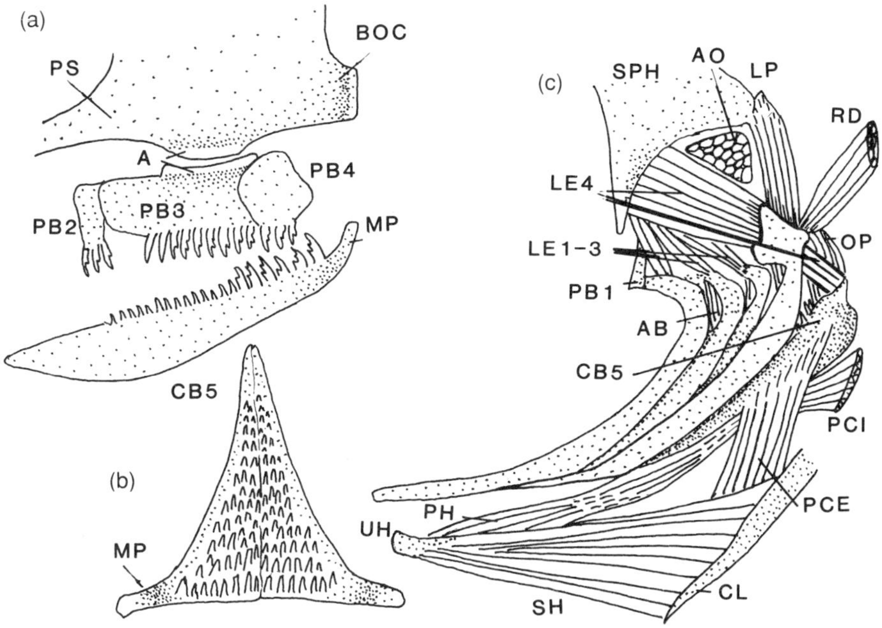

Fig. 6.7 Major anatomical components of the pharyngeal jaw apparatus in a cichlid. (a) The upper and lower pharyngeal jaws of a piscivorous cichlid. Note the basipharyngeal joint (A) between the basicranium and the upper jaw, which is made up of the pharyngobranchials (2–4). (b) The lower pharyngeal jaw composed of the sutured left and right fifth ceratobranchials. (c) Lateral aspect of the musculature associated with the pharyngeal jaws. Note the muscle sling composed of the fourth levator externus muscle (LE4) which extends from the sphenotic region of the neurocranium to the muscular process of the lower pharyngeal jaw (CB5). Abbreviations: A, basipharyngeal joint; AB, adductor branchialis; AO, adductor operculi; BOC, basioccipital; CB5, fifth ceratobranchial; CL, cleithrum; LE 1–3, levator externus 1–3; LE4, fourth levator externus; LP, levator posterior; MP, muscular process of lower jaw; OP, obliquus externus; PB 1–3, pharyngobranchials 1–3; PB4, toothplate of fourth pharyngobranchial; PCE, pharyngocleithralis externus; PCI, pharyngocleithralis internus; PH, pharyngohyoideus; PS, parasphenoid; RD, retractor dorsalis; SH, sterno hyoideus; SPH, sphenotic; UH, urohyal.

pharyngeal jaws which are provided with large molariform teeth and highly pinnate hypertrophied muscles; (2) the papilliform morph, with relatively slender pharyngeal jaws which are operated by mostly parallel-fibred smaller muscles (Liem and Kaufman, 1984). The two morphs occur in natural populations and tend to segregate along the food axis only during ecological bottlenecks (Liem and Kaufman, 1984).

The different morphological interspecific variants in the pharyngeal jaw apparatus clearly reflect adaptations to preferred diets (Liem, 1973; Hoogerhoud, 1986a).

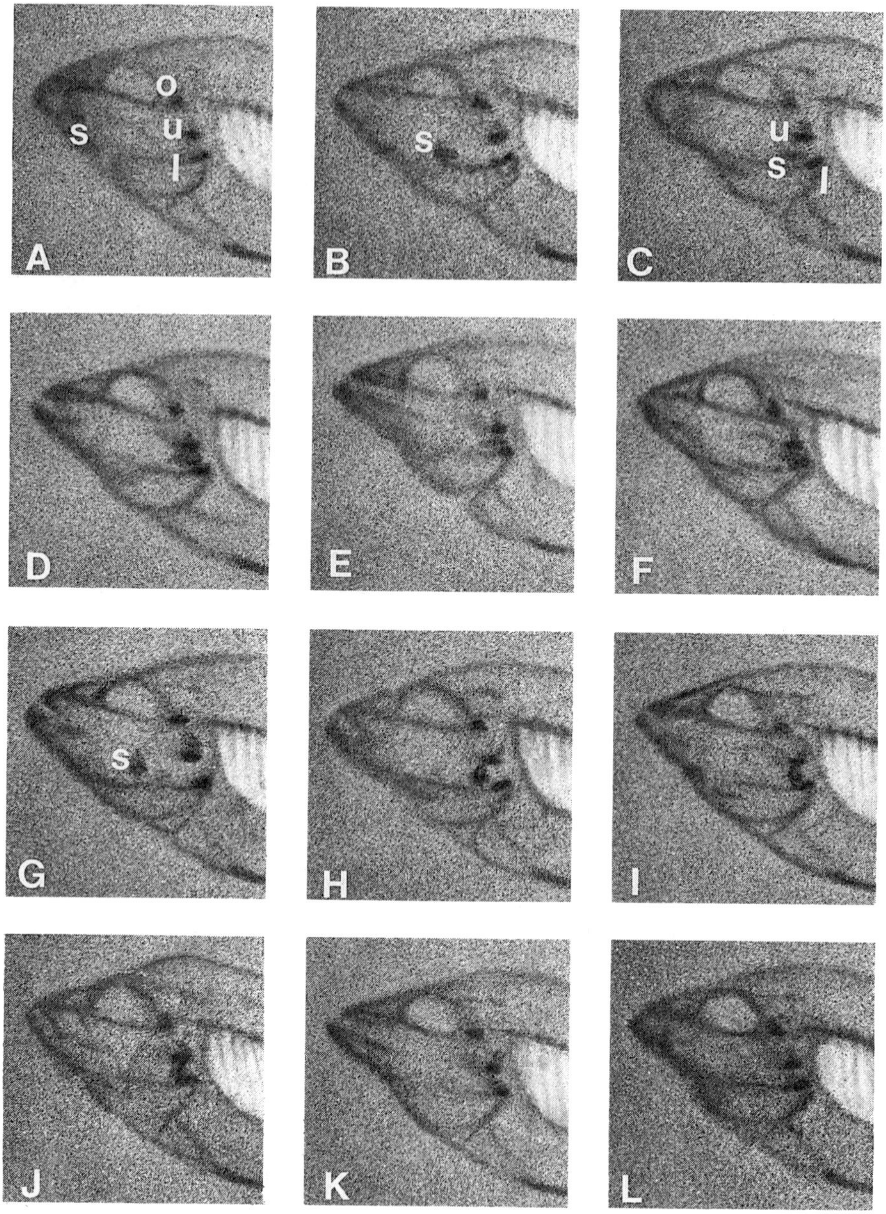

Fig. 6.8 Kinematics of the pharyngeal jaws and the action of the hydrodynamic tongue of *Astatoreochromis alluaudi*. Selected prints of a cineradiographic film taken at 100 frames s^{-1} while the fish was feeding on a snail. The kinematic sequence begins with the captured snail (s) being in the front of the buccal cavity (a). The snail is transported by a water current toward the pharyngeal jaws (b). The upper jaw (u) moves forward (c) and meets the dorsally held lower jaw (1), thereby exerting a

Kinematics and muscle function

The kinematics and muscle actions of the cichlid PJA have been studied by cineradiography and electromyography (e.g. Liem, 1973, 1986). Both the pattern of pharyngeal jaw movements and the electromyographic profile of muscle activities are conserved throughout the evolutionary radiation of the cichlids (Liem, 1986; Aerts *et al.*, 1986). For example, when *Astatoreochromis* eats a snail, both the prey and pharyngeal jaws can be seen on an X-ray film (Fig. 6.8). The prey is a snail which is clearly discernible on the X-ray film. Prey is transported to the pharyngeal jaws by water currents. Once it has reached the jaws, the upper pharyngeal jaw pivots forward, while the lower jaw moves upward (Fig. 6.8(c) and (d)) to exert a shearing action on the prey by the simultaneous retraction of the upper and lower jaws (Fig. 6.8(e) and (f)) which approach each other. Electromyography has shown that shearing is accomplished by action of the levator externus 4 and levator internus muscles, while crushing is the result of action of the levator externus 4, levator posterior and retractor dorsalis muscles (Liem, 1973, 1978, 1986; Liem and Kaufman, 1984).

The kinematic and electromyographic patterns are conserved throughout the radiation of cichlids. However, structural diversity is quite spectacular. In general, piscivores have elongate, slender and mostly parallel-fibred, strap-like levator muscles and elongate jaws (Liem, 1978; Witte and Barel, 1976). Durophagous cichlids have shorter, stouter pharyngeal jaws with massive, pinnate-fibred, shorter levator muscles (Hoogerhoud and Barel, 1978; Witte and Barel, 1976). Invertebrate feeders exhibit intermediate morphologies (Anker, 1978).

Functionally the cichlid pharyngeal jaw apparatus is very versatile. While maintaining the basic kinematic and electromyographic patterns, the jaws can move in such a way as to crush or shear prey, lacerate food items such as fish, compact filamentous algae prior to swallowing, or arrange scales on top of each other like rolls of coins. These functions are facilitated by specialized dentition (Liem, 1973).

Hydrodynamic tongue

It is generally believed that the cichlid pharyngeal jaw apparatus is capable of manipulating and positioning prey items. However, recent evidence obtained by high-speed cineradiography (Fig. 6.8) has shown otherwise. In *Astatoreo-*

---

shearing force on the snail. Crushing ((d) to (f)) is effected by strong retraction of both the upper and lower jaws while the latter is pulled dorsally. In this sequence the fish fails to crush the snail on its first attempt. The hydrodynamic tongue is used to reposition the snail back in the buccal cavity (g). Subsequently, the snail is moved between the jaws (h) and crushed ((i) to (k)) and the broken shells are expelled (l). Abbreviations: l, lower pharyngeal jaw; o, otolith; s, snail; u, upper pharyngeal jaw. Cycle lasts 450 msec.

chromis the positioning of the snail prey item can be observed directly with high-speed cineradiography. The snail is sucked in and propelled by a water current from the front of the buccal cavity toward the pharynx (Fig. 6.8(a) to (c)). The current is created by an observable expansion of the pharynx (the base in the truncated cone model). At first the fish attempts to shear and crush the snail (Fig. 6.8(d) to (f)). As the crushing attempt fails, the anterior part of the buccal cavity can be seen to expand, thereby creating a forward water current carrying the snail with it (Fig. 6.8(g)). It is in this forward compartment of the buccal cavity that the snail is positioned and turned by water currents. Once the snail has been turned it is again drawn backwards by lifting of the anterior floor of the buccal cavity. The snail is then carried by water currents to a new position between the pharyngeal jaws (Fig. 6.8(h)) and the fish exerts shearing and crushing forces (Fig. 6.8(i) and (j)). In the new position the snail shell is weaker and fails under the crushing forces (Fig. 6.8(k)). This is followed by the expulsion of the broken shell pieces by means of water jets (Fig. 6.8(l)).

Bemis and Lauder (1986) have proposed the term **hydrodynamic tongue** for the positioning of prey items within the buccal cavity of the lungfish by means of water currents. It is clear from cineradiographic analysis of many species of cichlids that prey repositioning is not done by the pharyngeal jaws but by water currents and jets. The hydrodynamic tongue in cichlids is responsible for prey transport, positioning and repositioning. Hydrodynamically, repositioning is accomplished by creating selective local distortions of the truncated cone of the buccal cavity. Local expansion of the base of the cone, or a local compression of the anterior cone (lifting of the anterior part of the buccal floor) will result in a posteriorly directed current carrying the prey with it. Localized expansion of the tip of the cone will create a forward-directed current. Asymmetrical distortions of the side walls of the cone can produce whirlpools, turning the prey. Localized distortions of various parts of the truncated cone are produced by symmetrical or asymmetrical actions of the levator arcus palatini, adductor arcus palatini, geniohyoideus, sternohyoideus, dilator and adductor operculi muscles.

The hydrodynamic tongue appears to play a key role in cichlids that have to reposition prey items. Mollusc-crushing cichlids often have to turn or reposition the snails or bivalves until a weaker area of the shell can be acted upon by the shearing and crushing forces of the pharyngeal jaw apparatus. In these molluscivorous cichlids the form and function of the buccal apparatus may reflect adaptations related to the hydrodynamic tongue rather than prey capture. The capture of snails and bivalves does not require special adaptations; however, the configuration and degree of movements of the side walls of the buccal cavity floor made up by the hypobranchial apparatus, the geniohyoideus and sternohyoideus muscles do influence or determine the efficiency of the hydrodynamic tongue. Other cichlids engaging in extensive repositioning of prey items within the buccal cavity are those with specialized feeding preferences, e.g. filamentous algae and arthropods, which are respectively compacted and masticated prior to swallowing. All these feeding types are characterized by highly differentiated levator arcus palatini, adductor arcus palatini, dilator operculi, adductor operculi, geniohyoideus and sternohyoideus muscles, and palatopterygoid arch and hyoid morphology conducive to a variety of extreme movements.

The role of the hydrodynamic tongue may also be quite significant in oral brooding cichlids. However, no experimental analysis has been made of the functional role of the hydrodynamic tongue during oral brooding in a cichlid.

6.3 POLYPHYLETIC FUNCTIONAL DIVERSIFICATION OR ADAPTIVE RADIATION?

Adaptive radiation is a pervasive and influential concept in macroevolutionary theory. **Adaptive radiation** denotes the diversification into different ecological niches by species derived from a common ancestor. The cichlid fishes of the Great Lakes of Africa are often discussed in recent books on evolution (e.g. Grant, 1986; Futuyma, 1986) as an impressive and classical example of adaptive radiation. It is generally accepted that body form has remained relatively conservative, but that the trophic apparatus has undergone a spectacular adaptive radiation, especially among the haplochromines in Lake Victoria, the cichlids of Lake Malawi and in Lake Tanganyika (Fryer and Iles, 1972; Greenwood, 1984).

Functional morphology has demonstrated unequivocally that the trophic apparatus of cichlids in Lake Malawi and Lake Tanganyika has undergone a functional diversification unparalleled by any other family of vertebrates (Liem, 1974, 1978, 1979, 1980a; Liem and Osse, 1975). The functional diversification of Lake Tanganyika cichlids is summarized in Fig. 6.9 in the same way as that of Lake Malawi cichlids by Fryer and Iles (1972, p. 488, Fig. 333), which has been reproduced in many textbooks (e.g. Moyle and Cech, 1988, pp. 402–3). A comparison of the two figures reveals numerous similarities. The functional and morphological diversification of Lake Tanganyika cichlids can be explained in the same way as that of Lake Malawi cichlids. Generalized insectivorous ancestors capable of living among rocks and over sand, and not specialized for life in either situation, would gradually be displaced from rocky habitats by forms better adapted for life in rocky habitats. Such displaced generalists would not find any refuge on sandy habitats where in the meantime a group of well-adapted specialists would have become dominant inhabitants. The result would be extinction and the community would consequently be composed of specialists which outcompeted the jack-of-all-trades generalist forms. This process would eventually lead to gaps in the phylogeny of Lake Tanganyika cichlids as great as those between Lake Malawi taxa. In Lake Tanganyika the ancestral cichlids can be hypothesized to have

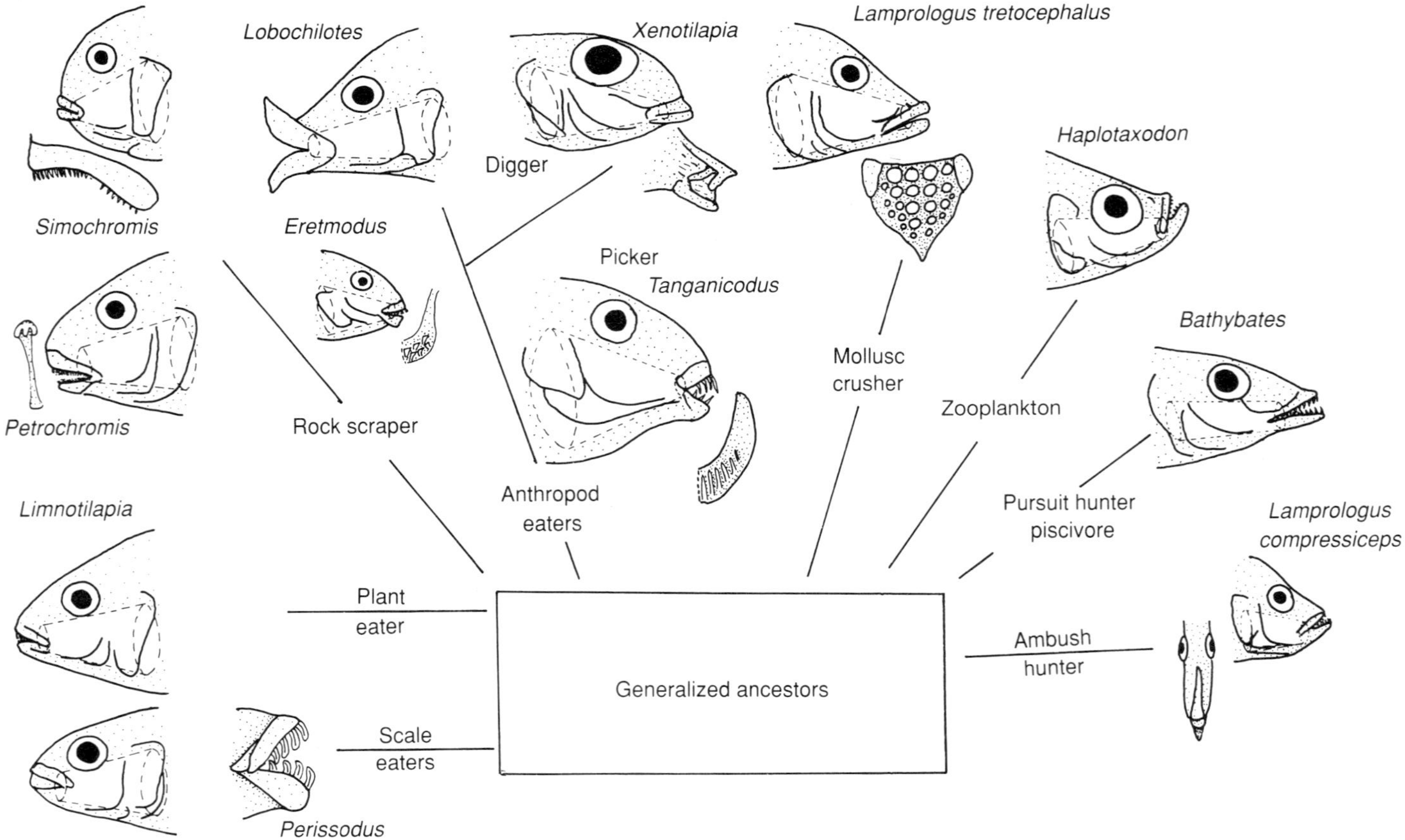

Fig. 6.9 Polyphyletic functional diversification of Lake Tanganyika cichlids. The variations in the shape of the buccal cavity, ranging from cylinders to attenuated cones, are drawn in the head profiles.

given rise to the following major trophic specialists (Fig. 6.9): scale eaters (*Perissodus*) with highly specialized leaf-like teeth; higher plant herbivores (*Limnotilapia*); rock scrapers (*Petrochromis*) feeding on various algae, diatoms, etc.; various arthropod eaters, including pickers (*Tanganicodus*) and those digging in the sand (*Xenotilapia*) with their protrusible, ventrally directed mouths; durophagous forms (*Lamprologus tretocephalus*) with very heavy pharyngeal jaws and molariform teeth; zooplankton feeders (*Haplotaxodon*); pursuit-hunting piscivores (*Bathybates*) and ambush-hunting carnivores (*Lamprologus compressiceps*).

As is the case in Lake Malawi, it can be argued that piscivory and zooplanktivory are based on very similar functional designs, with a more cylindrical buccal cavity, shallow palatopterygoid and large mouth (compare *Haplotaxodon* and *Bathybates* in Fig. 6.9). Scale eaters (*Perissodus*) also possess a design with the features encountered in open-water planktivores and piscivores. Several of the scale eaters are also able to switch to zooplankton (Liem and Stewart, 1976). Strikingly-similar specializations are found among the species flocks of the Mbuna in Lake Malawi and the rocky-habitat cichlids of Lake Tanganyika (*Petrochromis, Simochromis, Eretmodus*). Because the specializations involve the attenuated, truncated cone design of the buccal cavity, with deep palatopterygoid shapes, small mouths, large adductor mandibulae complexes and often multicusped, stalked teeth, it is often argued that rock-inhabiting cichlids such as the Mbuna are monophyletic (Fryer and Iles, 1972). Unfortunately the phylogenetic evidence for monophyly of the Mbuna is still very tentative, since the hypothesis is based on similar specializations which are typically associated with biting and may well represent convergent characters. In Lake Tanganyika the rocky-habitat cichlids belong to the *Ophthalmotilapia* assemblage (Liem, 1981a; Greenwood, 1983b), the *Petrochromis* lineage, and the *Tanganicodus* group (Liem, 1979). That is they are polyphyletic.

In Lake Tanganyika the functional diversification of the cichlids (Fig. 6.9) is of a polyphyletic nature (Liem and Stewart, 1976; Liem, 1979, 1981a; Greenwood, 1983b; Stiassny, 1981a, b). The pattern of diversity in the trophic apparatus of Lake Tanganyika cichlids (Fig. 6.9) should not be called adaptive radiation as defined on p. 145. We are dealing with polyphyletic functional diversification.

Hypothesized trophic adaptive radiations expressed in Fig. 6.9 for Lake Tanganyika and by Fryer and Iles (1972: Fig. 333) for Lake Malawi do not reflect accurately either the evolutionary process or its pattern. These hypothesized patterns imply that the historical factors contained in various lineages do not play a significant role in the evolutionary process. Instead omnipotency is ascribed to the 'generalized' ancestor(s), with the extrinsic determinants playing a dominant role in producing the dramatic diversification in the cichlid trophic apparatus in a way analogous to that in the Hawaiian honeycreepers and Darwin's finches (Grant, 1986). Comparative

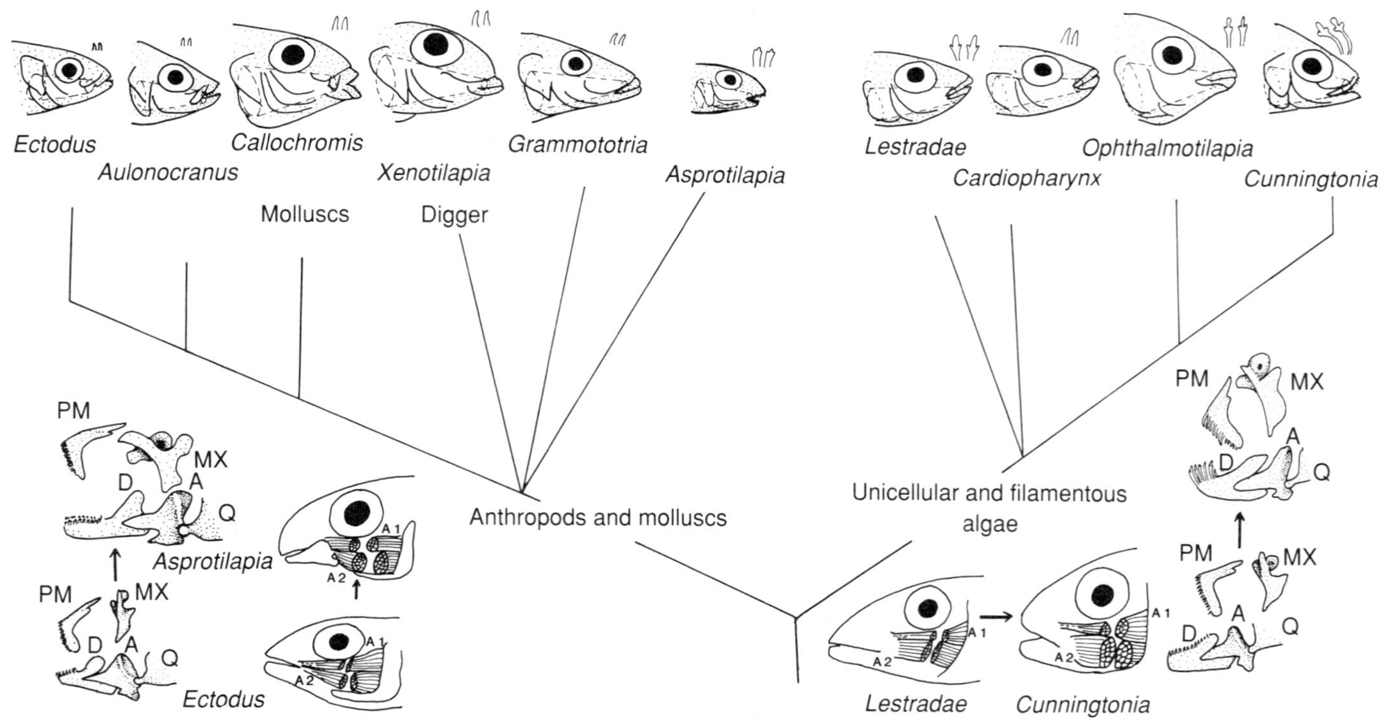
Ectodus
Aulonocranus
Callochromis
Xenotilapia
Grammototria
Asprotilapia
Lestradae
Cardiopharynx
Ophthalmotilapia
Cunningtonia
Molluscs
Digger
Anthropods and molluscs
Unicellular and filamentous algae
PM
MX
D
A
Q
A 1
A 2
Asprotilapia
Ectodus
Lestradae
Cunningtonia

functional studies based on a phylogenetic foundation of the cichlid trophic apparatus have revealed that the similarities in the patterns of diversity of cichlids, the Hawaiian honeycreepers and Darwin's finches are more apparent than real. Cichlid functional design allows specialists to be jacks of all trades at the same time (Liem, 1980a), enabling them to engage in extensive prey switching (McKaye and Marsh, 1983; Ribbink *et al.*, 1983a). Witte (1984a) has also shown that morphologically-similar species differ more in feeding niches than was believed. Trophic competition is thought to have been the driving force in the radiation of Darwin's finches, and character displacement is well documented (Grant, 1986). In cichlids, character displacement in trophic structures has never been shown, indicating that the diversification of cichlids was not triggered solely by competitive interactions within the great ecological richness of the East African lakes.

An analysis of adaptive radiation must begin with the phylogeny of the group (Stiassny, 1981b). I have chosen the *Ophthalmotilapia* assemblage of Lake Tanganyika because its phylogeny is better known than that of any other major group of cichlids (Liem, 1981a; Greenwood, 1983b). The assemblage is monophyletic; it has undergone an evolutionary or adaptive radiation, with the emergence of ten lineages which are generally recognized as genera. As one can observe from the variations in the shape of the buccal cone, jaw structures and adductor mandibulae muscle configurations (Fig. 6.10), the radiation is canalized. Within this radiation, herbivore, arthropod and mollusc feeders evolved on a common basic body plan. No invertebrate pickers, zooplanktivores, scale eaters and piscivores evolved. It is clear that the radiation is constrained by historical factors.

It is more accurate to view the spectacular trophic diversification of cichlids in Lake Tanganyika, not as a single adaptive radiation within a monophyletic assemblage, but as multiple radiations originating from multiple ancestors. Multiple origins would favour the differentiation of new lineages with very different morphological and ecological directions. Each lineage underwent a canalized radiation. Thus, the radiation of the *Ophthalmotilapia* assemblage is distinctly different from that of the *Lamprologus* assemblage. For example, the

Fig. 6.10 Adaptive radiation in the monophyletic *Ophthalmotilapia* assemblage. The feeding specializations are superimposed upon an independently derived cladogram. One lineage is characterized by arthropod and mollusc feeders, while the second lineage is represented by predominantly herbivorous taxa. The more primitive members in each of the two lineages are characterized by more-cylindrical buccal cavities. Specialization in feeding is accomplished by the development of buccal cavities in the shape of truncated cones. At lower left, the transformation from generalized (*Ectodus*) to specialized (*Asprotilapia*) is summarized in respect to jaw structure and adductor muscles. Similarly, at lower right, such transformations are summarized for *Lestradea* (the primitive member) and *Cunningtonia* (the derived member). Abbreviations: A1, A2, parts A1 and A1 of the adductor mandibulae muscle complex; A, angular; D, dentary, MX, maxilla; PM, premaxilla; Q, quadrate.

mollusc crusher of the *Ophthalmotilapia* assemblage (represented by *Callochromis*) differs significantly from the mollusc crusher of the *Lamprologus* group (represented by *L. tretocephalus*) (Figs. 6.9, 6.10). The differences in morphological design reflect differences in ancestry (historical or intrinsic determinants of form) rather than subtle differences in feeding habits.

If the diversification of the trophic apparatus of cichlids in Lake Tanganyika can be explained as having originated from multiple ancestors, each of which had a canalizing effect on the subsequent radiations, many unsolved problems can be approached in novel ways. For example, two major areas may need a complete reassessment.

First, the diversity in functional design of the cichlid feeding apparatus cannot be explained exclusively by the long-favoured mechanisms of competition, differential feeding efficiencies and character displacement, none of which has been experimentally demonstrated. The feeding plasticity, as demonstrated in the laboratory (Figs. 6.3 to 6.6) and in the field (e.g. McKaye and Marsh, 1983; Ribbink *et al.*, 1983a; Witte, 1984a), is a potentially powerful mechanism to avoid competition. The trophic diversity is to a large extent a consequence of the historical factors contained in the multiple ancestors. Functional studies showing the extraordinary capabilities for prey switching in cichlids do not conflict with the notion of the importance of historical factors underlying the multiple canalized radiations.

Second, the evolutionary history of the cichlids in each of the East African Great Lakes needs to be examined with the assumption that multiple, differently designed ancestors may have been involved. Greenwood (1983b) has criticized the widely accepted notion that the major lakes are faunistically closed basins, and have been so almost since their inception, with the result that the cichlids of a lake are presumed to have evolved from one or a few ancestral species originally trapped there. Greenwood offers some preliminary evidence for close phylogenetic relationships between members of the *Ophthalmotilapia* assemblage of Lake Tanganyika and *Lethrinops* of Lake Malawi. He also hypothesized relationships between other cichlid species from lower Ewa, Lake Malawi and Lake Tanganyika. It is also possible that multiple colonizations of the lakes have taken place.

In short, comparative functional morphology has demonstrated the shortcomings of our current and generally favoured interpretations of cichlid diversifications, as monophyletic in origin and driven by competition in faunistically closed basins.

Chapter seven

Feeding relationships

K. Yamaoka

7.1 INTRODUCTION

It is well known that an extraordinary adaptive radiation has occurred in the feeding habits of cichlid fishes, especially in the East African Lakes (Poll, 1956; Greenwood, 1974a; Lowe-McConnell, 1987). However, previous work focused mostly on stomach content analysis to infer what cichlids took, and there was no comprehensive study on feeding relationships except Fryer's (1959) pioneering work in Lake Malawi.

One of the most interesting ecological problems about cichlids is how numerous allied species sharing similar foods coexist, because it has been widely accepted that related species with similar food requirements cannot easily do so (Hori, 1983, 1987; Ribbink *et al.*, 1983a; Takamura, 1984; Lowe-McConnell, 1987).

The problem of coexistence cannot be approached only by stomach content analysis. The importance of feeding behaviour in resource partitioning has been stressed in marine fishes (Jones, 1968; Smith and Tyler, 1973) and in cichlids (van Oijen, 1982; Yamaoka, 1982, 1983a; Hori, 1983, 1987; Takamura, 1983).

Recently, Japanese and Zairean workers have jointly conducted ecological studies on the feeding habits of cichlids in Lake Tanganyika (Kawanabe, 1981; Yamaoka, 1982, 1983a; Hori, 1983, 1987; Takamura, 1983, 1984; Nshombo *et al.*, 1985; Gashagaza and Nagoshi, 1986; Mbomba, 1986; Kondo, 1986; Gashagaza, 1988). The feeding ecology of the endemic cichlids of Lakes Malawi and Victoria has been studied mainly by South African and Dutch workers, respectively (Marsh *et al.*, 1981; Lewis, 1982; Ribbink *et al.*, 1983a, b; McKaye and Marsh, 1983; Marsh and Ribbink, 1985; Witte, 1981, 1984a; Witte and Witte-Maas, 1981; van Oijen *et al.*, 1981; van Oijen, 1982). Almost all results by these research teams suggest that when examined in detail, species that have been regarded as sharing the same trophic requirements show slight but distinct interspecific differences in feeding ecology: feeding behaviour, feeding sites and habitat.

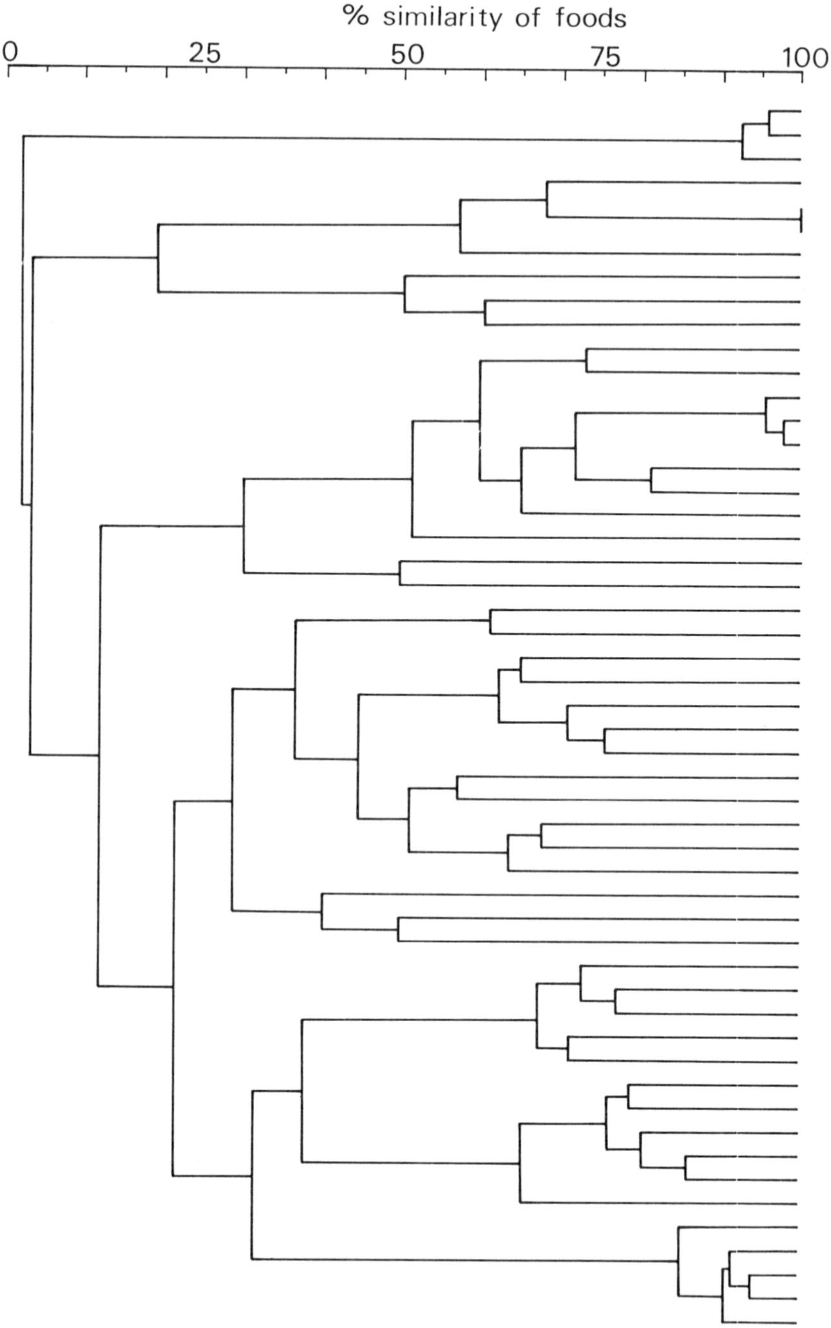

Fig. 7.1 Cluster analysis (using Mountford's average linkage method) of similarity of food for each species pair, and feeding-habit groups of fishes, including non-cichlids, from Lake Tanganyika. Values for feeding-habit groups are mean percentage

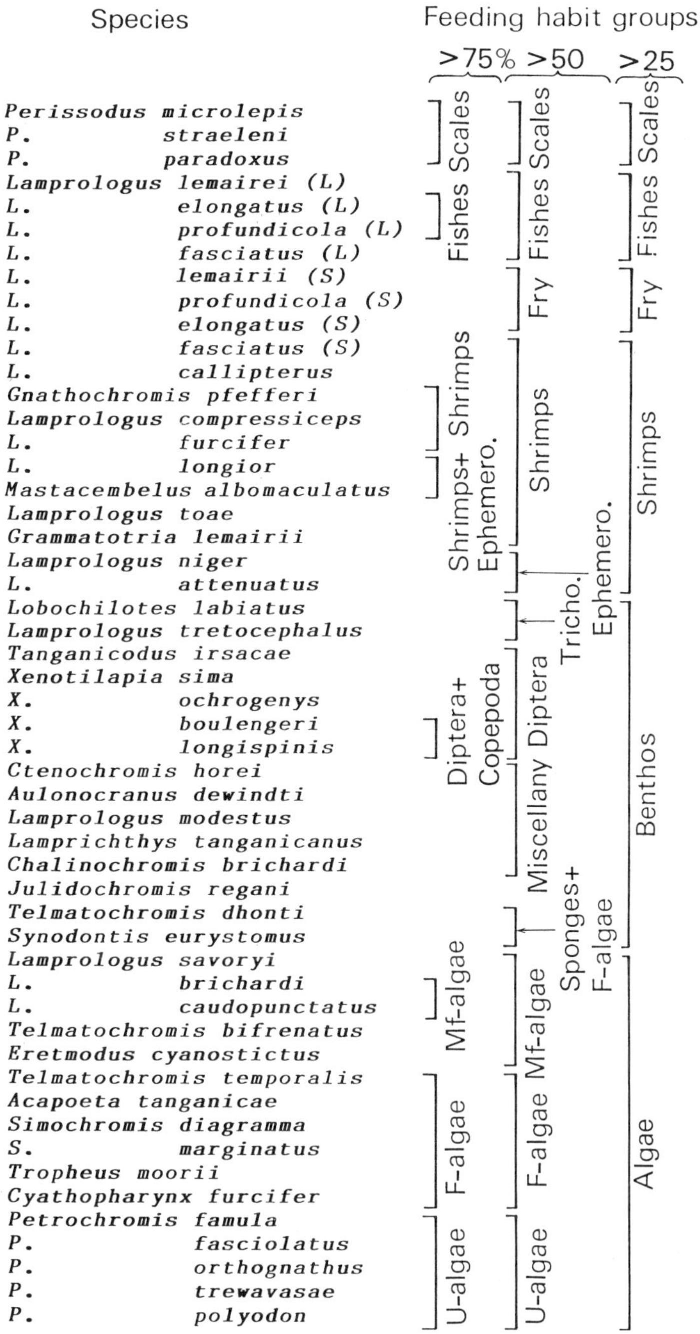

similarity of food. U-algae, unicellular algae; F-algae, filamentous algae; Mf-algae, filamentous algae ⩽ 5 μm in width, probably blue green algae; L, large specimen; S, small specimen (modified with permission from Hori, 1987).

In this paper, I do not intend to be concerned with the overall feeding habits of cichlids, world-wide, but will discuss the importance to coexistent species of interspecific differences in feeding behaviour. Because observations on feeding behaviour have been conducted mainly in Africa, especially in the Great Lakes region, emphasis will be on cichlids from that region.

7.2 FEEDING BEHAVIOUR

Before discussing feeding behaviour, some information about classification of the feeding-habit groups must be given. Figure 7.1 shows various feeding-habit groups found in the rocky habitat of Lake Tanganyika and demonstrates that the category of feeding habits differs when the percentage value of food similarity changes (Hori, 1987).

I shall deal mainly with the feeding behaviour of epilithic algal feeders and some carnivorous groups which have been thoroughly studied (Fryer, 1959; Fryer and Iles, 1972; Holzberg, 1978; Yamaoka, 1982, 1983a; Ribbink *et al.*, 1983a; Takamura, 1983, 1984; Hori, 1983, 1987).

Epilithic algal feeders (aufwuchs feeders)

This group has been well known since Fryer's (1959) work on the Mbuna in Lake Malawi. According to Keenleyside (1979), three general types of herbivorous fishes can be recognized: grazers, browsers and phytoplanktivores. The first two of these form major groups among epilithic algal-feeding cichlids.

Grazers

According to Jones (1968), grazers are fishes that pick up large quantities of the substrate while feeding. Keenleyside (1979) also defined them as fishes that remove algae from the substrate by scraping or rasping at the surface and sucking the loosened materials into the mouth, or by biting pieces out of the substrate. I combine these two definitions in this paper and use the term to designate fishes that comb unicellular algae from filamentous algae that are attached firmly to rocks, together with loosened materials including fine sand particles, by pressing the tooth band against the rock surface. They take foods with repeated mouth openings and closings, just like the 'sucking' of *Petrotilapia tridentiger* in Lake Malawi (Fryer, 1959). This definition is partly in accord with the description of loose aufwuchs feeders in Fryer (1959). Ribbink *et al.* (1983a) use the term 'brushing' instead of 'grazing' to designate this mode of food collection.

Grazers are one of the commonest components in the rock-frequenting cichlid communities of Lakes Tanganyika and Malawi. Six *Petrochromis* species

Table 7.1 Behavioural characters in five species of the genus *Petrochromis**

Species	*Grazing speed*† (s^{-1}) (*mean* ± *S.D.*)‡	*Bites per bout* (*mean* ± *S.D.*)‡
Petrochromis polyodon	1.54 ± 0.30 (149)	4.63 ± 1.87 (121)
P. trewavasae	1.82 ± 0.32 (131)	3.61 ± 1.50 (108)
P. famula	2.65 ± 0.50 (87)	1.66 ± 0.77 (166)
P. orthognathus	2.90 ± 0.63 (110)	2.74 ± 1.28 (115)
P. fasciolatus	4.10 ± 0.44 (100)	6.29 ± 2.30 (117)

*Source: Yamaoka (1982).
†Number of mouth opening and closing cycles per second.
‡Number of bouts observed is shown in parentheses.

(Yamaoka, 1982, 1983a, b, 1985, 1987; Takamura, 1984; Hori, 1987) and *Limnotilapia loocki* (Yamaoka, in prep.) in Lake Tanganyika and 17 *Petrotilapia* species (Ribbink *et al.*, 1983a) in Lake Malawi typically belong to this category.

There are interspecific differences in methods of grazing unicellular algae among *Petrochromis* spp. (Table 7.1). Yamaoka (1982, 1983a, b) suggested that *P. polyodon* showing slow grazing speed may comb diatoms from filamentous algae gently and selectively; in contrast, *P. fasciolatus* showing much higher grazing speed may roughly comb off particles, including diatoms and sand grains attached to filamentous algae, rather nonselectively. These slight differences in grazing behaviour might affect resource partitioning and permit the coexistence of closely related species.

In Lake Malawi, besides the *Petrotilapia* species, which have been regarded as morphological and ecological equivalents of *Petrochromis* species from Lake Tanganyika (Fryer and Iles, 1972; Ribbink *et al.*, 1983a), there is another grazing species, *Pseudotropheus zebra*. This species shows similar dental form to that of *Limnotilapia loocki* (Fryer, 1959; Fryer and Iles, 1972) and, according to Fryer (1959) and Holzberg (1978), feeds by grazing. The stomach contents of *P. zebra* consist mainly of loose aufwuchs (Fryer, 1959; Ribbink *et al.*, 1983a).

One epilithic algal-feeding species, *Neochromis nigricans*, has been reported from Lake Victoria (Greenwood, 1956, 1974a, 1980; Fryer and Iles, 1972; Witte, 1984a). *N. nigricans* nibbles algae from a rigid substrate (Greenwood, 1974a). Its ingested material shows a preponderance of diatoms (Greenwood, 1956). Jaw teeth on the outermost row are bicuspid, tall, slender but robust, movably implanted and very closely set (Greenwood, 1956, 1980). Judging by these features, *N. nigricans* might be classified as a grazer. This species is probably an ecological equivalent of *Pseudotropheus zebra*, though Fryer and Iles (1972) considered it to be a morphological and ecological equivalent of members of the genus *Pseudotropheus* in Lake Malawi and *Tropheus* spp. in Lake Tanganyika, which are browsers.

Browsers

I mostly follow Jones' (1968) definition of browsers: browsers are herbivores that bite and tear off bits of multicellular benthic algae without (or at least rarely) ingesting any of the inorganic substrate. Epilithic algae have been classified as filamentous, microfilamentous or unicellular (Takamura, 1984; Hori, 1987). Therefore, the multicellular benthic algae included in the category of epilithic algae are filamentous and microfilamentous algae. Browsers repeat several bites regularly in each feeding bout, but they do not press the tooth band against the rock substrate (Yamaoka, 1983a).

Browsers are also one of the main components along rocky shores of Lakes Tanganyika and Malawi. In Lake Tanganyika *Tropheus* (six species), *Simochromis* (five species), *Pseudosimochromis curvifrons* and *Limnotilapia dardennii* (Yamaoka, 1983a; Poll, 1986), and in Lake Malawi *Pseudotropheus* spp. (Fryer, 1959; Ribbink *et al.*, 1983a), except the *P. zebra* species complex, can be classified as typical browsers. Browsers in Lake Tanganyika clearly ingested less inorganic matter than grazers (Takamura, 1984).

Yamaoka (1983a) has shown that interspecific differences in browsing behaviour can be found among *T. moorii*, *P. curvifrons*, *S. marginatus*, *S. babaulti*, *S. diagramma* and *L. dardennii* from Lake Tanganyika. The variable measured in the analysis of feeding behaviour was the interval between bites in one bout. A series of short repeated bites makes a browsing bout. The interval between bites is regarded as handling time and fish with the shortest handling times are considered to be the most adapted to browsing. The interval of *T. moorii* is shortest and that of *L. dardennii* is longest, therefore *T. moorii* can be regarded as the most specialized and *L. dardennii* the least specialized for browsing filamentous algae by shearing. Hori *et al.* (1983) treated the latter species as an omnivore. *Pseudosimochromis curvifrons* browsed with the second shortest interval (Yamaoka, 1983a).

There are differences in feeding behaviour among *Pseudotropheus* browsers in Lake Malawi (Fryer, 1959; Fryer and Iles, 1972; Ribbink *et al.*, 1983a). For example, *P. tropheops* differs in feeding behaviour and dental morphology from *P. fuscus* and so there are differences in the food collected: *P. tropheops* removes even the most firmly attached algae which are not nibbled free by *P. fuscus* (Fryer and Iles, 1972).

In addition to grazers and browsers, other modes of food collection among epilithic algal feeders in Lake Tanganyika are recognized: scrapers, tappers and suckers (Yamaoka, 1987, pers. obs.; Yamaoka *et al.*, 1986).

Scrapers

Scrapers are fishes that scrape off epilithic algae (Yamaoka *et al.*, 1986; Yamaoka, 1987). *Eretmodus cyanostictus* and *Spathodus marlieri* inhabiting the rocky or stony shores of Lake Tanganyika belong to this category. They collect epilithic algae from the rock surface by repeating several scrapes in quick succession and when they are feeding, the scraping noise can be heard distinctly underwater (Yamaoka *et al.*, 1986). This suggests that friction between the spatulate tooth crown and the rock substrate is strong, causing wear of the crowns. To compensate for this wear, jaw dentition consists of several tooth groups aligned in a single row, each group consisting of three or four equal-sized teeth arranged in a tier (Yamaoka *et al.*, 1986).

Liem (1979) suggested on the basis of laboratory work that both *Eretmodus* and *Spathodus* can visually select their diverse prey individually and that they are invertebrate pickers. Fryer and Iles (1972) also thought that the remarkable spatulate teeth of *E. cyanostictus* are used for the extraction of small insects and crustaceans from among the algal covering of the rocks. However, ecological studies including field surveys and ecomorphological analyses clearly show that they are epilithic algal scrapers, not invertebrate pickers (Yamaoka *et al.*, 1986).

According to Takamura (1984), *E. cyanostictus* took mostly filamentous algae (69%) and a small amount of unicellular algae (15%). In contrast, Yamaoka *et al.* (1986) demonstrated that *E. cyanostictus* fed mainly on microfilamentous algae (47.6%) and *S. marlieri* on unicellular algae (42.2%). Therefore, the scraping technique seems to produce a lower selectivity of epilithic algae.

From its dental morphology (Greenwood, 1973b), *Haplochromis limax* from Lake George seems to take aufwuchs by scraping. *Gobiocichla ethelwynnae*, a goby-like cichlid species inhabiting rapids of the Cross River in Cameroon (Roberts, 1982), may be ecologically and morphologically an intermediate between scrapers and browsers.

Tappers

Tappers are fishes that take mainly epilithic algae, keeping the body axis parallel to the rock surface as they feed. The mouth is ventrally situated and the jaw margin is straight. The jaw teeth do not form a tooth band . Tappers pull out the epilithic algae by entangling them in the sharp tooth edge, and jerking the upper and lower jaws.

When feeding, *Asprotilapia leptura* in Lake Tanganyika approaches the rock closely. A series of very quick, repeated taps (bites) makes a feeding bout. During a bout, *A. leptura* moves forward gradually (Yamaoka, 1990). The jaw teeth are not tightly implanted on the alveolar; this can be considered as an adaptation for absorbing the shock caused by repeated, hard tapping on the surface with prominent protrusion of the mouth. Since *A. leptura* takes filamentous and unicellular algae almost equally (Takamura, 1984), the tapping behaviour cannot discriminate filamentous algae from unicellular algae as foods.

Two species of the genus *Labeotropheus* in Lake Malawi are morphologically similar to *A. leptura* (Fryer, 1959). However, members of *Labeotropheus* are different from *A. leptura* in the following features: (1) the jaw teeth are fixed (Trewavas, 1935); (2) food is collected by a series of 'nibbles'; (3) the mouth is kept pressed to the rock surface throughout the process and (4) the fish moves forward very slightly while feeding (Fryer, 1959). This suggests that *Labeotropheus* spp. are not strictly tappers.

Some ecological differences were found between *Labeotropheus fuelleborni* and *L. trewavasae* (Ribbink *et al.*, 1983b). The former feeds predominantly from upper surfaces of rocks while the latter favours lower surfaces. In this way, closely related species may share food resources by utilizing different feeding sites at the same rock.

Suckers

Suckers mainly take what is called loose aufwuchs. They also tend to ingest organic debris from the substrate. When feeding, they appear to use the bucco-opercular cavity as a suction pump at the moment of mouth opening. Trophic specialization of the jaw apparatus in suckers is at a lower level than that of grazers, browsers, scrapers and tappers.

Yamaoka (1987) classified *Ophthalmotilapia nasuta* and *Cyathopharynx furcifer* from Lake Tanganyika as suckers. The teeth of these two species do not have a sharp cutting edge or deep cusp gap which can be regarded as useful in herbivory. The stomach contents consist mainly of fine sand, unicellular algae and debris (Poll, 1956; Takamura, 1984). This might suggest that the two species are both epilithic algal feeders and deposit omnivores. This notion is supported by the fact that *Ophthalmotilapia* and *Cyathopharynx* are phylogenetically closely related to omnivorous species inhabiting the sandy bottom (Liem, 1981a; Greenwood, 1983b; Yamaoka, 1985) and the omnivorous species exploit a spectrum of food resources (Liem, 1981a).

In Lake Malawi, two species of the genus *Gephyrochromis* appear to be suckers. They probably form a link between rock scrapers, such as some of the species of *Pseudotropheus*, and deposit feeders (Fryer and Iles, 1972). They take mainly loose aufwuchs with sediment materials (Fryer, 1959; Fryer and Iles, 1972; Ribbink *et al.*, 1983a).

Piscivores

According to van Oijen (1982), piscivorous species are those for which fishes (from eggs to adults) form the major part by weight of the adult diet. The term 'fishes' includes whole organisms as well as parts of fishes. Therefore scale eaters and paedophages are, in general, covered by this definition, but they are discussed separately here (pages 160–3).

Piscivores form one of the main trophic groups in terms of species number, especially in Lake Victoria (Greenwood, 1974a, 1981b; van Oijen, 1982). Over 80 piscivorous species, i.e. *c.* 40% of the total number of cichlids, have been taken from the Mwanza Gulf of Lake Victoria (van Oijen *et al.*, 1981).

However, observations on the feeding behaviour of piscivores in nature are much less common than observations on the epilithic algal feeders, because prey capture by piscivores is a less frequent event (Fryer and Iles, 1972).

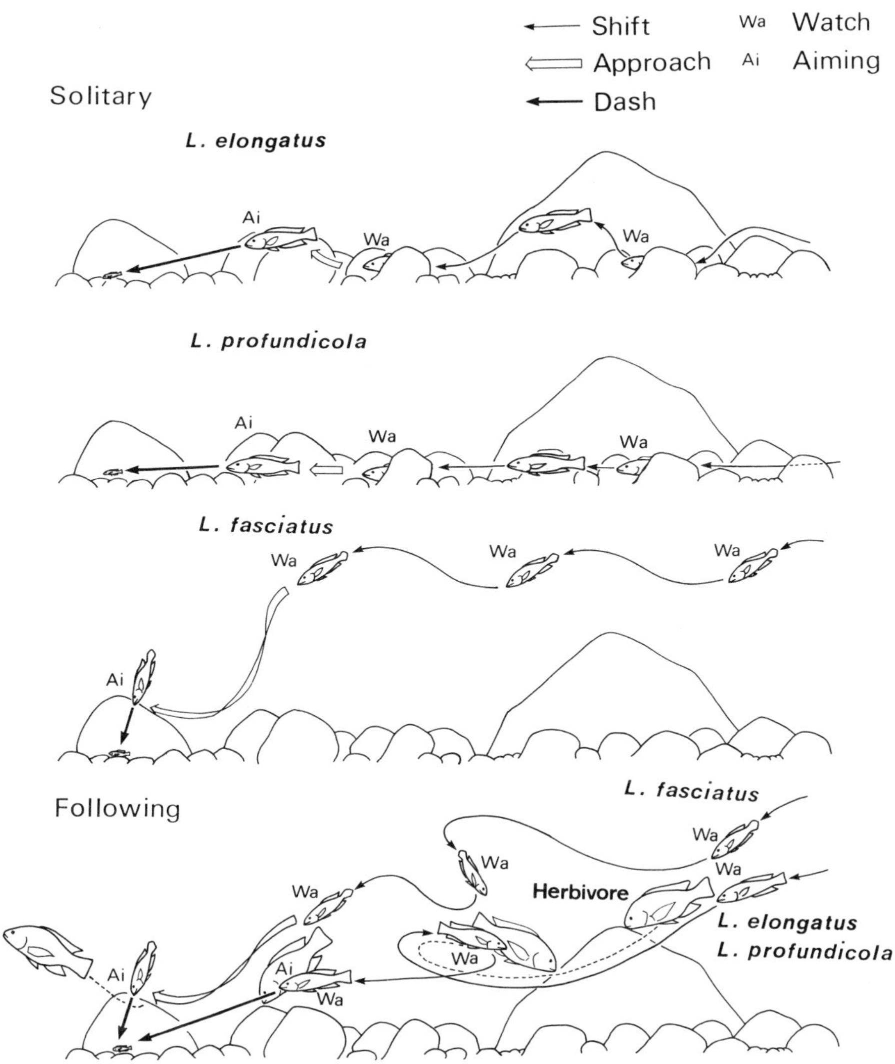

Fig. 7.2 Sequence of hunting behaviour of three piscivores, *Lamprologus elongatus*, *L. profundicola* and *L. fasciatus*. The three upper drawings depict solitary hunting and the bottom drawing shows what happens when a large fish is used as cover. 'Shift' is a behaviour that occurs before prey is detected; 'approach' behaviour occurs after prey is detected (Hori, unpublished).

Hori (1983, 1987) studied the feeding behaviour of three sympatric piscivorous species of the genus *Lamprologus* in Lake Tanganyika: *L. elongatus*, *L. profundicola* and *L. fasciatus*. All three species hunted for small fishes and *c.* 60% of prey species were common to them, but their hunting techniques differed (Fig. 7.2). In general, *L. elongatus* searches for prey along the bottom and dashes slightly downward, while *L. profundicola* weaves through rocks and stones, and dashes horizontally. *L. fasciatus* searches for prey from mid-water, approaches it from above and behind, and then dashes downward. Occasionally individuals (mainly of *L. elongatus* and *L. profundicola*) follow a large herbivorous fish such as *Petrochromis polyodon*. Using the large fish as cover, the fish approach and then dash suddenly from cover to attack their prey; which is also employed by marine fishes in the Aulostomidae and Fistulariidae (Keenleyside, 1979).

The giant cichlid *Boulengerochromis microlepis*, which has characteristic morphological features facilitating rapid locomotion (Fryer and Iles, 1972), has been observed feeding in nature. Kuwamura (1986c) described solitary juveniles (13–14 cm TL) hiding near the bottom and abruptly darting towards small (cichlid?) fish nearby, whereas adults of 40–50 cm TL swam rapidly into the coastal boulder zone, chasing and attacking a school of *Stolothrissa tanganicae*.

Matthes (1961) reported different hunting behaviour for *B. microlepis*; juveniles often followed shoals of clupeids, together with *Lates*, trying to herd them into shallow bays, whereas mature adults caught small cichlids by surprise after a stealthy approach or by cornering them between rocks. These observations suggest that *B. microlepis* can change its hunting behaviour, depending upon the mode of existence of its prey.

Van Oijen (1982) studied ecological differentiation among the piscivorous haplochromine cichlids of Lake Victoria, using specimens collected by several fishing methods. He concluded that each species has its own unique combination of food preferences, habitat and feeding behaviour. Besides specialization on prey species, prey size partitioning, often correlated with habitat restrictions, plays an important role in coexistence. Van Oijen *et al.* (1981) stressed that the haplochromine species, rather than being specialized predators on certain organisms, are specialized in particular predation techniques.

Scale eaters

Scale eaters (lepidophages) are fishes that subsist almost entirely on fish scales. Of all the sources of food to have been exploited by African cichlid fishes, the scales of other fishes are perhaps the least expected (Fryer and Iles, 1972). Scale-eating cichlids have been reported only from the East African Great Lakes: seven species of the genus *Perissodus* in Lake Tanganyika (Liem and Stewart, 1976), *Allochromis welcommei* in Lake Victoria (Greenwood, 1974a), and *Docimodus evelynae* (Eccles and Lewis, 1976), *Geniochromis mento* (Fryer, 1959; Ribbink *et al.*, 1983a), *Melanochromis* sp. (Ribbink *et al.*, 1983a) and two species of *Corematodus* (Fryer and Iles, 1972) in Lake Malawi.

In *Perissodus microlepis*, scales which are tightly piled up like a pack of cards in the stomach are nevertheless digested; the calorific value of scales obtained by one attack is approximately 50.21 joule, suggesting a comparable feeding efficiency to feeding on zooplankters. Since scales are easily renewable, they may be one of the most stable food resources in habitats where many fishes occur together (Nshombo *et al.*, 1985). *G. mento* also digests fish scales (Fryer *et al.*, 1955; Fryer, 1959).

Hori (1987) compared the feeding behaviour of *Perissodus microlepis* and *P. straeleni*. The hunting behaviour of the two species is composed of a search in mid-water, a stealthy approach, and a dash at the prey (Fig. 7.3). In general, *P. microlepis* approached and dashed from outside the range at which the prey could detect it. On the other hand, *P. straeleni* generally dashed from within the range at which the prey could detect it. *P. microlepis* was more successful in approaching than *P. straeleni*, but less successful in dashing. These two scale eaters thereby overcome the prey's vigilance in different ways.

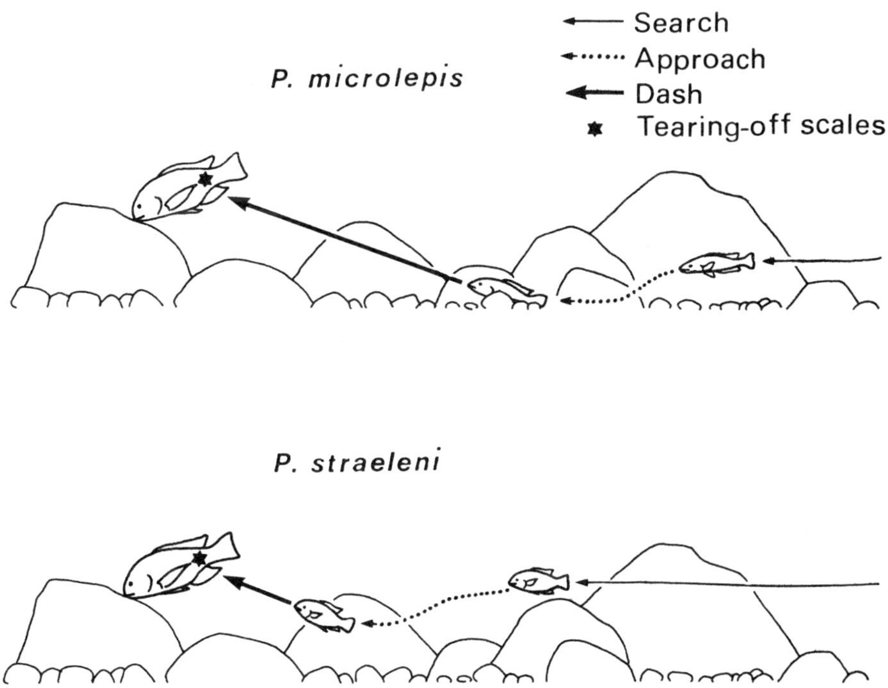

Fig. 7.3 Sequence of hunting behaviour of two scale eaters, *Perissodus microlepis* and *P. straeleni* (modified with permission from Hori, 1987).

Nshombo *et al.* (1985) suggested that the success of scale eating in *P. microlepis* was dependent upon whether the scale eater could take advantage of the prey's unguarded moments. Hori (1987) did not detect a difference in the situations of prey fishes attacked by *P. microlepis* and *P. straeleni*. Both species mainly attacked when their prey were feeding or swimming in mid-water, and their feeding efficiency was almost identical (Hori, 1987).

P. paradoxus was also found in shallow waters in Lake Tanganyika and ingested fish scales (Marlier and Leleup, 1954; Hori, 1987). The feeding behaviour of this species has not been observed. However, since the adults swam together with *Haplotaxodon microlepis* at Mbemba in Zaire (Yamaoka, pers. obs.) and Myako in Tanzania (Kuwamura, pers. comm.), and since the two species are very similar in body coloration and in the presence of a dark spot on the caudal peduncle, *P. paradoxus* may be acting as a mimic of *H. microlepis* and feeding on its scales.

Such a feeding relationship would be similar to that found in *Corematodus* spp. in Lake Malawi (Trewavas, 1947; Fryer and Iles, 1972). *C. shiranus*, the larger species, mimics mainly *Tilapia squamipinnis* (Trewavas, 1947) on whose scales it feeds. Individuals of *C. shiranus* live in ones and twos among the shoals of their model and host (Fryer and Iles, 1972). In contrast, *C. taeniatus*, a smaller species, seems to be adapted for taking scales from smaller species than the *Tilapia* victimized by *C. shiranus* (Fryer and Iles, 1972).

All species of *Perissodus* except *P. microlepis* have remarkable recurved jaw teeth (Fryer and Iles, 1972; Liem and Stewart, 1976). This characteristic form of jaw teeth makes one imagine that the cusp edge of recurved teeth is inserted under the edge of a scale, and that when the mouth is closed, scales are bitten off. However, this seems to be incorrect. Field observations on *P. straeleni* suggest that when they strike against the prey after a dash, the mouth is fully opened and the wide plane of each tooth with longitudinal ridges, not the cusp edge, is pressed strongly against scales. At that moment, the lepidophagous fish wrenches its body in a complex manner. It would appear that strong friction between teeth and scales removes the scales from the prey's body. Sometimes, just after this process, the fish snaps up dislodged scales as they drift in the water (Yamaoka, pers. obs.).

Paedophages

Wilhelm (1980) suggested that the term 'paedophage' should be given to species feeding only on buccal-stage offspring of mouthbrooding cichlids. However, since his definition would exclude species that feed on progeny of substrate brooders, I mainly follow the definition of Greenwood (1967, 1974a): fish feeding exclusively (or principally) on the embryos and yolk-sac larvae of other cichlids.

Eight species of paedophages have been reported from Lake Victoria: *Lipochromis obesus, L. maxillaris, L. melanopterus, L. cryptodon, L. microdon, L. parvidens, Haplochromis cronus* and *Astatotilapia barbarae* (Greenwood, 1959a, 1967, 1974a, 1980). *Lipochromis taurinus* from Lakes Edward and George is also a paedophage (Greenwood, 1973b, 1980). From Lake Malawi, McKaye and Mackenzie (1982) and McKaye and Kocher (1983) reported that at least three cichlids, *Cyrtocara orthognathus, C. liemi* and an undescribed species, are paedophages.

These African paedophages all feed on the young of mouthbrooding species. By contrast, one substrate-brooding Asian cichlid in Sri Lanka, *Etroplus maculatus*, is paedophagous, feeding on eggs and embryos of a congener, *E. suratensis* (Ward and Wyman, 1977). The interrelation between these two species is complex; on the basis of aquarium observations, Wyman and Ward (1972) first described the two species engaging in a cleaning symbiosis, with *E. suratensis* (the larger species) as the host, and *E. maculatus* the cleaner. In fact, the advantage of cleaning to *E. maculatus* is more likely the survival and reproductive success of adult *E. suratensis*, which in turn provide eggs and embryos for the predatory habit of *E. maculatus* (Ward and Wyman, 1977).

Three species of *Cyrtocara* engage in a unique head-ramming behaviour, forcing targeted females to expel eggs, embryos or larvae from their mouth (McKaye and Kocher, 1983). However, they each employ a different attacking method. *C. orthognathus* approached a female from underneath and behind, swimming until it was underneath the belly of the brooding female *c.* 1–2 fish lengths behind and 45° below, and then darted up, ramming the female in the hyoid region. *C. liemi* attacked from 0.5 to 2 m below the female, at a 70–90° angle, and struck the same body region as *C. orthognathus*. By contrast, the undescribed species attacked from above and hit *C. pleurotaenia* females on the snout near the premaxillary pedicel. On the basis of these observations, McKaye and Kocher (1983) concluded that snout-engulfing behaviour to extract the brood, which was described in a Lake Victorian haplochromine paedophage in aquaria by Wilhelm (1980), is not prevelant, and reflects laboratory confinement. They regarded the ramming or butting behaviour as important methods for these paedophagous cichlids to obtain food.

A controversy has existed over how paedophagous cichlids take eggs, embryos or larvae from brooding females. Greenwood (1959a, 1974a, 1981b) insisted that on the basis of anatomical data and aquarium observations, snout-engulfing was a possible technique employed by the large-mouthed species. Fryer and Iles (1972) suggested that the eggs and embryos are voluntarily released and that this might be one of the devices which enable cichlids to regulate their own population densities. However, since field observations made by McKaye and Kocher (1983) are all on the paedophages from Lake Malawi, the possibility of snout-engulfing by Lake Victoria species remains unanswered. The suggestion proposed by Fryer and Iles (1972) is fascinating, but now tends to be ignored, because it is generally accepted that behaviour decreasing the fitness of individuals cannot evolve.

Insectivores and other benthic arthropod feeders

Three methods of prey collection are recognized among Malawian insectivores (Fryer and Iles, 1972). The first is exhibited by *Labidochromis maculicauda* (Lewis, 1982; Ribbink *et al.*, 1983a) (=*L. vellicans* in Fryer (1959) and Fryer and Iles (1972)), which picks out individual insects and crustaceans from the algal mat by a single nip, using its long, recurved teeth and a quick twist of the body. In Lake Tanganyika, *Tanganicodus irsacae*, which is closely related to *Eretmodus* and *Spathodus* (Liem, 1979; Yamaoka, 1987), preys mainly on chironomid larvae with its elongate, forceps-like rostral teeth, and shows similar behaviour to that of *L. maculicauda* (Yamaoka *et al.*, 1986). *T. irsacae* picks at the prey item, while maintaining contact with the substrate by its ventral fins.

The second method is shown by fishes of the genus *Lethrinops* inhabiting the sandy shore in Lake Malawi. *L. furcifer* feeds almost exclusively on chironomid larvae by plunging its snout into the sand, taking up a mouthful of sand and after withdrawing its head, discharging the collected material through a gill raker sieve via the opercular openings (Fryer and Iles, 1972).

Grammatotria lemairii, which is common in the littoral region in Lake Tanganyika (Kawabata and Mihigo, 1982; Kuwamura, 1987), shows very similar feeding behaviour to that of *L. furcifer* (Yamaoka, pers. obs.). It takes mainly shrimps (Hori, 1987) or diatoms, insects and molluscs (Brichard, 1978), by plunging its head into the sand and ejecting sand through the gill openings. *Xenotilapia flavipinnis* preying mainly on harpacticoid and cyclopoid copepods and ostracods shows similar feeding behaviour. However, it spits sand not only from the gill opening but also from the mouth, and also picks up prey from the surfaces of rocks and gravel (Yanagisawa, 1986). *X. sima* ejects the sand either from the mouth only (Fryer and Iles, 1972) or from both the mouth and gill opening (Hori, pers. comm.).

The third method is shown by '*Haplochromis*' *euchilus*, which has fleshy lobes on its lips. Fryer and Iles (1972) inferred that these hypertrophied lips are sensitive structures which are used to detect insects by touch, because when feeding it places the lobes against a rock, presumably 'feeling' for food.

Hypertrophied lips are also well developed in *Lobochilotes labiatus* in Lake Tanganyika. According to Hori (1987) it takes mainly Trichoptera, though Poll (1956) and Hori *et al.* (1983) regarded it as an omnivore. Fryer and Iles (1972) viewed it as a morphological and ecological equivalent of '*H.*' *euchilus* in Lake Malawi and *Paralabidochromis chilotes* in Lake Victoria. However, when the feeding behaviour of *L. labiatus* was observed, its hypertrophied lips did not appear to function as sensitive structures. It did not place the lobes against a rock, but protruded the mouth abruptly against interstices between rocks and the gravel bottom (Yamaoka, pers. obs.). At the time of abrupt protrusion, the fleshy lobes of the lips may function to absorb the shock caused by striking the mouth against the substrate. Hori (1987) described *L. labiatus* taking prey by vigorously sucking out each insect from the interstices between rocks. This indicates that very similar forms of enlarged lips may fulfil different functions for feeding. Histological study is needed to elucidate the functions of the hypertrophied lips.

Besides these three feeding methods of insectivores, variations in hunting behaviour have been observed by Hori (1983) in five sympatric benthic arthropod feeders in the genus *Lamprologus* (*L. callipterus*, *L. compressiceps*, *L. furcifer*, *L. leleupi* and *L. modestus*) in Lake Tanganyika (Fig. 7.4). *L. compressiceps* showed species-specific hunting behaviour. It cruised at 5–15 cm above the bottom, stopped to hover just above a crevice in the rocky substrate, gradually assumed a vertical position with head downwards and then rushed into the crevice to snatch shrimps. This behaviour resembles that of '*H.*'

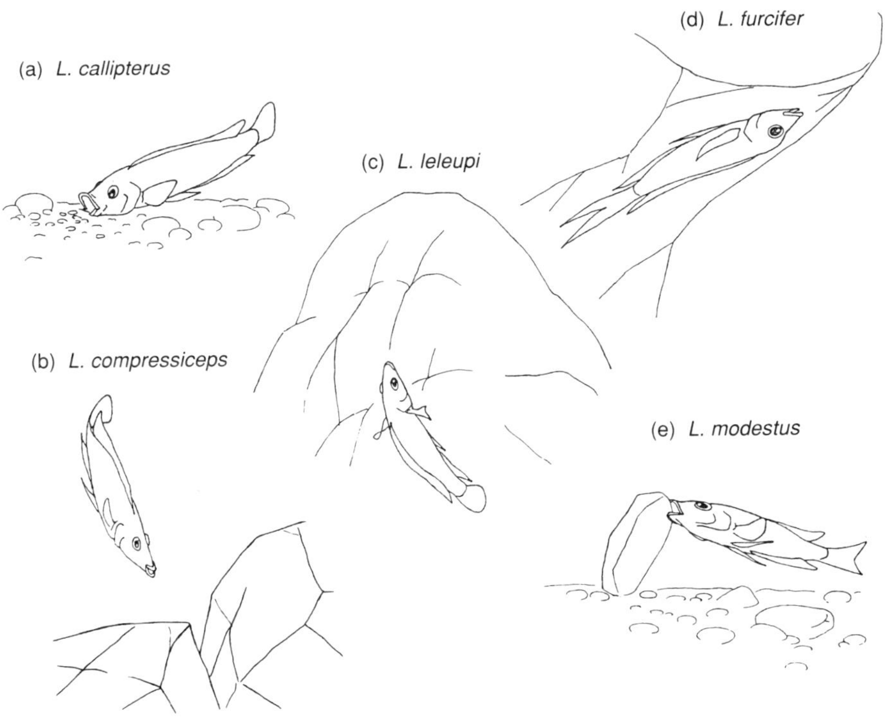

Fig. 7.4 Feeding behaviour of five benthic arthropod feeders of the genus *Lamprologus* (modified with permission from Hori, 1983).

compressiceps in Lake Malawi (Fryer and Iles, 1972) and the piscivorous *L. fasciatus* (Hori, 1983). Another shrimp-eating species, *L. furcifer*, inhabited the undersides of overhanging rocks and the vertical plane of (usually smooth) rocks and obtained food there (Hori, 1983; Yanagisawa, 1987). When a pair of the cyprinodontid *Lamprichthys tanganicanus* spawned near its habitat, *L. furcifer* snatched their eggs (Hori, 1983). *L. callipterus* continually engulfed quantities of sand and debris, spitting out inedible materials. Its hunting behaviour was classified into two types: picking up easily (probably eating chironomid larvae), and pecking after a short dash (probably eating shrimps) (Hori, 1987). *L. leleupi* often swam slowly around an uneven site or a fine crevice, keeping the snout almost in touch with the surface, and snatched prey from the surface. *L. modestus* showed four types of behaviour: (1) swimming slowly just above various substrates and picking up prey with a sudden dart, (2) turning over a flat piece of rubble and picking up exposed prey (Fig. 7.4(e)), (3) wriggling its body vigorously on the substrate to disturb gravel and then snatch exposed prey, and (4) following just behind a pair of *Lamprichthys tanganicanus* on a rock surface beyond the boundary of its usual cruising area and snatching their eggs.

These five species of *Lamprologus* share a daytime diet consisting of shrimps and aquatic insects, but each species has one or more characteristic and elaborate hunting techniques, and the preferred substrate or type feeding sites are different for each species (Hori, 1983).

7.3 VERSATILITY IN FOOD AND FEEDING BEHAVIOUR

Liem (1978, 1980a) has suggested on the basis of detailed laboratory experiments on feeding repertoires of cichlids from the Great Lakes of Africa that the taxa most specialized phylogenetically and morphologically are not only specialists, but also jacks of all trades. This paradoxical view contradicts the alternative view that the most morphologically specialized taxa are remarkable specialists for certain feeding habits. If specialists are simultaneously jacks of all trades, the belief that the trophic morphology in cichlids is adaptive should be reconsidered (Liem, 1980a, b).

Van Oijen (1982) and Yamaoka (1983a) have criticized Liem's conclusions as being based on laboratory artefacts. Van Oijen (1982) stressed the importance of ecological aspects in the natural environment in functional morphological studies, as suggested by Bock (1977, 1980) and Motta (1985). McKaye and Marsh (1983) tested Liem's conclusions in the field and agreed with them, showing that *Petrotilapia tridentiger* and *Pseudotropheus zebra*, which have been considered as specialized aufwuchs-feeding species in Lake Malawi (Fryer and Iles, 1972), are facultative in their feeding repertoire.

Ribbink *et al.* (1983a) also supported Liem through extensive field observations on the Mbuna, including these two species.

I think it premature to draw a conclusion about the feeding versatility in African Great Lakes cichlids only from these facts gained by laboratory and field studies. Each Great Lake has its own characteristic level of specialization. If we view *P. tridentiger* and *P. zebra* in Lake Malawi from the level of specialization in Lake Tanganyika equivalents, can these two species be truly regarded as specialized in their feeding? Yamaoka (in prep.) suggests that epilithic algal feeders in Lake Malawi are at a lower level of specialization. In *Pseudotropheus spp.* there are many potential taxa to be developed and diversified into Malawian equivalents of *Tropheus, Simochromis, Pseudosimochr-*
experiments on feeding repertoires of cichlids from the Great Lakes of Africa
omis or Limnotilapia. Petrotilapia spp. are, strictly speaking, not equivalent to *Petrochromis* spp., but only to *Petrochromis fasciolatus* (*Yamaoka*, in prep.), which has been supposed from a functional morphological point of view to be the most generalized member within its congeners (Yamaoka, 1982). It seems impossible to regard *P. tridentiger* and *P. zebra* from Lake Malawi as most specialized at the level of specialization in Lake Tanganyika.

Pseudotropheus zebra and *Petrotilapia tridentiger* were placed in the category showing the widest variation of feeding repertoires, and were also considered highly specialized (Liem, 1980a, b). This contradicts my view presented above. Liem considered that bilateral asymmetry in the firing patterns of head muscles in specialized cichlids from the African Great Lakes may represent an important component of the continuously modulated jaw mechanism, contributing to the widest variation of feeding repertoires and adjustments of the tooth band against irregular rock substrates while rock grazing. I consider the bilateral asymmetry of muscle firing in *P. zebra* and *P. tridentiger* to be a secondary adaptation for algal grazing on rough rock surfaces, because *Petrochromis, Eretmodus* and *Spathodus* (in which strong friction occurs between tooth crown and rock substrate, necessitating adjustments by a modulated jaw mechanism (Yamaoka, 1982; Yamaoka *et al.*, 1986)) show bilateral asymmetry of firing (Liem, 1979, 1980a).

Limnotilapia dardennii was regarded as one of the most specialized taxa on the basis of laboratory observations (Liem, 1980a). However, field observations suggest that *L. dardennii* is a generalized algal-feeding species. Of browsing species, this is the least dependent upon epilithic algae (Takamura, 1984) and it is considered an omnivore by Hori *et al.* (1983). Behaviourally, it sometimes scrapes quickly with a few repeated bites, like grazers on rock surfaces and reeds (pers. obs.), and of browsing species is the least adapted for browsing (Yamaoka, 1983a). Concerning its habitats, Kawabata and Mihigo (1982) reported that *L. dardennii* inhabits not only rocky shores but also the Ruzizi estuaries. Morphologically, *Limnotilapia* most nearly resembles a single ancestral form (Regan, 1920). These observations suggest that this species has wider ecological versatility and generalized features for browsers in Lake Tanganyika.

It may be necessary to apply electromyographic analyses to the most specialized browsers, *Tropheus moorii* and *Pseudosimochromis curvifrons*, to test Liem's paradox that specialists are simultaneously jacks of all trades. These specialized browsers open the mouth weakly while feeding. This is implied by the fact that *Petrochromis polyodon's* grazing marks were more than 370 times as large as *T. moorii's* browsing marks (Takamura, 1983). This suggests that specialized browsers need biting power concentrated on a small area, rather than the delicate control of jaw movements found in grazers and scrapers. If so, complicated muscle activities, including pronounced asymmetrical firing of muscles, become of little importance in specialized browsers. In *T. moorii* and *P. curvifrons*, therefore, there may be less pronounced, or no, asymmetrical firing of muscles while feeding, compared with grazers.

The evolutionary success of the cichlids in the African Great Lakes has been attributed to their trophic versatility and flexibility (Liem, 1978, 1980a; McKaye and Marsh, 1983; Ribbink *et al.*, 1983a). This hypothesis treats the cichlid faunas of the three Great Lakes as one evolutionary unit. However, it must be recognized that the cichlid fauna in Lake Tanganyika seems somewhat different phylogenetically from those in Lakes Malawi and Victoria (Yamaoka, 1985; Nishida, in prep.), and interspecific relationships in feeding in Lake Tanganyika are much advanced (Hori, 1987). If the phylogenetically and morphologically most specialized taxa in Lake Tanganyika cichlids are not only specialists, but also jacks of all trades as suggested in Lake Malawi cichlids by Liem (1978, 1980a), the mutualisms and facultative commensalisms prevailing in Lake Tanganyika cichlids (Takamura, 1983; Hori, 1987) would not have evolved. Each species must be able to recognize the species-specific behaviour of other species in order to develop such interrelationships.

7.4 HOW TO COEXIST?

In general, it has been accepted that the coexistence of many closely related species of cichlids in the African Great Lakes is attributable to the explosive adaptive radiation in their feeding habits (Fryer and Iles, 1972; Greenwood, 1974a). However, Ribbink *et al.* (1983a) concluded that how and where they feed appear to be more important for resource partitioning than the foods they take. The importance of slight differences in feeding behaviour for resource partitioning was also suggested for Lake Tanganyika grazers and browsers by Yamaoka (1982, 1983a). Hori (1987) insisted that differences in food preference or habitat use among coexisting related species are merely a condition that permits coexistence. Takamura (1983, 1984) and Hori (1983, 1987) tried to find direct evidence that facilitates the coexistence in

Lake Tanganyika of predatory and epilithic algal-feeding cichlids. They found that commensal and mutualistic relationships prevail in the rocky shore cichlid community.

A typical commensal relationship was seen among benthic animal feeders that sometimes feed together, forming a mixed-species school: *Lamprologus callipterus*, *Lobochilotes labiatus* and *Gnathochromis pfefferi* (Hori, 1987). The feeding rates of these species were examined in various feeding situations. *L. callipterus* and *G. pfefferi* caught more shrimps when they fed by following *L. labiatus*, or in conspecific or mixed-species schools, than when they were solitary. *L. labiatus* did not show any significant difference in feeding rate between the various situations. The presence of *L. labiatus*, a caddis-fly eater, therefore provides a situation favourable to the two species of shrimp eaters.

A relation resembling mutualism between *Petrochromis polyodon*, a highly specialized grazer, and *Tropheus moorii*, a highly specialized browser, was studied by Takamura (1983, 1984). *T. moorii* preferred to browse on rock surfaces where *P. polyodon* and other *Petrochromis* had already grazed and had left little sand. *P. polyodon* feeding mostly on unicellular algae appears to benefit *T. moorii*, which feeds mostly on filamentous algae, by decreasing the particles attached to the filamentous algae. In addition, a *T. moorii* was observed nibbling on the lateral body surface of an inclining *P. polyodon*, suggesting cleaning behaviour by *T. moorii*, but since that behaviour was observed only once, more data are needed to ascertain the degree of mutualism between these highly specialized grazers and browsers.

Examples of mutualism were also found between scale eaters in the field (Hori, 1987). When observing how feeding changed in *Perissodus microlepis* and *P. straeleni*, he found that for each species, when a scale eater of the other species was present within a radius of 50 cm, the success ratio was significantly higher than that in solitary hunting (Fig. 7.5). No significant difference was observed, however, when the second individual was of the same species as the first. This suggests that vigilance against one lepidophagous species makes a prey fish less wary of attack by another lepidophagous species.

Three piscivorous *Lamprologus* species (*L. elongatus*, *L. profundicola* and *L. fasciatus*) exhibit a similar phenomenon to that found in the scale eaters (Hori, 1987). By following large, herbivorous fishes, *L. elongatus* and *L. profundicola* achieved a much higher success ratio than in solitary hunting. When a piscivore of another species was near by, *L. profundicola* and *L. fasciatus* showed a significantly higher success ratio than when hunting alone (Fig. 7.6). This suggests that *L. profundicola* and *L. fasciatus* make use of the presence of congeneric piscivorous species for hunting. By contrast, *L. elongatus* obtained little advantage from the presence of congeneric species, presumably because *L. elongatus* was by far the most abundant of the three in the study area and the prey fishes would probably take precautions against the most abundant piscivore. The above mutualism may be realized only under conditions in which each species has a specific hunting technique.

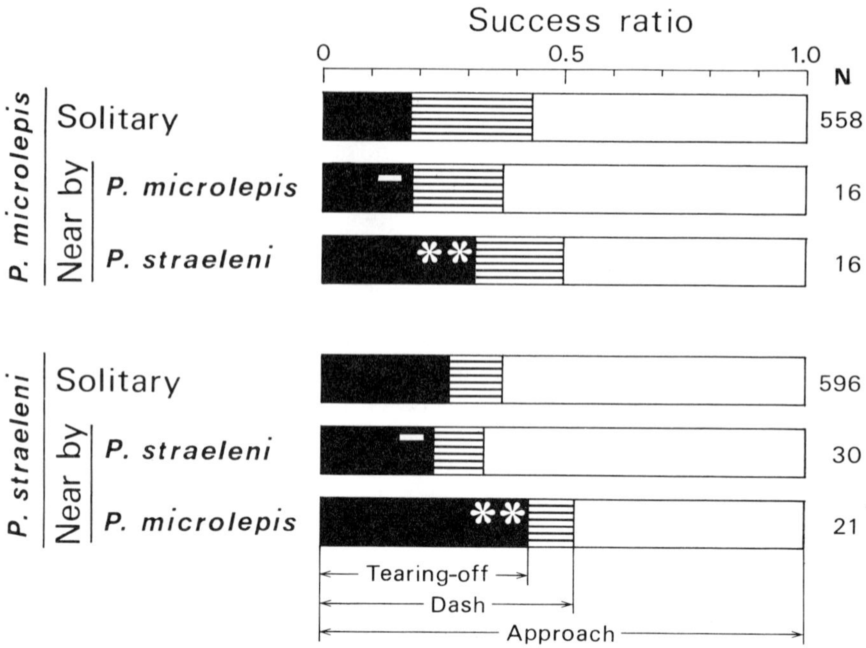

Fig. 7.5 Hunting success ratio of two scale eaters, *Perissodus microlepis* and *P. straeleni*, with their foraging situations. The success ratios of 'tearing-off' and 'dash' are shown as proportions of 'Approach', which is set at 1.0. **Indicates that the rates are significantly different from that in a solitary feeding situation at the 5% level.—Indicates that the difference is nonsignificant (modified with permission from Hori, 1987). Significance level was tested with approximation to normal distribution.

A similar relation was inferred in three paedophagous species of *Cyrtocara* in Lake Malawi by KcKaye and Kocher (1983). Although these species share the behaviour of ramming the head of mouthbrooding females, they each employ a different method (page 163). The variation in the attack pattern of the paedophages may make it difficult for females to develop effective countermeasures. McKaye and Kocher suggested that these different methods of utilizing the same food resource probably contribute to the ability of the three species to coexist.

In this way, the commensalism among fishes of different feeding habit groups and the mutualism among species of the same feeding habit group found in Lake Tanganyika may be factors that positively facilitate their coexistence, and thus raise the species richness by operating on different parts of the community. Mutualism also should be studied from the viewpoint that it raises the species richness of each feeding-habit group without any resource partitioning, if their feeding behaviour can diverge (Hori, 1987). Hori stressed

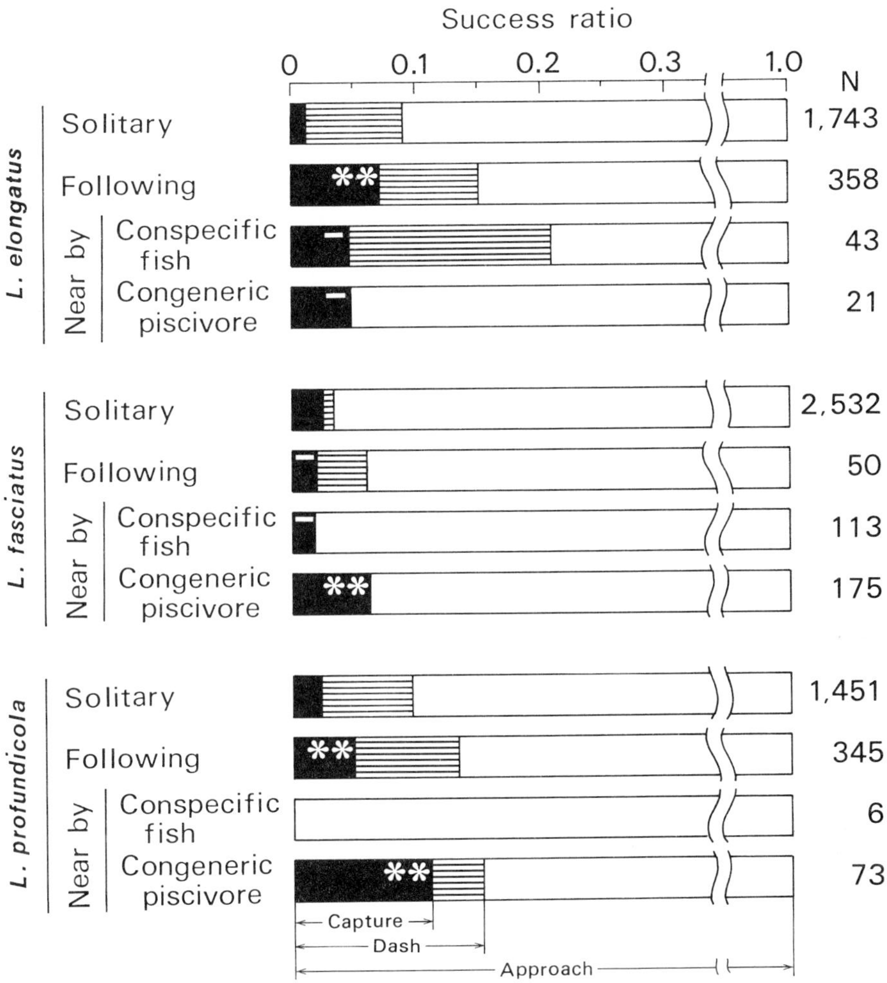

Fig. 7.6 Hunting success ratio of three piscivores, *Lamprologus elongatus*, *L. fasciatus* and *L. profundicola*, with their foraging situations. Results of statistical analysis are shown as in Fig. 7.5 (modified with permission from Hori, 1987).

the importance of studying the connection among three kinds of interspecific relationships in feeding – commensalism, mutualism and competition – in ecological communities within tropical ecosystems. More intensive field study is needed and special attention should be paid to individual differences, to say nothing of intra- and interspecific relationships in the community, towards obtaining an understanding of coexistence mechanisms and adaptive radiation in feeding habits of cichlid fishes.

ACKNOWLEDGEMENTS

I wish to express my heartfelt thanks to many people whose generosity, encouragement, advice and/or critical comments on earlier drafts of the manuscript have made this work possible. Those deserving special recognition are T. Iwai, H. Kawanabe, M. Nagochi, Y. Yanagisawa, M. Hou, T. Kuwamura and K.Takamura.

Chapter eight

Mating systems among cichlid fishes

G. W. Barlow

8.1 INTRODUCTION

This essay describes mating systems among cichlid fishes and factors that appear to have shaped them. Mating systems and patterns of parental care influence one another to such a degree that a discussion of one inevitably involves the other (Baylis, 1981). Consequently, this chapter and that by Miles Keenleyside on parental behaviour should be read as a pair because they are so interrelated.

Conventionally, cichlids are divisible into two groups (Baerends and Baerends-van Roon, 1950; Kuwamura, 1986a; Lowe-McConnell, 1959; Wickler, 1962a). The substrate guarders are monogamous, relatively isomorphic species with adhesive eggs and helpless larvae with large yolk sacs which are guarded in a nest and are termed wrigglers. The adults form pair bonds such that a different individual cannot readily be substituted for the mate (Wickler, 1962a). The other group of species, the mouthbrooders, are polygamous and usually dimorphic with females carrying the eggs and wrigglers in their mouths. Mating is brief, typically minutes, so no pair bond is formed. In both groups, when the young start to swim they are called fry.

The evolution of mating systems in cichlid fishes has evidently proceeded from monogamy with biparental care to polygamy with maternal care (Barlow, 1963, 1964; Gittleman, 1981). Further, substrate-guarding species evidently gave rise to mouthbrooders (Balon, 1977; Breder, 1934; Fryer and Iles, 1972; Lowe-McConnell, 1959; Myers, 1939; Oppenheimer, 1970; Wickler, 1966); cichlids have evolved a number of variations on these two themes.

8.2 AN ACCOUNT OF MATING SYSTEMS

Monogamy

Fishes engage in two types of monogamy (Barlow, 1984, 1986); (1) monogamous marine fishes of tropical reefs, with one exception, consistently spawn with the same partner almost daily over long periods, but without care of young; (2) monogamous pairs of freshwater species, such as cichlids, stay together and care for the fry until they are independent.

Substrate guarders

Originally, monogamy was thought almost synonymous with substrate guarding (Baerends and Baerends-van Roon, 1950). The primitive Etroplinae of Asia are monogamous substrate-guarding species (Barlow, 1970; Ward and Wyman, 1977); so too are most of the cichlids of Central America, such as *Cichlasoma sensu lato*, *Herotilapia*, and *Neetroplus* (Barlow, 1974). Many cichlids in South America provide further cases, e.g. *Aequidens*, *Astronotus*, *Cichla*, *Cichlasoma sensu stricto*, *Pterophyllum* and *Symphysodon*, as do several riverine and some lacustrine species in Africa, e.g. *Hemichromis*, *Lamprologus* and *Tilapia* (Keenleyside and Bietz, 1981; Loiselle, 1985a).

Breeding in those and similar cichlids starts with pair formation, which lasts about an hour and grades smoothly into courtship lasting a few to several days (Baerends and Baerends-van Roon, 1950). The male of a pair is almost always larger than the female but sexual dimorphism is otherwise slight. Rarely, the female is bigger than the male, e.g. among the larger species of *Julidochromis* in Lake Tanganyika (Konings, 1988).

Early studies reported that the male establishes a breeding territory to which the female comes. *Cichlasoma maculicauda* fits this pattern in nature (Perrone, 1978a), as does *Cichla ocellaris* (Zaret, 1980). Sometimes this applies to *Cichlasoma citrinellum* (Barlow, 1976); at other times the pair jointly acquires a territory (McKaye, 1977, McKaye and Barlow, 1976a). The firemouth of Mexico, *C. meeki*, is likewise labile; either a male or a pair founds a territory. Apparently, when competition for sites is low, males acquire a territory alone, but when competition increases, a pair may be required (Neil, 1984; compare Barlow, 1976 with McKaye and Barlow, 1976a, McKaye, 1977).

Keenleyside and Bietz (1981) observed pairs of *Aequidens vittatus* in Surinam founding a territory together, as did McKaye (1977) for several species of *Cichlasoma* in Nicaragua and Loiselle (1977) for the African cichlids *Hemichromis fasciatus* and *Tilapia zillii*. Aquarium observations on *Etroplus maculatus* suggested a similar process (Barlow, unpublished).

The parents share in parental care though males and females usually invest differently (Barlow, 1974; Kuwamura, 1986a; Neil, 1984; Smith-Grayton and Keenleyside, 1978; Townshend and Wootton, 1985). The female is usually more closely tied to the fry than is the male, who repels intruders, ranges further away and flees first when endangered (Keenleyside and Bietz, 1981; Kuwamura, 1986a; McKaye and Barlow, 1976a; Neil, 1984; Perrone, 1978a).

The territories are small, often *c.* 1 m^2 of substrate, depending on size of the species, and typically are not used for feeding (e.g. Barlow, 1976; McKaye and Barlow, 1976a; Meral, 1973; Kuwamura, 1986a; Perrone, 1978a). Pairs sometimes breed in aggregations (Loiselle, 1977; McKaye and Barlow, 1976a). Commonly, however, pairs breed apart from other conspecific fish and wander slowly with their mobile fry (Barlow, 1976; Meral, 1973; Neil, 1984).

Mouthbrooders

Polygamy predominates among mouthbrooders, but several species are monogamous (Apfelbach, 1969; Kuwamura, 1986a). Monogamous South American mouthbrooders are known in the genera *Aequidens*, *Geophagus*, *Gymnogeophagus* and *Satanoperca* (Loiselle, 1985a; Reid and Atz, 1958; Timms and Keenleyside, 1975). These species, typically isomorphic but with the male larger than the female, start the parental phase as substrate brooders. They spawn adhesive eggs on a firm substrate. At about the time of hatching, 1–2 d later, both parents take the larvae into their mouths (termed primitive mouthbrooding). Later the fry flee into their parents' mouths when endangered. In *Geophagus altifrons* and *G. megasema*, the eggs are taken into the mouth immediately after spawning (Loiselle, pers. comm.). The evolutionary derivation from monogamous substrate guarders is apparent.

A monogamous genus of Africa rivers, *Chromidotilapia*, parallels the South American 'earth eater' (Linke and Staeck, 1981b; Myrberg, 1965). *C. batesii* delays picking the eggs up for 3 d but in most species loosely-adhesive eggs are taken up toward the end of spawning, which lasts about an hour. The female (e.g. *C. finleyi*, *C. linckei*) or the male (*C. guntheri*) may brood the eggs first, or the female may give the larvae to the male to brood (*C. linkei*). Both parents later care for the free-swimming fry.

Other monogamous mouthbrooders in Africa may have evolved along a parallel path. In the simplest condition, illustrated by *Perissodus microlepis* of Lake Tanganyika, the female broods the eggs and wrigglers and the pair then jointly defends the fry (Yanagisawa, 1986). In a further step in the same lake, the female *Eretmodus cyanostictus* gathers the loose eggs into her mouth at spawning. Later the brood, presumably as wrigglers (Yanagisawa, 1986), is transferred to the father (Kuwamura, 1986a). Such prolonged biparental care of fry, hence monogamy, was seen in six species in Lake Tanganyika by Kuwamura (1986a) and *Xenotilapia flavipinnis* by Yanagisawa (1986; further examples in Konings, 1988).

Sarotherodon galilaeus is of special interest. It is often mentioned as a biparental mouthbrooder, hence a monogamous species. However, it appears to represent an intermediate situation (summarized in Trewavas, 1983). The male and female pair up briefly and dig a nest. When they spawn, each takes eggs into its mouth. They then go their separate ways. Such a mating system could give rise either to maternal or to paternal species (Barlow, 1974; Baylis, 1981; Ridley, 1978).

Paternal care

Exclusively-male care is known for only two cichlid species, *Sarotherodon melanotheron* (Trewavas, 1983) and *S. occidentalis* (Loiselle, 1985b). The male fertilizes the eggs, then broods them in his mouth. Given that paternal care is the general rule among bony fishes (Blumer, 1979), these cichlids deserve closer attention. The nature of their mating system, however, is clouded. Limited field observations (Trewavas, 1983) indicate that males of the blackchin tilapia, *S. melanotheron*, nest colonially, as do many polygynous tilapiine cichlids. In aquaria, in contrast, the isomorphic male and female form a pair and court from one to a few days (Aronson, 1949; Barlow and Green, 1969, 1970; Loiselle, 1985b); in this respect, they seem intermediate between substrate guarders and the brief encounter and spawning of most polygamous mouthbrooders.

Some aquarium studies suggest that the female departs after spawning (Aronson, 1949; Barlow and Green, 1970). On the contrary, if the female lays more eggs than the male can brood, she picks up the excess and broods them (Aronson, 1949); this indicates the potential for biparental care. Trewavas (1983) examined several blackchin tilapia in the field, however, and found that only males were ovigerous.

Polygamy

Substrate guarders

Aquarists have discovered that many substrate-guarding *Apistogramma* of South America are apparently polygynous, forming harems (see also Burchard, 1965). Sexual dimorphism is often extreme. In aquaria, large males exclude or dominate smaller males and preside over the lesser territories of synchronously breeding females. Further haremic species from South America are found in *Acarichthys*, *Crenicara*, *Guianacara*, *Nannacara* and *Gymnogeophagus* (Loiselle, 1985a). African haremic species occur in *Nanochromis*, *Teleogramma* and in lake-dwelling lamprologines, about which most is known. Haremic cichlids are regularly diminutive species though the genera *Acarichthys* and *Guianacara* contain large haremic species (Loiselle, 1985a).

The lamprologines are substrate guarders. Many are monogamous, but several are haremic (Kuwamura, 1986a; Konings, 1988; Loiselle, 1985a; Wickler, 1965a). Males of *Lamprologus* have territories or domains encompassing the territories of two or more females. Kuwamura (1986a) noted that males defend small ($< 5\ m^2$) exclusive breeding territories or dominate in large (up to $20\,m^2$) domains, depending on the species. Snail-dwelling haremic males have territories less than $1\,m^2$ (Konings, 1988).

Lamprologus furcifer is the extreme case. The domains of males reach up to $100\,m^2$ and may contain the territories of up to 16 females (Yanagisawa, 1987). The females reproduce out of phase with one another. The dominant male continually visits females, usually spending less than 30 s with each but occasionally staying up to 20 min. Yanagisawa thought this indicated sequestering by the male of females about ready to spawn.

So far, no haremic mouthbrooders have been discovered in the field. In aquaria, however, the South American mouthbrooder *Gymnogeophagus balzanii* forms harems (Loiselle, 1985a).

Mouthbrooders

Only in recent years have polygamous mouthbrooding cichlids been studied closely in nature. Males are usually, but not always (e.g. Konings, 1988), larger and more colourful than females. Polygynandry prevails in that males mate with several females, and females may apportion their eggs among more than one male when males are close together. Mating typically lasts only a few minutes and then the female swims off with the eggs (termed advanced mouthbrooding). Four different situations have been reported: without territories, transient territories, persistent display sites and lekking; they are fuzzy sets that grade into one another.

Aterritorial breeding has been reported for rift-lake cichlids in Africa. In some species of *Tropheus* in Lake Tanganyika, the male and female move across the rocky bottom, stopping now and then to spawn one or a few eggs (Brichard, 1978). The consequences of this arrangement deserve closer scrutiny.

Transient spawning territories are also infrequent. Some species, e.g. within *Tropheus*, alternate between aterritoriality and transient territories (males of *T. moorii* move about in groups whereas females are solitary; Konings, 1988). Additional genera from Lake Tanganyika said to show both these types of behaviour are *Eretmodus*, *Simochromis*, *Spathodus* and *Tanganicodus* (Brichard, 1978). Kuwamura (1986a) reported temporary defence of breeding sites by males in 11 species of Tanganyikan mouthbrooders, e.g. *Gnathochromis pfefferi* and *Lobochilotes labiatus*.

Transient spawning territories are also known for Lake Malawi cichlids but are again said to be exceptional. Among the Mbuna, for instance, males of *Pseudotropheus williamsi* defend transient territories (Ribbink *et al.*, 1983a). Of 70 species that McKaye (1984) observed to spawn over sand in Lake Malawi, only two did so while moving about, *Cyrtocara moori* and *Lethrinops* sp.

More common is the persistent defence of a display site by a male who is then visited by females. This spawning mode prevails among the Mbuna of Lake Malawi (Ribbink *et al.*, 1983a) and their counterparts in Lake Tanganyika, e.g. *Petrochromis*, *Simochromis* and *Cyphotilapia*. Persistent-defence breeding is distinguished from lekking by three features: breeding sites are not clustered, males defend their sites continuously for a long period, and males feed on their territories (Kuwamura, 1986a; Ribbink *et al.*, 1983a).

Lekking, or arena-type breeding, among fishes was first described by Fryer and Iles (1972) and later reviewed and delimited by Loiselle and Barlow (1979). Lekking males gather in colonies, each on a small territory. A female visits and spawns with one or more males in the nests built by them. The males of some species construct a volcano-like structure, or ring the nest with turrets, the detailed features of which are species-specific (Konings, 1988; Lowe (McConnell) 1956, 1959; McKaye, 1983). So far, among cichlids, lekking is known only for African mouthbrooding species that nest in open areas. There it is common among lake and riverine species of haplochromine and tilapiine cichlids (e.g. Fryer and Iles, 1972; Konings, 1988; Lowe (McConnell), 1956, 1959).

McKaye (1983) investigated lekking in detail in the Lake Malawi species *Cyrtocara eucinostomus*. A vast ribbon-shaped colony forms at Cape McClear that varies from 5000 to 50 000 breeding males, depending on the breeding season, which has two peaks. Within a season, breeding is synchronized. Of the males' nests, 90% are confined to a strip of bottom 3–9 m deep. The males occupy their nests during the morning and feed on plankton near by during the afternoon, returning to their nests to sleep. The females descend from the nearby school in the morning, and each may spawn with up to 12 males.

Some South American 'earth eaters,' *Geophagus sensu lato*, are polygynous mouthbrooders but are not haremic (Loiselle, 1985a). *Geophagus hondae* (= *steindachneri*) is apparently a maternal mouthbrooder, as noted in passing by Keenleyside and Bietz (1981; see also Sprenger, 1971), as are *G. pellegrini* and *G. crassilabris* (Loiselle, pers. comm.). Unfortunately, none of these species has been studied in the field.

Plasticity of mating systems

Substrate guarders

Biparental care effectively removes both parents from further reproduction during care. Each parent should therefore be tempted to desert in order to invest in additional matings (Trivers, 1972). The female invests more in eggs than does the male in sperm, and she is the limiting sex because of the relatively small number of eggs compared with sperm. Consequently, the male is expected to desert more often than the female.

In the presumably primitive Etroplinae, the male may even invest more in parental care than the female (unpublished data), which is highly unusual for cichlids. Pairs observed in the field were persistently monogamous despite communal brooding with its opportunities for desertion (Ward and Samarakoon, 1981; Ward and Wyman, 1977). Another cichlid is tightly monogamous under aquarium conditions; neither male nor female *Cichlasoma citrinellum* deserted regardless of the sex ratio (Rogers, 1987).

Monogamous male cichlids tend to flee, and females to remain with the young, when danger threatens. In general, males spend more time away from the brood than do the females (Fryer and Iles, 1972; Barlow, 1974, 1976; McKaye and Barlow, 1976a; Neil, 1984; Townshend and Wootton, 1985). Evidence is accumulating for a tendency to desert among males of many presumably monogamous species. With a female-biased sex ratio in experimental ponds, some males of *Herotilapia multispinosa* deserted their first mate to mate with a second (Keenleyside, 1983). Males of *Cichlasoma tetracanthus* behaved the same way in aquaria (Loiselle, pers. comm.).

In the field, some males of *Aequidens caeruleopunctata*, *Cichlasoma panamense* and *Tilapia mariae* are faithful, but other males forsake their females to find new mates (Barlow, 1974; Townshend and Wootton, 1985; Schwanck, 1987b, respectively). Among *Cichlasoma tuba* in Costa Rica, the male schools with conspecific fish while the female cares for the eggs; when the fry start to swim, the male rejoins the family (Meral, 1973); some males of *T. mariae* do the same (Schwanck, 1987b).

Occasional males of the Central American convict cichlid, *Cichlasoma nigrofasciatum*, have two mates in the field (Meral, 1973). Under seminatural conditions, some male convict cichlids acquired second mates when the sex ratio was biased toward females (Keenleyside, 1985). In Panama, polygyny has been seen in *Aequidens caeruleopunctata* and *Cichlasoma panamense* (Barlow, 1974; Townshend and Wootton, 1985). *Lamprologus brichardi*, in Lake Tanganyika and in the laboratory, is variously monogamous or bigamous depending on the number of females the male can sequester (Limberger, 1983).

Apistogramma wickleri of South America is intermediate between haremic and monogamous species. In aquaria, the male bonded with a female and helped her to defend the breeding site, but only until spawning was completed (Burchard, 1965). Young, small males of the haremic *A. borelli* mated monogamously with females when no large males were present in aquaria; remarkably, the young males then showed the distinct yellow and black colour pattern of females (Burchard, 1965).

Small, excluded male *Apistogramma* may sneak (Loiselle, 1985a). Within the territories of reef-dwelling *Lamprologus* in Lake Tanganyika, smaller subordinate males avoid the dominant male and hide in crevices at his approach (Yanagisawa, 1987). These males probably behave as sneakers at opportune moments.

Aquarists have found that South American 'earth eaters' show a wealth of mating types in captivity, extending from monogamous through haremic to openly polygamous, and from substrate guarding through primitive and advanced mouthbrooding (Loiselle, 1985a: table on p. 231). Some of the species are noted for varying between monogamy and harem polygyny.

Mouthbrooders

Monogamy is uncommon in African mouthbrooders, probably because maternal mouthbrooding provides the opportunity for males to abandon their mates, e.g. *Tilapia mariae* (Schwanck, 1987a, b). Given the infrequency of monogamy, the options exploited by most maternal mouthbrooders typically range from aterritorial polygamy to lekking; a number of species are plastic in this respect (see page 177).

The sneaker-male tactic should be frequent. To illustrate, the school of fish over the lek of *Cyrtocara eucinostomus* in Lake Malawi contains small conspecific males that resemble females; when the territorial male chases other intruders, such as egg predators, these small sneaker males slip into the nest and try to spawn with the visiting female (McKaye, 1983).

8.3 CONSERVATISM OF MATING SYSTEMS

Especially effective behavioural adaptations may persist in the face of large differences in ecology. Once evolved, they may be stable under changing ecological conditions (Fischer, 1988). Thus substrate guarding has persisted among African, Asian and Neotropical cichlids despite long periods of separation and radical differences in ecology in each lineage. Several lines, however, have produced mouthbrooders.

In some places, most notably in Lake Tanganyika, mouthbrooders and substrate guarders persist side by side, subject to similar ecological conditions, and have identical mating systems. Whatever ecological circumstances led to the origin of mouthbrooding, they are not great enough to impose a single mode of caretaking on coexisting cichlids.

Differences in mating systems often derive from differences in feeding habits (Wittenberger, 1981). Consequently, one could expect mating systems of cichlids to vary in some predictable fashion with changes in feeding habits. The feeding behaviour of Central American cichlids ranges from alga-scraping herbivory through detritus feeding to gleaning invertebrates to piscivory. Their mating systems, however, have diverged little (Barlow, 1974). All are substrate guarders and most are monogamous; the exceptions, which are probably more frequent than currently known, are facultatively polygynous. Perhaps this conservatism of mating systems is simply a consequence of the relative recency of the cichlids there (Miller, 1966; Myers, 1966) compared with the Amazon Basin.

The cichlids of the African rift lakes suggest that recency may not be the only issue. Lake Malawi is estimated to be 1–2 m y old. The diversity of feeding behaviour is impressive (Ribbink *et al.*, 1983a). Yet all the species are maternal mouthbrooders, save one, the biparental mouthbrooder *Tilapia rendalli*; however, *T. rendalli* is not endemic to Lake Malawi (Ribbink *et al.*, 1981). The absence of substrate guarders could reflect nothing more than a historical accident, their absence among the original colonizers. Nonetheless, since the cichlids there have radiated so incredibly, one might have expected the rediscovery of substrate guarding.

Lake Tanganyika is even older and has both maternal mouthbrooders and substrate guarders, as a result of different founder stocks. Again, the diversity of mating systems within each lineage is impressive (see page 177), but the mouthbrooders have not given rise to substrate guarders, or vice versa.

The form of parental care limits the mating systems. By definition, substrate guarders can neither lek nor have male-display sites. Haremic systems are frequent among substrate guarders but are apparently absent among mouthbrooders. Furthermore, mouthbrooding appears to have originated a number of times from substrate guarding.

8.4 FACTORS SHAPING MATING SYSTEMS

Mating systems, whether of substrate guarders or mouthbrooders, results from the interaction of ecology and fundamental aspects of reproductive biology. Ecology determines the time and place of breeding and shapes reproductive behaviour through such factors as the temporospatial distribution of food, competitors and predators, and the nature and availability of breeding sites (Baylis, 1981; Williams, 1966; Emlen and Oring, 1977). Intrinsic factors constrain the mating system in concert with the ecological setting.

Intrinsic factors

Conflict between the sexes arises from anisogamy and how this relates to differential investment by the sexes in gametes and the succeeding offspring (Williams, 1966, 1975; Trivers, 1972; Maynard Smith, 1977; Baylis, 1981). The critical factor may be which sex limits the reproductive rate of the other (Baylis, 1981; Trivers, 1972). In cichlids, females invest more in gametes than do males. Females are also the limiting sex because their rate of producing eggs is slower than the males' rate of producing sperm. Thus males try to maximize lifetime reproductive success by fertilizing the eggs of as many females as possible whereas females seek to promote their fitness by enlisting males in parental care.

If the period of essential male care exceeds that needed for the female to produce another clutch of eggs, then the male becomes the limiting sex and sex-role reversal could be expected. The male blackchin tilapia, *Sarotherodon melanotheron*, broods the eggs and larvae for *c.* 15 d (Shaw and Aronson, 1954), as does *S. occidentalis* (Loiselle, 1985b). The young are not taken back into the mouth, but it is not known how long, if at all, the male stays with the young in the field. Nor is it known how long the female requires to produce a second clutch of eggs. In aquaria, cichlids commonly produce a second clutch in less than two weeks (see page 188). Even if the male ceases care after 15 d, he still needs a period of feeding to make up for the fasting during mouthbrooding (pers. obs.; Trewavas, 1983). The male may well be the limiting sex in these species.

Intrinsic factors are regularly confounded by extrinsic ones. Ultimately, differences in mating systems among cichlids which experience the same intrinsic effects must be attributed to extrinsic factors.

Extrinsic factors

Differences in the physical environment and seasonality affect mating systems. Wet and dry seasons have profound effects on freshwater fishes in the tropics (Lowe-McConnell, 1964, 1969b; Marsh *et al.*, 1986; McKaye, 1977, 1984; Neil, 1984; Zaret, 1980). Breeding among cichlids often occurs sporadically year-round but has a distinct peak just before or at the onset of the rainy season.

The major seasonal effect is on food supply and amount of water available (Baylis, 1981). Low water levels concentrate fish, making feeding and survival difficult for small species while favouring large piscivores. Fluctuating water level, and changing stream velocity, may have led to a few species spawning on leaves which they can move (Barlow, 1974; Timms and Keenleyside, 1975; see also Lowe-McConnell, 1964).

It is often unclear, however, which is the primary factor because physical environment, feeding, absolute size and exposure to predation interact. For instance, feeding on plankton is often associated with schooling in open water, and spawning may occur there (see page 183). However, some plankton feeders hover over the reef and refuge and spawn there as monogamous pairs, e.g. *Neolamprologus elongatus* in Lake Tanganyika (Konings, 1988). Nonetheless, the following discussion is organized around the physical environment, predation and feeding.

The physical environment

Two examples illustrate the effect of the physical environment. The first concerns spawning in cichlids adapted to living in the water column in Lake Tanganyika. Most such cichlids must return to a firm substrate to spawn. In *Paracyprichromis* the male sequesters a female who points her head straight down next to, but not touching, the vertical face of a rock and releases her eggs. The male emits a cloud of sperm, fertilizing the falling eggs which the female catches in her mouth (Konings, 1988). Thus while the substrate serves only to orientate the response, it still appears necessary.

The closely related *Cyprichromis* has gone a step further. The male holds a 'static' lek territory in the open water near shore to which the female comes. The head-down female releases and catches the eggs before fertilization. The male arches his body and holds the trembling yellow tips of both pelvic fins toward her. She mouths the fin tips and thereby receives sperm (Konings, 1988). *Limnochromis* also spawns in the water column.

The plankton-feeding *Haplochromis chrysonotus* in Lake Malawi has undergone parallel evolution. The female releases and catches falling eggs, which are probably fertilized in her mouth. The males hold lekking stations in the open water (Eccles and Lewis, 1981). Thus selection has produced emancipation from the substrate for spawning in open-water species: that facilitated lekking through aggregating behaviour and high mobility (Loiselle and Barlow, 1979).

The second example is provided by a number of small species of *Lamprologus* and *Neolamprologus* in Lake Tanganyika (Konings, 1988). Lamprologines are typically reef dwellers, but they have occupied the open, sand bottom by taking refuge in snail shells such as those of *Neothauma*. The abundance of shells, and the small size necessary to reside in them, have had profound effects.

Where the shells accumulate in piles, *N. multifasciatus* live in high densities and display little intraspecific aggression, but further details are lacking (Konings, 1988). Typically, shells are scattered about, and a monogamous male and female may occupy a shell together. In another species, each mate has its own shell. Shell size constrains sexual dimorphism in size.

L. callipterus is different. The male is much larger than the female. He holds a territory *c.* 50 cm in diameter encompassing a number of shells, each occupied by a female that is the sole caretaker of the young. Neighbouring territorial males steal shells from one another, presumably with their female occupants. To prevent such thievery, Konings (1988) believes, the monogamous species of ostracophilic lamprologines bury their shells, leaving just the openings exposed.

Clearly, the abundance of shells is an important factor shaping the mating system of these cichlids. But we need to know much more about differences in feeding behaviour and the like to understand why haremic and monogamous lamprologines coexist in the face of the same abundance of shells.

Some small maternal mouthbrooders in Lake Malawi, e.g. *Pseudotropheus livingstoni*, also live over the sand and occupy empty snail shells. Ribbink *et al.* (1983a) reported that they are not haremic, but apparently they are (McKaye, pers. comm). A comparison of the shell dwellers of the two lakes would be rewarding.

The effect of predation

Exposure to predation should influence the mating system. The risk of predation on adults can limit their activities, whereas predation on the fry can select for biparental care, or its absence for uniparental care.

Substrate guarders. The eggs are immobile, concentrated in one place and, when laid in a cavity, relatively easy to defend. (Predation on fry can be so great that parents can protect them only in a cavity (Ribbink *et al.*, 1981).) One parent should suffice to guard eggs, but no field data are available to test this assumption: if true, it could account for the lower investment by males in care of eggs. However, two parents are often needed to obtain a territory as well as to protect the school of fry, and that may have driven the evolution of monogamy (Barlow, 1974, 1984, 1986).

Predation on fry may have selected for monogamous behaviour among previously haremic males in the open-bottom-dwelling lamprologines of Lake Tanganyika. *Lamprologus congolensis*, a riverine and haremic species (Wickler, 1965a), is probably representative of the stem stock (Loiselle, 1985a). The founding populations would have encountered higher levels of predation on their fry in these more exposed lacustrine biotopes, favouring monogamy (Loiselle, 1985a; see also Konings, 1988).

Adults of large species are themselves little constrained by predators, so they can afford to nest in the open away from shelter. Nevertheless, that exposes their fry to small predators, necessitating biparental guarding. Those species also move large distances with their young, further exposing them to predators. Examples include *Boulengerochromis microlepis* in Lake Tanganyika (Kuwamura, 1986c), *Cichla ocellaris* in South America (Lowe-McConnell, 1969b; Zaret, 1980) and *Cichlasoma managuense* in Nicaragua (pers. obs.).

To test whether two parents are necessary to defend fry from predation, Keenleyside and Bietz (1981) removed one parent, male or female, from *Aequidens vittatus*. The young were quickly taken by predators (see also Keenleyside, 1978; Perrone, 1978b; Zaret, 1980). When Yanagisawa (1985) removed either parent in the monogamous maternal mouthbrooder *Perissodus microlepis* in Lake Tanganyika, the rate of predation on young rose.

At times, however, one parent may suffice. *Cichlasoma panamense* were biparental when predators on the fry were abundant. Exclusively maternal care was more frequent when predators on fry were few. At the same time, however, the food supply fell and the sex ratio rose to two females per male, confounding the interpretation (Townshend and Wootton, 1985). Such observations, nonetheless, suggest how haremic mating systems may have evolved when mating with a second female did not reduce the survivorship of the first female's offspring.

Haremic dwarf cichlids tend to live in close association with complex substrates. The intricate structure of reefs, snail shells and the like afford the tiny adults protection from predators, reducing a predator-imposed constraint on the evolution of conspicuous coloration (Barlow, 1974).

Close use of the habitat also places the fry of such cichlids in situations less exposed to predation (Loiselle, 1985a; McKaye, 1984). Reduced risk of predation on fry permits their care by one parent and promotes the evolution of serial or harem polygyny. However, selection must also operate on the behaviour of the fry to complete the process.

The fry of haremic *Apistogramma* in South America do not school up in the water (i.e. are not nektonic) in the typical cichlid fashion. Instead, they hug the bottom (i.e. are benthic), are cryptically coloured, move only in small darts, and only move one at a time; they are difficult to detect (Barlow, unpublished). Nagoshi (1983) and Kuwamura (1986a) described parallel behaviour among the haremic lamprologines of Lake Tanganyika. Kuwamura ranked 17 species on a gradient from biparental to maternal care. As the species became more maternal the fry became less nektonic and more benthic.

Kuwamura (1986a) added that food for the fry is critical and could even drive the system. Young of biparental cichlids typically school up in the water and feed on microplankton continuously renewed by moving water, even in lakes. Benthic fry of haremic cichlids, in contrast, often seem to glean prey from the bottom and thereby run the risk of depleting local food. They may need a stable supply of benthic prey to permit uniparental care.

But even in some biparental cichlids, such as *Papiliochromis* and *Pelvicachromis* of Africa, the fry hug the bottom. They risk being swept away in the strong current where they are raised. Benthic fry may be a necessary but not a sufficient adaptation for uniparental care.

Another effect of predation on eggs and young may be to select for aggregations of breeding cichlids. Pairs of *Etroplus maculatus* in Sri Lanka were more successful when grouped than when isolated (Ward and Samarakoon, 1981). When the level of predation rose in a Panamanian stream, the *Cichlasoma panamense* there bred in aggregations (Townshend and Wootton, 1985). Pairs of *Tilapia zillii* on the edge of a cluster of 53 pairs experienced attacks from predators whereas more-centrally located pairs did not (Loiselle, 1977). The breeding *T. zillii* were not in phase. Pairs of *T. mariae* in their native Africa settled together and synchronized their breeding in relation to the phase of the moon. Most fry were lost to nocturnal predators. Care of the fry peaked at full moon, enabling the parents to defend their fry at night (Schwanck, 1987a). In Florida, introduced *T. mariae* breed synchronously (Baylis, pers. comm.).

Mouthbrooders. Lowe (McConnell) (1956) found that lekking male *Oreochromis* in African swamps often went to the surface for air because the dissolved oxygen concentration in the water was so low. She also pointed out that mouthbrooding, with its accelerated spawning act, would be favoured under such conditioning (see also Lowe-McConnell, 1959). The question is whether this represents the circumstances in which mouthbrooding arose, or whether mouthbrooding preadapts those fish to breeding in oxygen-poor water.

Predation on eggs laid in open nests has also been proposed as the driving force in the evolution of maternal mouthbrooding (Fryer and Iles, 1972; Lowe-McConnell, 1975; Oppenheimer, 1970; Wickler, 1962a). Consider the situation above leks of *Cyrtocara eucinostomus*. Lurking there are dangerous cichlids, such as *Cyrtocara labifer*, that dive into the nest at the moment of spawning; these egg eaters are the objects of frequent attacks by the nesting males (McKaye, 1983).

Comparative studies indicate a progression from a delay in taking the eggs into the mouth to picking them up instantly (e.g. Trewavas, 1983). Thus mouthbrooding may not have been triggered by predation on the eggs but, once it started, predation may have accelerated the behaviour.

Egg predators may have affected the mating system of mouthbrooding cichlids in another way: by breeding in phase, as when lekking, mouthbrooders may reduce predation through the well-known dilution effect (McKaye, 1983).

Predation on fry could have promoted monogamy among mouthbrooders in African lakes. The biparental *Xenotilapia flavipinnis* are territorial over sand-and-rock bottoms. Both parents stay with the fry and retrieve them into the mouth. Yanagisawa's (1986) observations led him to believe that two parents are required to protect fry, which accounts for the lack of desertion by either parent.

Predation on adults is also effective. Male haplochromines are conspicuously coloured and are taken by diving birds much more often than are females (Trewavas, 1938). Differential predation on males skews the sex ratio toward a surplus of females and promotes polygyny.

The lekking grounds of *Cyrtocara eucinostomus* in Lake Malawi are confined to a specific range of depths owing to predation on the adults. The upper depth is limited by wave action in the shallows, and to a small degree by piscivorous cormorants. The lower limit is set by the catfish, *Bagrus meridonalis.* At night catfish move up from deeper water to prey on the sleeping adult *C. eucinostomus* and other cichlids that share the lek. The lower depth corresponds roughly to the upper limit of the catfish's nightly excursion (McKaye, 1983). This compression of the breeding population contributes to the dense packing of male nests and hence lekking.

Feeding behaviour

The behaviour of seeking and capturing food, when studied in relation to the distribution of food in space and time, has proved valuable in comprehending the adaptiveness of mating systems in other kinds of animals (e.g. Jarman, 1974; Emlen and Oring, 1977; Wittenberger, 1981). Typically, females array themselves to maximize energetic intake and males match the distribution of females while minimizing time spent obtaining food (Hoffman, 1983; Schoener, 1971). Students of cichlids, however, have made little attempt to relate mating systems to feeding even though food habits have been analysed (e.g. Lowe-McConnell, 1959).

It was once supposed that mouthbrooding is causally related to buccal morphology, especially the fine gill rakers associated with filtering or sifting fine particles of food (Fryer and Iles, 1972). The tilapia of Africa and the 'earth eaters' of South America are examples. However, mouthbrooders of Lakes Malawi and Tanganyika eat an incredible variety of food and have radically different trophic morphology, showing that feeding behaviour alone does not account for mouthbrooding.

Hori (1983) described the feeding behaviour of 13 *Lamprologus* in Lake Tanganyika, and Kuwamura (1986a) observed 52 cichlids in the same lake (see also Takamura, 1984). Hori's account hints at possible effects of feeding behaviour. The piscivores were in general the largest species. Large species are more inclined towards biparental care, and thus monogamy. Piscivores, at the top of the food chain, should also be infrequent and might therefore be more likely to stay with a mate, once found. The planktivores were among the smallest, most abundant species, and hence more usually polygynous.

Feeding on plankton in open water in Lake Malawi promotes aggregating and regular short migrations between the spawning ground and the feeding site, both of which may contribute to lekking (Loiselle and Barlow, 1979). Therefore while predation may constrain the depth distribution of lekking there, proximity to feeding sites and suitable substrates for nests may be decisive factors in positioning the lek (McKaye, 1983), though this does not explain why the fish lek in the first place.

Ribbink *et al.* (1983a) catalogued the habits of a large number of Mbuna in Lake Malawi, mostly of reef-dwelling species, into seven trophic groups. The feeding habits are diverse and include planktivores, scale and egg eaters, gleaners of invertebrates, and scrapers of algae or aufwuchs. The Mbuna's mating systems, however, are relatively conservative and polygamy prevails, usually centred on a breeding territory.

Particularly intriguing is the finding that some Mbuna of both sexes hold feeding territories, and in some others males defend feeding territories independent of breeding (Ribbink *et al.*, 1983a), a condition rarely reported among tropical freshwater fishes (Barlow, 1984, 1986). This is apparently a secondary condition derived from the maintenance of male breeding territories.

To make more sense of the situation, one would have to work with the species in the field. The data are sparse and lack detail.

Energetic considerations

The rate of obtaining food, and the costs of mating and raising young, can also influence the mating system. For instance, how quickly are energetic stores replenished to permit the production of another set of gametes? If the prospect of finding another mate is poor, or if food is scarce, it may pay to stay with the mate and invest further in care of the young, rather than risk loss of the current brood and wasted energy trying to start a new cycle.

Abundance of food often fluctuates seasonally. The onset of the rainy season may wash in large amounts of exogenous food, triggering reproduction (McKaye, 1977). Phasic trophic stimuli, or short breeding seasons, thus promote synchrony. Synchronous breeding militates against male desertion because the chances are small of finding a new, unmated female. Depending on other factors, synchronous breeding can favour monogamy (Knowlton, 1979) or lekking (Loiselle and Barlow, 1979).

Parental cichlids often feed little or not at all (Townshend and Wootton, 1984), so breeding depletes energetic stores. This is ameliorated if the parents take turns guarding the fry, and so can feed alternately, as in the Asian cichlid *Etroplus maculatus* (Ward and Wyman, 1977; Ward and Samarakoon, 1981). That is possible, however, only if one parent suffices to protect the fry. Pairs of the sympatric *E. suratensis* constantly attend their fry, apparently because of depredations by the abundant *E. maculatus* (Ward and Wyman, 1977).

In monogamous mouthbrooders in Lake Tanganyika, e.g. *Xenotilapia flavipinnis*, first the female broods the eggs, then the male broods the wrigglers. Yanagisawa (1986) believes that male care is favoured because it permits the female to feed at a normal rate and thus to shorten the interval between broods. The male benefits because the pair remains together on the combined feeding and breeding territory and produces more broods within a season.

Amount of food has been shown under experimental conditions to affect recovery time. Female *Cichlasoma nigrofasciatum* were fed at three different levels. When the ration was decreased, the number of eggs per clutch went down and the interval between clutches increased (Townshend and Wootton, 1984). The effect was so large that small, well-nourished females laid more eggs than large ones fed poorly (contrary to the general rule that number of eggs laid increases with female size). The interval between clutches ranged from 9 to 40 days.

In contrast, male *C. nigrofasciatum* were much less influenced by the amount of food. Thus as food decreases, the difference in reproductive potential between male and female increases (Townshend and Wootton, 1984). That should drive the behaviour toward monogamy as ovigerous females become less abundant, skewing the operational sex ratio toward an excess of males. Then males should sequester females. On the other hand, with less food the female produces fewer eggs, which may tempt the male to desert.

Townshend and Wootton (1984) went on to compare fecundity and size of *C. nigrofasciatum* in captivity and in a Nicaraguan lake. The lake fish were smaller and had lower fecundity, suggesting that food is a major factor constraining reproductive behaviour. Unfortunately, too few data of this type are available to relate richness of feeding to type of mating system.

One is tempted to generalize from the foregoing that, given enough food, a cichlid fish should be able to repeat endless reproductive cycles. Some species do indeed breed repeatedly in a season (e.g. Yanagisawa, 1986), but this may not be general. Consider *Etroplus maculatus* in Sri Lanka which breeds but once each season (Ward and Samarakoon, 1981). Because parental mates alternate caretaking and feeding, they ought to be able to spawn repeatedly at short intervals. To explore why they do not, pairs were fed a constant diet in aquaria

(Lamon and Ward, 1983). For three successive spawnings, the eggs were removed right after spawning, and each fish received a new mate.

Number and weight of eggs remained the same, and the males were able to fertilize them. However, the fat content of the eggs fell progressively *c.* 30%, suggesting a decrease in quality, and the days to spawning increased steadily for females, doubling from 5 d to the first spawning to 10 d to the third one. Reproductive effort decreased *c.* 70%, according to a derived index. Both sexes showed increasing signs of declining motivation to reproduce. Finally, the growth of both males and females was arrested, a cost to future reproduction.

Some species recover faster than others, even when food is ample. Small species have a higher metabolic rate than do large ones, and so can return to breeding condition relatively quickly, given sufficient food. Large species may thus be more likely than small ones to breed only once during a breeding season (Barlow, 1981). That, together with the tendency of large species to breed in the open, and to be infrequent because they are at the top of the food chain, ought to contribute to a stronger tendency toward monogamy among large cichlids. Unfortunately, no clear relationship emerges when large cichlids are tabulated without regard to ecological differences (Loiselle, pers. comm.). Finally, the huge *Boulengerochromis microlepis* of Lake Tanganyika may be semelparous (Konings, 1988); if so, that would help explain their monogamy, because both sexes should invest maximally in parental care.

The study of energetics in relation to mating systems is in its infancy. We need more information about the flow of energy from the ecosystem to the fish and how that is translated into offspring, together with data on the costs of reproduction. Then we can better understand the conditions under which monogamy and polygamy have evolved.

8.5 COMMENTARY

This review of mating systems has served more to show how little we know than to reveal cichlid mating systems as adaptations. It has been instructive, nonetheless, in three ways. First, I have presented a fuller spectrum of differences among species than was previously available, thanks to the growing literature. Thus arguments about evolutionary sequences become more persuasive as the steps shrink in size. More importantly, it has served to illustrate how cichlids differ so fundamentally from other teleost fishes.

Most teleosts provide no parental care of eggs or young. When caretaking occurs, it is typically the male alone that provides it, whether a substrate guarder or a mouthbrooder (Blumer, 1979). In contrast, all cichlids invest in eggs or young. Exclusively paternal care, however, has been reported for only two mouthbrooders: when just one parent is the caretaker, it is otherwise always the female. Among mouthbrooders, maternal care is prevalent, whereas substrate guarders are predominantly biparental. Biparental monogamy is infrequent amongst teleosts (Barlow, 1984, 1986), making cichlids even more exceptional. The most difficult question about the evolution of mating systems among cichlids concerns the origin of monogamy, which appears to be the stem condition. The definitive answer can never be known.

A second theme that emerged is consistent with the general picture of intrinsic differences between the sexes in other animal groups. Repeatedly, it is the male that is apt to desert. That also contrasts with the typical situation in teleosts.

Third, the connection between ecology and mating system is becoming more evident. This is clearest with regard to predation on the offspring and the nature of the environment. The roles of energetics and mode of feeding, however, must be important, though these have not yet been worked out. Above all, the many factors of the environment need to be regarded in an interactive framework. Size of fish, feeding behaviour, physical environment, seasonality and predation are all of a piece.

In the future, we need studies more focused on this interconnection. The research should concentrate on a given species, work outwards to embrace related species whose mating systems differ, and then jump to parallel lines of evolution in other cichlids. Thus we require ecological data such as feeding and predation at all stages of the life history. To that we need to add demographic studies, employing life tables that reveal age-specific mortality, growth and fecundity in relation to the timing of breeding and reproductive effort. Given enough information, we should have the power to predict the mating system.

Some groups of cichlids offer outstanding opportunities. First and foremost are the lamprologines of Lake Tanganyika. They combine different mating systems with the opportunity to observe them in the field and in the laboratory. Likewise, a comparison of key Mbuna in Lake Malawi with their counterparts among the *Tropheus* types of reef fishes in Lake Tanganyika would be rewarding. If good field conditions can be found in South America, *Apistogramma* and the related 'earth eaters' would be excellent subjects (if any reader knows of good sites for such work in South America, I would appreciate hearing from you).

ACKNOWLEDGEMENTS

This essay has benefited greatly from the critical comments of Jeffrey R. Baylis, Paul V. Loiselle, Miles H. A. Keenleyside and David L. G. Noakes.

Chapter nine

Parental care

M. H. A. Keenleyside

9.1 INTRODUCTION

If there is one life-history feature that characterizes the family Cichlidae it is the long period of parental care of the young after spawning. In all the more than 1000 species currently recognized (see page 1), the progeny are protected by one or both parents from spawning until independence, a period lasting at least 3–4 weeks, and in many species considerably longer. This prolonged and close association between parents and offspring, which involves active two-way communication between the generations, is highly unusual among fishes (Breder and Rosen, 1966; Keenleyside, 1979; Blumer, 1982; Gross and Sargent, 1985). It is the feature that makes cichlids so popular among tropical fish aquarists and investigators of fish reproductive behaviour and ecology.

In this chapter I review the known parental care patterns in the family. Emphasis is on describing the range and variety of general patterns of care, and the associated variation in specific behaviours towards their young shown by brood-guarding adults. I conclude by discussing some ideas about the evolution of cichlid parental care. Throughout I use the terms egg, wriggler and fry for the three behaviourally distinct stages in the early life of cichlids. **Wrigglers** are the newly-hatched, non-free-swimming young; when they begin free swimming they are **fry**. These labels are so familiar to behaviourists, aquarists and others that I am reluctant to replace them here with technically more suitable names used by specialists in developmental biology of fishes (e.g. Balon, 1975; Noakes, this volume).

9.2 GENERAL PATTERNS OF PARENTAL CARE

Traditionally the patterns of parental care in cichlid fishes have been classified on the basis of two criteria (Baerends and Baerends-van Roon, 1950; Breder and Rosen, 1966; Fryer and Iles, 1972). One is the physical relationship between parents and their offspring. **Substrate brooders** are species in which the parents maintain their brood on or close to the substrate from spawning until the young become independent, although they may briefly carry them around in the mouth as they transfer wrigglers or retrieve scattered fry. **Mouthbrooders** are species in which the parents carry their offspring in their mouth until independence. Some mouthbrooders have adhesive eggs which stick to the spawning substrate; mouthbrooding begins when the eggs hatch. Others have non-adhesive eggs and mouthbrooding begins at spawning. In this chapter I refer to these types respectively as **delayed** and **immediate** mouthbrooders. Equivalent labels used by other workers are larvophilic and ovophilic (Peters and Berns, 1982) and primitive and advanced (Loiselle, 1985a).

The second general criterion for classifying parental care patterns is the sex of the care-giving adult; thus there are **biparental**, **maternal** and **paternal brooders**. The variety of parental care systems created by the interaction of these two sets of descriptive criteria is summarized in Fig. 9.1 and will now be described.

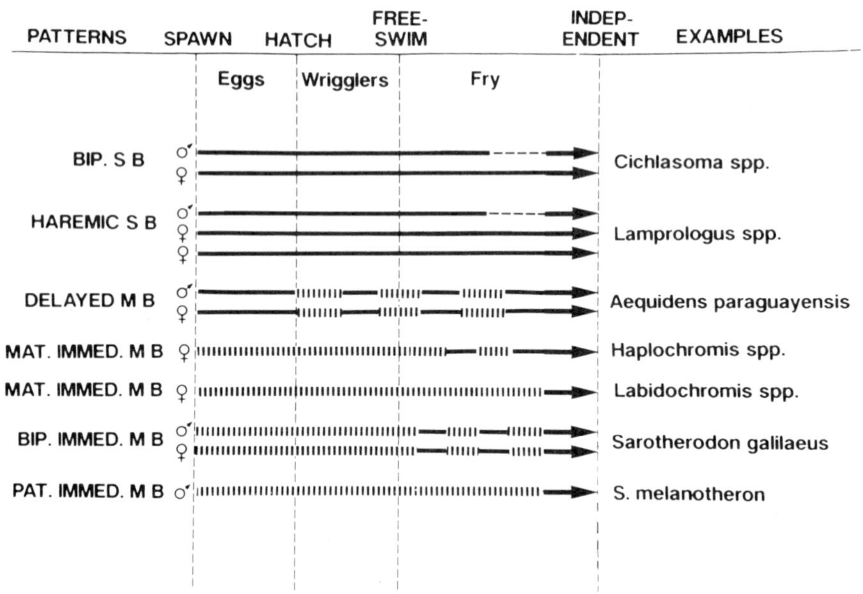

Fig. 9.1 Variation in parental care patterns among cichlid fishes, with one or more examples of each. Horizontal axis is generalized time scale from spawning to independence of young from parents. Solid line, young on or near substrate; hatched line, young in parent's mouth; broken line, facultative desertion by male parent. BIP., biparental; IMMED., immediate; MAT., maternal; MB, mouthbrooding; PAT., paternal; SB, substrate brooding.

Substrate brooding

In this system adhesive eggs are deposited on a firm surface, fertilized there, and after several days of intensive care they hatch into wrigglers. These are sucked or chewed out of the egg membranes by one or both parents, who carry them in the mouth and spit them into a pit excavated in the substrate, where care continues. The wrigglers are attached to the lining of the pit by sticky mucous secretions produced by six head glands (Brinley and Eulberg, 1953; Jones, 1972; Peters and Berns, 1982). In some species the wrigglers are spat onto leaves or among rootlets of aquatic vegetation, where they hang suspended from the plants or other wrigglers by the mucous threads (Breder and Rosen, 1966; Barlow, 1974; Courtenay and Keenleyside, 1983). When the young develop into free-swimming fry, the family unit moves slowly about; the fry feed on particles on or near the substrate and the parents guard them.

Typically this pattern of parental care is associated with cooperative custodial care of the brood by a monogamously mated pair. However, in some species the male may desert his mate before their brood reaches independence, leaving her to continue the care alone; in others the male spawns and remains bonded with more than one female and helps each to raise a brood, i.e. he is a harem polygynist. These variations in monogamous pair bonding are discussed below (page 20).

Immediate mouthbrooding

This can be divided into three subcategories, according to which parent orally carries the brood; hence there are maternal, paternal and biparental mouthbrooders.

Maternal mouthbrooding

This is the most common and best-known system. Spawning occurs on the substrate, usually at a site prepared by the male. The non-adhesive eggs are laid singly or in small groups, and the female immediately turns and picks them up in her mouth. The male releases sperm, as she is picking up the eggs, or in response to her nibbling at his anal fin or genital region; in either case, some or all of the eggs are probably fertilized inside her mouth (Mrowka, 1987b). In some species the male attacks and drives away other fishes that try to enter the nest and eat the eggs before they are picked up (McKaye, 1983, 1984). The female then leaves the spawning area and carries the brood orally until they have developed into free-swimming fry. In most species the mother releases her fry periodically to forage, and takes them back into her mouth when danger threatens and at night (Fryer and Iles, 1972; Trewavas, 1983). In some species the young are carried by the mother until they reach independence; after their first release they have no further contact with her. Examples of the latter are *Labidochromis vellicans* and *Labeotropheus* sp., both of which are members of the Lake Malawi Mbuna group (Wickler, 1965a; Balon, 1977) and *Oreochromis grahami* (Coe, 1966).

Paternal mouthbrooding

Very few cichlid species are known to practise mouthbrooding by the male parent only. *Sarotherodon melanotheron* is the only well-documented species, and its reproductive behaviour has been extensively studied in the laboratory (see Trewavas, 1983 for references). After several batches of eggs have been released and fertilized, the male picks them all up and carries them in his mouth. The female occasionally picks up some eggs, but in Aronson's (1949) study this only occurred when the male was very slow to pick them up, or when the male was smaller than the female, suggesting that she had released too many eggs to fit into his mouth. Male parental mouthbrooding in this species has been confirmed in the field (Pauly, 1976; Trewavas, 1983).

Biparental mouthbrooding

The oral brooding of the young by both parents from spawning until independence is also a rare system in cichlids. It is best documented in *Sarotherodon galilaeus*, where it has been observed in aquaria and confirmed in the field (Ben-Tuvia, 1959; Fishelson and Heinrich, 1963; Iles and Holden, 1969; Apfelbach, 1969). The eggs are released onto the substrate in batches of up to 40, fertilized by the male, and when spawning is complete the mass of eggs is picked up by one or both adults and carried orally until the young reach independence. If the fry leave a parent's mouth early they may be taken in again, especially at night (Trewavas, 1983). A closely related species, *S. multifasciatus*, also appears to be a biparental mouthbrooder (Whyte, 1975).

A variation of this parental system occurs in *Chromidotilapia guntheri* (Myrberg, 1965). When all the eggs have been released and fertilized they are picked up by the male only, who carries them until they have become free swimming. After this, when disturbed they are taken into the mouth of both parents. Whyte (1975) confirmed this unusual system in free-living *C. guntheri*.

In a comparative study of cichlid reproduction in Lake Tanganyika, Kuwamura (1986a) found that 28 of 35 mouthbrooding species practised typical maternal mouthbrooding. The other seven had one of two variations on this system. In some the eggs and wrigglers were carried by the female, but the older fry appeared to be carried only by the male. In the others, the egg batches were picked up by the female as they were being fertilized, but the fry entered the mouth of either parent for protection.

Delayed mouthbrooding

This system combines elements of substrate and mouthbrooding. Adhesive eggs are released onto a prepared site, fertilized and then guarded by both parents until hatching. The wrigglers are then picked up and orally brooded by one or both parents. In *Aequidens paraguayensis* the initial retrieval of hatching wrigglers is always by the female, but she soon releases them onto the substrate, after which each parent carries some of the brood (Timms and Keenleyside, 1975; Mrowka, 1984). As the fry develop they may be released to forage independently, but quickly return to the parents' mouths for protection. This parental system is accompained by an extended monogamous pair bond between male and female, as in substrate brooders. Delayed mouthbrooders are found among several Neotropical genera, including *Geophagus* (Reid and Atz, 1958), *Gymnogeophagus* (Loiselle, 1985a) and *Aequidens* (Timms and Keenleyside, 1975), and at least one West African riverine species, *Chromidotilapia batesii* (Linke and Staeck, 1981a).

9.3 BEHAVIOURAL PATTERNS ASSOCIATED WITH PARENTAL CARE

The two main functions of parental care behaviour are to protect the young from predators and to promote the normal patterns of growth and development of the young.

Protection

The protective role begins as soon as the first eggs are shed and ends when the link between parents and young is broken as the juvenile fish begin independent life. Among substrate brooders and delayed mouthbrooders (until oral transport begins), the parents attack and try to drive away any animals that approach their brood too closely. Direct evidence of successful predation on defended young is not easy to obtain in the field, except when individually recognized parents are seen attacking predators, and their broods decline in size over successive days (McKaye, 1977; Yanagisawa, 1987). Knowledge of predation on adult cichlids comes from gut analysis of presumed predators, including fishes, birds and Man (Fryer and Iles, 1972). However, predators of eggs, wrigglers or small fry may themselves be small fishes (including juvenile conspecifics) not taken by fishermen. Therefore, indirect evidence is commonly obtained by recording what the brood-guarding parents attack (e.g. Cichocki, 1977; Townshend and Wootton, 1985; Keenleyside *et al.*, 1990).

The spawning location can influence the success of parental guarding. For example, many substrate-brooding cichlids spawn in a cave or tunnel which they have excavated under rocks or in a soft substrate. Potential egg predators can be repelled by one or both parents stationed at the entrance (Barlow, 1984; McKaye, 1977; Nagoshi, 1983). Another method is to use a movable object as the spawning platform. Several Neotropical species spawn on a leaf chosen from the substrate litter (Barlow, 1974; Timms and Keenleyside, 1975; Keenleyside and Bietz, 1981). In a study of *Aequidens paraguayensis*, the parents selected a relatively easily-moved leaf when given a choice, and after spawning moved it away from a predator model presented near the leaf (Keenleyside and Prince, 1976). A mobile spawning site also allows the parents to move their clutch to safety in response to sudden changes in water level following heavy rains (Lowe-McConnell, 1964).

Mouthbrooding could be considered the ultimate in protection of defenceless young, because as long as the brooding parent is not itself eaten or seriously injured its brood should be safe. However in Lakes Victoria and Malawi, where mouthbrooding is the most common pattern among cichlids, several haplochromine species are paedophagous; their main diet is the eggs and young of other mouthbrooding cichlids. The mechanism by which they obtain this food has been disputed (Fryer and Iles, 1972; Greenwood 1974a; Wilhelm, 1980). The best evidence has come from underwater observations of three paedophagous species in Lake Malawi (McKaye and Kocher, 1983). These fish used their mouths to ram the opercular region and snout of mouthbrooding females. The attack caused the release of some young, which were quickly swallowed by the attackers. (See page 162 for further details.)

When a brood has reached the free-swimming fry stage, parental protection (in substrate brooders) is continued by constant vigilance and attacks at potential predators. Individual fry that stray from the brood are orally retrieved by a parent and spat back with the others. In addition, the parents may signal the young, using special body and fin movements termed **calling behaviour.** Two conspicuous components of calling are quick, lateral jerking or twitching movements of the head and body, called **jolting** (Baerends and Baerends-van Roon, 1950), and **pelvic fin flicking**, in which the pelvics are quickly snapped open and shut. The fry respond to these movements by approaching the signalling parents and settling onto the substrate. Among those mouthbrooding species in which the fry leave and re-enter the parent's mouth, calling includes jolting and pelvic fin flicking, and also the parent pitches slightly head-down and slowly backs up. In response, the fry rapidly approach, swarm around the head and soon enter the mouth (Baerends and Baerends-van Roon, 1950).

Promotion of growth and development

The second general function of parental care behaviour is the promotion of normal growth and development. For those species that maintain eggs on the substrate the most common parental behaviours are fanning and mouthing of the eggs. **Fanning** is also called ventilating and aerating because its primary function is to create a current of water past the eggs to remove metabolic waste products and increase the oxygen level near the eggs. The fanning parent holds position close to the clutch and moves the water with large-amplitude beats of the pectoral fins. Zoran and Ward (1983) distinguished between active and passive fanning in *Etroplus maculatus.* The former occurs in short bouts of rapid, vigorous pectoral fin beats, with the caudal fin also beating to compensate for the pectoral fin action. In passive fanning, the pectorals beat more slowly and for longer bouts; compensatory caudal fin movements are absent. Similar variation in fanning vigour occurs in *Herotilapia multispinosa* (Baylis, 1974; Smith-Grayton and Keenleyside, 1978) and probably in most substrate brooders.

Mouthing the eggs is a cleaning in pattern in which the parent fish places its mouth gently against the eggs and sucks away loose particles. Dead or fungus-ridden eggs are removed by more vigorous mouth contact, often called snapping, which is also used to remove larger objects such as gravel or snails (Bergmann, 1968).

During the egg care period, the parents prepare a site for the newly-hatched wrigglers. They excavate one or more small pits in the substrate near the clutch, or clean nearby leaves by vigorous mouth-picking movements. Newly-hatched wrigglers are carried in the parents' mouths to the pit or leaf. The choice of location for wrigglers may depend on ambient oxygen levels, with near-surface vegetation being used during hypoxic conditions (Courtenay and Keenleyside, 1983).

During the wriggler stage, fanning and mouthing of the brood continue. Occasionally a few wrigglers may be picked up, rolled about in the mouth and returned. The entire brood is commonly moved by the parents from one pit to another, or from vegetation to a pit, or vice versa. The proximate causes of these brood transfers are not well understood, but disturbance by predators may be important.

Major differences between substrate brooders and immediate mouth-brooders are that the eggs of the latter are not adhesive, and ventilation and cleaning of the eggs and wrigglers occurs inside the buccal cavity of the parent(s). Thus fanning and mouthing are replaced by **churning**, a conspicuous oral action also called chewing, mumbling and rolling (Baerends and Baerends-van Roon, 1950; Oppenheimer and Barlow, 1968; Fryer and Iles, 1972). It is a vigorous action, clearly distinct from ordinary respiratory movements, and it moves the brood around inside the parent's mouth. Churning is performed by all mouthbrooding cichlids while carrying their progeny in the buccal cavity; delayed mouthbrooding species begin churning as soon as they pick up their wrigglers from the spawning site. A decline in churning rate as the young approach independence strongly suggests that the primary function of churning is to meet the brood's respiratory needs, which the young increasingly assume as they develop their locomotor abilities shortly before leaving their parents (Oppenheimer and Barlow, 1968).

Finally, several actions performed by cichlid parents while guarding free-swimming fry are related to feeding by the young. **Leaf-lifting** is one of these. The parent grasps the edge of a leaf in its mouth, lifts it off the substrate and releases it. If it turns over while falling, the fry (and often the parent) forage on the newly exposed face. If it does not turn over, the adult repeats the action until successful. Clearly the parent is exposing a food supply for its young. Leaf-lifting has been recorded for *Cichlasoma nigrofasciatum* and *C. panamense* in streams in Costa Rica and Panama, respectively (Meral, 1973; Keenleyside *et al.*, 1990; Townshend and Wootton, 1985) and has been seen in other species held in aquaria (Barlow, 1974).

A second parental behaviour which makes food more readily available for fry is **findigging**, described in detail by Williams (1972). The adult settles onto the substrate and with vigorous, rapid beats of the pectoral fins, stirs up loose materials. In response, the fry immediately rush in and begin feeding. This behaviour has been described for several Neotropical cichlids.

The most unusual method of providing food for offspring occurs in species where the fry ingest mucus from the body of the parent. This was first described for *Symphysodon discus*, where the parents' epidermal mucus production is increased during the fry-brooding period, and the young fish regularly bite at the adult's body and swallow the mucus (Hildemann, 1959). This action, now often termed **micronipping**, has been described for many cichlid species, although parent-contacting by fry without feeding is also common (Quertermus and Ward, 1969; Noakes, 1979). Micronipping, with apparent ingestion of mucus, can be induced in some tank-held cichlids by starving the fry (Noakes, 1979), but field observations of *Cichlasoma citrinellum* (Noakes and Barlow, 1973a) and of *Etroplus maculatus* and *E. suratensis* (Ward and Wyman, 1977) show clearly that in those species it is a regular, common feature of foraging by the fry, even in the presence of other foods.

9.4 MALE–FEMALE ROLES IN PARENTAL CARE

Among species with biparental substrate brooding, the two parents do not usually share all parental duties equally. Although both sexes seem able to perform all the parental behaviours, a common generalization is that the female concentrates on activities directly associated with the brood (fanning, mouthing, snapping), whereas the male mainly patrols the brood-rearing territory and repulses predators (Fryer and Iles, 1972; Barlow, 1974, 1984; Keenleyside, 1979; Loiselle, 1985a). This separation of parental roles is most apparent during the egg and wriggler stages. When the young have become fully mobile fry, the main parental duty is protecting them from predators, and the parents share this task more or less equally.

Another measure of role-sharing is the difference in total time spent close to the brood by each parent. Typically one parent stays within one or two body lengths of the brood while the other is further away, chasing potential brood predators, foraging or simply staying alert. The two may exchange positions as the further parent returns to the brood; the exchange often includes a mutual 'greeting ceremony' in which the fish briefly display to each other before switching roles. In general, the female is said to spend more time close to their brood than the male does.

Data testing the universality of these parental role-sharing 'rules' are summarized in Table 9.1. Where fanning and mouthing of eggs and time spent close to eggs and wrigglers were measured, the female parent scored higher than the male in 16 of the 21 comparisons, and the parents were not different in the other five. This clearly supports the contention that females are more actively involved than males in direct care of their non-free-swimming young. In contrast, active defence of eggs and wrigglers (by attacking potential predators) was not primarily a male-biased activity. There are nine such comparisons in Table 9.1 and in only two were male parents consistently more aggressive than female parents. In the other seven cases, either the parents were equally aggressive, or females were more hostile to predators than were males. Table 9.1 also includes 14 comparisons of the parents' behaviour while guarding fry. In only one of these did males score higher than females (calling fry in *T. mariae*). In the other 13 cases females were more active (four) or the parents were equal (nine).

Thus, these quantitative studies (on species from Africa, Asia and the Neotropics) support the general 'rule' that in biparental cichlids the female is more involved than the male in direct care of immobile young (eggs and wrigglers). However, even though males spend less time than females close to the brood at these stages, this does not imply more active brood defence by the males. Parental attack frequency may not be the best way of assessing antipredator effectiveness. Males are typically larger than their mates in biparental species, and the presence of a relatively large, actively patrolling male may inhibit potential predators from coming within the male's attack distance. Finally, the parents' behaviour while guarding mobile fry does support the generalization that such broods need the active protection of both male and female.

9.5 PAIR-BOND STABILITY DURING BROOD CARE

Pair bonding is defined as any close and long-lasting association between a female and a male (Wilson, 1975). Thus, the brief association between the members of a spawning pair of immediate mouthbrooding cichlids (followed by uniparental care of the young) is not a pair bond. Among species with biparental brood care, the prolonged male–female association clearly indicates the existence of a pair bond. My interest here is in reviewing existing knowledge about intraspecific variation in stability of this bond, and about the causes of this variation.

The best evidence comes from studies in the field and with fish held in large outdoor pools. In laboratory aquaria, brood-rearing pairs are subject to

Table 9.1 Results of quantitative comparisons of male–female parental behaviours in some biparental cichlids. E, eggs; W, wrigglers; F, fry; Fd, field study; all others are aquarium studies

Species	Fanning eggs	Mouthing eggs	Time near brood			Defence of brood			Calling fry	Authority
			E	W	F	E	W	F		
Tilapia mariae	♀>♂									Apfelbach (1969)
T. guineensis	♀=♂									Apfelbach (1969)
T. mariae	♀=♂								♂>♀	Baldaccini (1973)
Pterophyllum scalare	♀>♂*	♀=♂							♀=♂	Chien and Salmon (1972)
Etroplus maculatus					♀=♂				♀=♂	Cole and Ward (1969)
Herotilapia multispinosa	♀=♂	♀=♂				♂>♀	♀=♂	♀=♂		Smith-Grayton and Keenleyside (1978)
Cichlasoma cyanoguttatum	♀>♂	♀>♂	♀>♂	♀>♂	♀>♂	♀>♂	♀>♂	♀=♂		Itzkowitz and Nyby (1982) (Fd)
C. cyanoguttatum			♀>♂	♀>♂	♀=♂					Itzkowitz (1984)
C. maculicauda	♀>♂					♀=♂	♀=♂	♀>♂		Perrone (1978a) (Fd)
C. panamense	♀>♂				♀>♂			♀=♂		Townshend and Wootton (1985) (Fd)
C. citrinellum						♂>♀	♀=♂	♀=♂	♀>♂	Rogers (1988)
Aequidens vittatus	♀>♂	♀>♂	♀>♂	♀>♂	♀=♂		♀>♂			Keenleyside and Bietz (1981) (Fd)
A. paraguayensis	♀>♂	♀>♂								Timms and Keenleyside (1975)

*Day 1 of 3.

unnatural restrictions on space and on shelter for their offspring, and to a biotic community that is less complex than the natural one. These conditions must lead to pressures on the parental pair bond that are different from those under natural conditions.

Instability in a pair bond can take the form of **desertion** of mate and brood by one partner, or establishment of additional pair bonds by one member, i.e. simultaneous **polygamy**. In biparental species, only males desert: evidence comes from field studies of three species in Central America (Meral, 1973; Carlisle, 1985; Townshend and Wootton, 1985), from outdoor pond work with two Central American species (Keenleyside, 1983; Keenleyside *et al.*, 1989), and from Lake Tanganyika, where several *Lamprologus* species have been observed with broods being guarded either by both parents or by the female only, implying male desertion (Nagoshi, 1983; Kuwamura, 1986a).

Field workers also report that the male of a brood-guarding pair flees before the female does, and takes longer to return, when a human observer approaches (Barlow, 1976; Zaret, 1980; Neil, 1984; Carlisle, 1985). This suggests that the male is less firmly attached to its brood than is the female, and that under conditions appropriate for permanent mate desertion, the male is more likely to desert.

His desertion appears facultative and seems associated with reduced predation pressure or an unbalanced sex ratio in the local breeding population, or both. Male desertion by the substrate-brooding *Cichlasoma panamense* in Panama was common only where brood predators were scarce and females strongly outnumbered males (Townshend and Wootton, 1985). Presumably, males were more prepared under such conditions to leave the guarding of their brood to the females and to seek extra breeding opportunities.

In outdoor pond experiments with *Herotilapia multispinosa* (another biparental substrate brooder), in which the adult sex ratio was systematically varied, mate desertion was only by males and was common only in the female-surplus condition (Keenleyside, 1983). Most deserting males spawned again within 6 d of deserting, suggesting that the opportunity to breed with an unattached female was the cause of desertion.

The developmental stage of the brood may also influence a male parent's decision either to remain and help the female guard, or to desert and search for another breeding opportunity. The fry of substrate-brooding species become less vulnerable to predation as they grow (Perrone, 1978b; Neil, 1984; Carlisle, 1985), presumably because they become stronger, more agile swimmers, and learn where to hide. In a field study of *Cichlasoma nigrofasciatum*, deserted females guarding broods were often found, but always with their young at a relatively advanced free-swimming stage, suggesting that younger broods are more vulnerable and therefore the male parents do not desert them, or that if they do, the broods are lost to predators (Keenleyside *et al.*, 1990).

Simultaneous polygamy is the other major variant on the monogamous mating system of substrate-brooding cichlids. I know of no evidence for polyandry among pair-bonded cichlids, but polygyny has been documented in several groups (see Barlow, this volume, page 176).

9.6 BROOD-MIXING

Whether a brood of young cichlids is guarded by one or by both parents, an obvious change during the parental care period is a reduction in brood size. Predation is probably the most common cause of this decline, and predation pressure may be so intense that most breeding attempts fail (McKaye, 1977; Yanagisawa, 1987). However, there may be other explanations and there are also cases of increase in brood size. Some free-swimming fry may leave their parent(s) and join a nearby brood of the same or another species, or two or more entire broods may coalesce and be communally guarded by several adults, presumably their parents.

The possibility of accidental brood-mixing will occur to anyone watching families of substrate-brooding cichlids in the field, especially under crowded conditions. Attacks by predators and territorial fights between the parents of adjacent broods cause local turbulence, and displaced fry might readily approach or be retrieved by unrelated adults. Clear evidence of such mixing among the cichlids of Lake Jiloá, Nicaragua, was provided by McKaye and McKaye (1977). Experimental manipulations in the field (McKaye and McKaye, 1977; Perrone, 1978b; Carlisle, 1985) and in the laboratory (e.g. Noakes and Barlow, 1973a; Myrberg, 1975) have produced the generalization that fry-brooding adults will accept conspecific fry from other broods into their own broods if the fry sizes match. If the strangers are larger than the host's fry, they are repulsed or eaten; if they are much smaller they are usually eaten by either the host fry or their parents.

However, McKaye and McKaye (1977) saw active 'kidnapping' of strange fry by brood-guarding adults, showing that not all mixing is accidental. They also saw mixed-species broods being guarded by adults of one of the component species. Observations of mixed-species schools of fry being guarded by adults of one or both parent species were also made in Nigerian rivers by Burchard (1967) and Sjolander (1972).

In addition, field workers have reported large schools of fry of one species, containing many more individuals than a typical single brood, being guarded jointly by two or three pairs of adults. Communal parental care has been documented for *Etroplus suratensis* in brackish lagoons in Sri Lanka (Ward and Wyman, 1975, 1977), *Cichlasoma citrinellum* in Lake Jiloá, Nicaragua (McKaye and McKaye, 1977) and *Tilapia rendalli* in Lake Malawi (Ribbink *et al.*, 1981).

McKaye and McKaye (1977) concluded that the most likely explanation of both active incorporation and passive acceptance of strange fry into a brood is the reduced probability of capture of the host's young by a predator as brood size increases, i.e. the 'dilution effect'. However, this effect will work only if the predator attack rate does not increase in proportion to brood size increase. If it does increase, then the probability of an individual fish being captured will be the same, whatever its group size (Pitcher, 1986).

Another explanation is that since cichlid fry show strong schooling behaviour, individuals that become separated from their 'home' school may join any other nearby school. If this happens to many members of adjacent broods (especially conspecifics), the parents might be unable to distinguish their own young, and end up collaborating in guarding their combined brood.

So far the discussion of brood-mixing has been concerned only with substrate-brooding species. The upper limit on the number of fry that can be incorporated into a school of mixed parentage is probably set by the parents' defence capabilities or by the availability and size of shelters for the young at night. In contrast, maximum brood size in mouthbrooding species is limited by the size of the buccal cavity of the guarding parent, which in most species is the mother. One might expect mouthbrooding females to be reluctant to accommodate strangers, which would occupy space needed for their own offspring, yet brood-mixing, both intra- and interspecific, has been described for several mouthbrooding species (Ribbink *et al.*, 1980).

This phenomenon was first described by Ribbink (1977), who discovered three species of bottom-living cichlids in Lake Malawi carrying in their mouths, along with their own young, some fry of a more pelagic, open-water species, *Haplochromis chrysonotus*, which he called a 'cuckoo' by analogy with the avian brood parasite. Further work showed that at least 15 Lake Malawi species will carry heterospecific fry in their mouths. Some mixed broods contained young of one, two or three foreign species, and although most mixed broods had less than 5% foreigners, a few had as many as 50% (Ribbink *et al.*, 1980). These authors suggest that *H. chrysonotus*, and other species that remain in open water while carrying their broods, go to the bottom to release their fry. Their schooling tendencies may then lead the fry to join broods of bottom-living species which are still being guarded. When these mothers retrieve their young they may inadvertently take in some of the strangers.

An unusual example of the active transfer of fry from one brood to another has been described for *Perissodus microlepis* in Lake Tanganyika (Yanagisawa and Nshombo, 1983). This is a biparental mouthbrooder in which the eggs and wrigglers are brooded by the female only, but the male remains near by, and when the fry have become free swimming both parents guard them and occasionally shelter them in the mouth. If one parent deserts, or is experimentally removed, while the brood is at the fry stage, the remaining parent may take the fry into its mouth a few at a time and release them into a neighbouring conspecific brood. This has been called 'farming out' of the young, and its common occurrence in *P. microlepis* is suggested by the frequent presence of fry of two or more size groups within broods being guarded by one pair of adults (Yanagisawa, 1985). Yanagisawa suggests that this may be a single parent's best option for protecting its brood. The foster parents accept the strange young even though they are potential competitors for buccal cavity space in case of predator attacks.

'Helpers at the nest' are those members of a brood which do not leave the home territory as their parents begin the next breeding cycle, but stay and help to rear the next brood. This phenomenon is widespread among birds (Brown, 1987) and occurs in some mammals (Alcock, 1989), but has only been documented among fishes in a few Lake Tanganyikan substrate-brooding cichlids in the genera *Lamprologus* and *Julidochromis* (Taborsky and Limberger, 1981). By far the best known of these is *L. brichardi*, in which helping has been observed in the field (Brichard, 1978; Taborsky and Limberger, 1981) and intensively studied in the laboratory (Taborsky, 1984, 1985; Hert, 1985). Juveniles of from one to four size classes may be found in the breeding territory of their parents, not simply tolerated, but actively assisting in brood-rearing by chasing predators, cleaning and occasionally fanning eggs and wrigglers, and digging to keep the breeding shelter free of sand (Taborsky, 1984).

The ecological conditions that promote helping behaviour in place of direct reproduction in *L. brichardi* appear to be the high local population densities (of con- and heterospecifics) and resulting intense competition for shelters and suitable breeding sites on the rocky substrate. Taborsky's estimates of costs and benefits to helpers indicated that their growth rate was slower than that of non-helpers (because of effort expended in territory defence and brood rearing), but mortality was reduced (through access to the family shelter and protection by the breeding pair and other helpers) and the reproductive output of their parents was greater than that of adults without helpers (Taborsky, 1984).

9.7 EVOLUTION OF PARENTAL CARE

Parental care among cichlid fishes varies in two dimensions: the location of the young while being cared for, and the stability and duration of the parental pair bond during a single breeding attempt. In general, substrate brooders and delayed mouthbrooders have long-term monogamous pair bonds, whereas among the immediate mouthbrooders there is much variability in pair-bond duration and stability.

Substrate brooding is widely considered to be the ancestral pattern, while mouthbrooding is the more specialized and more recently evolved (Lowe-McConnell, 1959; Fryer and Iles, 1972; Barlow, 1974; Keenleyside, 1979; Peters and Berns, 1982). The main evidence supporting this generalization is the geographic distribution of the two patterns. Substrate brooding is much the most common form throughout the Neotropics, it is found among both lacustrine and riverine species in Africa, and it is the only parental pattern in the Asian species. Many substrate-brooding species have a wide range across northern and western Africa, indicating a long history on that continent. In contrast, almost all mouthbrooding species are found in Africa, and many are restricted to one of the African Great Lakes. Among the few exceptions are *Sarotherodon melanotheron*, a paternal mouthbrooder which is widely distributed in West African rivers and estuaries, and *S. galilaeus*, a biparental mouthbrooder with an even broader distribution across North-central Africa (Trewavas, 1983). On the American continent, only three *Geophagus* species are known to be immediate mouthbrooders; two of them are restricted to the extreme north-west of South America, and one is found near by in Panama (Gosse, 1975). Several other species are delayed mouthbrooders, and their distributions are not well known (Loiselle, 1985a).

A second line of evidence supporting the greater phylogenetic age of substrate brooding is the greater adhesiveness of their eggs and wrigglers. In most substrate brooders the eggs are highly adhesive and stick firmly to the spawning substrate. The eggs of immediate mouthbrooders are not adhesive, although in some species of *Sarotherodon* they have a vestigial adhesive layer (Iles and Holden, 1969; Trewavas, 1983). This difference continues after hatching: in substrate brooders, the wrigglers remain where they were placed, attached by mucous threads from three pairs of head glands; in the wrigglers of mouthbrooding species (immediate and delayed) these glands are vestigial or absent (Peters and Berns, 1982; Trewavas, 1983).

The most primitive parental care pattern in the Cichlidae, that of monogamous, biparental substrate brooding, has itself probably evolved from an earlier pattern of male-only care. The latter is recognized as the most primitive form of care in fishes (Blumer, 1979; Gittleman, 1981; Gross and Sargent, 1985). Williams (1975) thought the evolution of biparental from uniparental care a difficult problem because he assumed that one parent can rear a brood as effectively as two, and that selection should favour the male leaving the eggs with the female and searching for further breeding opportunities. Although uniparental care may be adequate for guarding a clutch of eggs and mass of wrigglers (especially if they are in a cave or pit), two parents may be needed to protect free-swimming fry. Evidence from the laboratory (Keenleyside, 1978) and the field (Keenleyside and Bietz, 1981; Nagoshi, 1987) shows that when one of a brooding pair is removed, the remaining parent cannot protect the brood effectively from predation. Other observers of biparental species in the field have been impressed by the need for both parents to protect the brood when predators are abundant (e.g. Barlow, 1974, 1984; McKaye, 1977). Male desertion followed by female-only care may be a facultative event that is common only when risks of predation on fry are low (Townshend and Wootton, 1985).

The question then arises: what selection pressures on early cichlids led to a parental system in which the active, mobile young stay close together near their spawning site, foraging on or near the substrate, and requiring the protection of both parents? In most other fishes with parental care, the male alone broods the eggs (usually at a prepared site on the substrate), and soon after hatching, the young disperse into the surroundings where they forage and seek protection more or less independently. Parental care ends with this dispersal. One advantage of such a system for freshwater species such as gasterosteids and many centrarchids is that both the male and the female (who played no part in parental care) can breed again in the same season. Barlow (1984) suggested that for cichlids the breeding season may be so short that the female lacks time to produce a second batch of eggs, and she can benefit more by helping the male to rear their first brood. However, it seems unlikely in tropical waters that female cichlids should be so limited by time and foraging opportunities that they could not produce more clutches and spawn more than once in a season if they were to leave their first clutch to the male. In captivity, females of biparental species can spawn again in less than 2 weeks if the first brood is lost early (Keenleyside, 1983, 1985).

Gross and Sargent (1985) argued that with the advent of male parental care in fishes, selection would favour fewer and bigger eggs because, since developmental time increases with egg size, larger eggs will spend more time in the "safe harbor" (Shine, 1978) of the egg stage and less time in the more vulnerable early juvenile, post-hatch stage. And with larger eggs, the hatchlings will also be larger and hence less vulnerable. Gross and Sargent also suggested that continued increase in egg size would raise their value to the female, eventually leading to selection pressures for care by the female, resulting in biparental care.

However, this does not explain why newly-mobile cichlids stay near the spawning site, requiring protection, usually by both parents, rather than dispersing and searching for food and shelter independently, as do the young of most species with male-only care. A possible explanation of this long-term dependence on parental care lies in the common observation of field workers that predatory attacks on guarded fry are often made by larger, sub-adult cichlids of the same or related species. When juveniles have become independent of their parents they often remain in or near the same shallow, shoreline areas, foraging on a variety of items including smaller, younger cichlid fry being guarded by parents (McKaye, 1977; Townshend and Wootton, 1985). It may be that habitat with enough food and shelter for many unguarded cichlid fry is simply not available in tropical waters that are rich in diurnally and nocturnally active piscivores. If the young cichlids are to survive the first few weeks of their free-swimming life they need both parents as guardians.

If the preceding argument is reasonable, one is then faced with the apparent contradiction of the evolution of uniparental care (usually maternal mouth-brooding) from the earlier biparental state. A possibly relevant feature of the biparental relationship is the apparent instability in the male–female bond.

This is to be expected on theoretical grounds, because males should try to maximize their lifetime reproductive output by fertilizing as many eggs as possible (Williams, 1975), and therefore even parental males should be ready to desert their mate and brood, and search for more breeding opportunities (Trivers, 1972).

However, such instability in the parental pair bond should not lead to male desertion unless the female can protect the brood as effectively as the two parents together. This seems to be where mouthbrooding entered the evolutionary picture. Two changes are necessary, however, if biparental substrate brooding is to evolve into uniparental mouthbrooding by the female: the brood must be small enough to be carried in the female's mouth, and the eggs must not be adhesive.

The Neotropical cichlids may provide the best material for exploring these evolutionary steps. Although biparental substrate brooding is most common, both biparental delayed and uniparental immediate mouthbrooding also occur (Keenleyside, 1979). Several authors have suggested that oral transport of newly hatched wrigglers to previously excavated pits set the stage for longer-term mouthbrooding. Parents carrying wrigglers might not release them if the predation risk is high (Myers, 1939; Lowe-McConnell, 1959; Collins, 1972; Baylis, 1981).

The obstacle in the progression to immediate mouthbrooding is the adhesive egg. *Aequidens paraguayensis* (a biparental delayed mouthbrooder) did not pick up the wrigglers early, even when brood predation pressure was high, but waited until the eggshells broke – a temperature-dependent event, not one that is influenced by parental reactions to predator density (Dupuis and Keenleyside, 1982). When egg adhesiveness is reduced, mouthbrooding can begin at spawning, and this form of parental care has been highly successful in the African Great Lakes. It is especially interesting that immediate mouthbrooding by the female occurs in three stream-living Neotropical species (*Geophagus steindachneri*, *G pellegrini* and *G. crassilabrus*). Little is known of the ecological conditions in which these species breed, and intensive field studies might produce valuable material for understanding the evolution of this pattern, which is so common in the African Great Lakes but so rare elsewhere among cichlids.

The advent of immediate maternal mouthbrooding provides the opportunity for male desertion when the female has her clutch fertilized and safely inside her mouth. This can be followed by male competition for further matings, which may lead to the development of lek-type assemblages of male spawning territories. In an extreme example, *Cyrtocara eucinostomus*, more than 50 000 males congregate on a 4 km-long breeding arena in Lake Malawi (McKaye, 1983). However, even in such a system, males may have a parental role beyond supplying sperm for fertilizing visiting females' eggs. At a breeding arena in Lake Malawi used by five mouthbrooding species, six other cichlid species attempted to eat the newly spawned eggs. Three of these were 'nest darters', which dashed into a nest and ate eggs before the female could turn and pick them up. The other three were 'mouth rammers', which struck mouthbrooding females in the head, causing them to lose some young, which the rammers promptly ate. The territorial breeding males frequently attacked both types of brood predator, thus playing an antipredator role in defence of their young (McKaye and Kocher, 1983; McKaye, 1984).

In conclusion, it will be apparent that in this review I have concentrated on field studies. This was not accidental. Carefully planned laboratory work will always be needed to solve most proximate problems. However, I believe that the combination of new facts based on direct observation of breeding fishes in the wild, and new insights into possible ultimate explanations that come from immersing oneself in the animal's world, is the surest way to advance our knowledge of fish behaviour in general and parental care in cichlid fishes in particular.

ACKNOWLEDGEMENTS

I am grateful to George W. Barlow and Kenneth R. McKaye for helpful comments on this chapter.

Chapter ten

Ontogeny of behaviour in cichlids

D. L. G. Noakes

10.1 INTRODUCTION

Studies of the complex parent–young interactions of cichlids have usually concentrated on the adults (Chapter 9). I believe that considerations of behavioural ontogeny are important, not just to redress this imbalance, but also to understand the behaviour of adults. The early life history and behavioural ontogeny of cichlids are best summarized from the perspective of adult reproductive behaviour. Cichlids are typically categorized on the basis of their reproductive habits as either substrate brooders (which display custodial care and guarding of fertilized eggs and developing young), or mouthbrooders (which carry the fertilized eggs and developing young within the buccal cavity) (e.g. Peters, 1937; Baerends and Baerends-van Roon, 1950; Lowe-McConnell, 1959; Wickler, 1966; Fryer and Iles, 1972; Barlow, 1974; Keenleyside, 1979; Peters and Berns, 1982; Huntingford, 1986; Kuwamura, 1986a). Although there is increasing evidence that this dichotomy is an over-simplification (e.g. Kuwamura, 1986a), the emphasis remains on reproduction as a key element in the biology and evolution of cichlids.

In a review of tilapiine cichlids, Noakes and Balon (1982) suggested that these show two alternative patterns of life history and ontogeny (see also Noakes and Godin, 1988). Guarder species (substrate brooders) have been described as **altricial**, and bearer species (mouthbrooders) as **precocial**, in their early development (Noakes and Balon, 1982). Important differences in fecundity, egg size and quality, and early development are summarized in Table 10.1. I will elaborate how these differences in reproductive style relate to both early life history and development of behaviour.

10.2 DEVELOPMENTAL STAGES AND THE TIMING OF HATCHING

I treat development as occurring in functional stages, and place emphasis on states and events that occur during relatively long intervals (Balon, 1975b),

Table 10.1 Summary of developmental and life-history features of guarder and bearer cichlids*

Characteristic	*Condition in*	
	Guarders	*Bearers*
Fecundity (number of eggs)	up to 7000	up to 1000
Egg diameter (mm)	1–1.5	up to 5.0
Yolk: % lipid	<25	up to 45
colour	clear, yellow	opaque, orange
Total length at first exogenous feeding (mm)	5–6	9–10
Age at first reproduction (y)	0.5–2	1–4
Juvenile mortality	relatively high	relatively low

*Table adapted with permission from Noakes and Balon (1982).

rather than at relatively brief transitions (Noakes, 1981; Noakes and Godin, 1988). I employ a general scheme of terminology of developmental intervals modified from Balon (1975) for reasons elaborated elsewhere (Noakes, 1978a, 1981), with the terms **period**, **phase** and **stage** used in the same hierarchical fashion, as explained in the next paragraph. The embryonic period begins with fertilization (or activation) of the egg, and ends with the transition from endogenous to exogenous feeding. The larval period follows, beginning with the transition to exogenous feeding and ending when the median fin-supports and vertebral centra are differentiated and calcified. The juvenile period begins at the end of the larval period and lasts until the first maturation of gametes. Gonadal maturation represents the transition to the adult period, which lasts until death (Fig. 10.1).

Each period is composed of one or more phases, and each phase consists of one or more stages. This scheme retains the generally accepted concept of the stage as the smallest, fundamental unit of ontogeny, while incorporating the two other commonly used terms (phase and period) into the overall hierarchical system. Every species must have at least the embryonic and adult periods, by definition, but either or both of the larval and juvenile periods can be eliminated in particular species (Noakes and Godin, 1988).

For example, the young of bearer cichlid species do not have a larval period. They undergo direct development, from fertilized egg to fully formed juvenile, while using their yolk as the sole (or at least primary) source of nutrition (endogenous nutrition). Meanwhile they are carried within the buccal cavity of one or both parents (Balon, 1986). By contrast, guarder cichlid species have all four periods in their life history. They undergo an indirect development, including a larval period (Youson, 1988; Flegler-Balon, 1989), after they exhaust their yolk material and become dependent upon exogenous nutrition

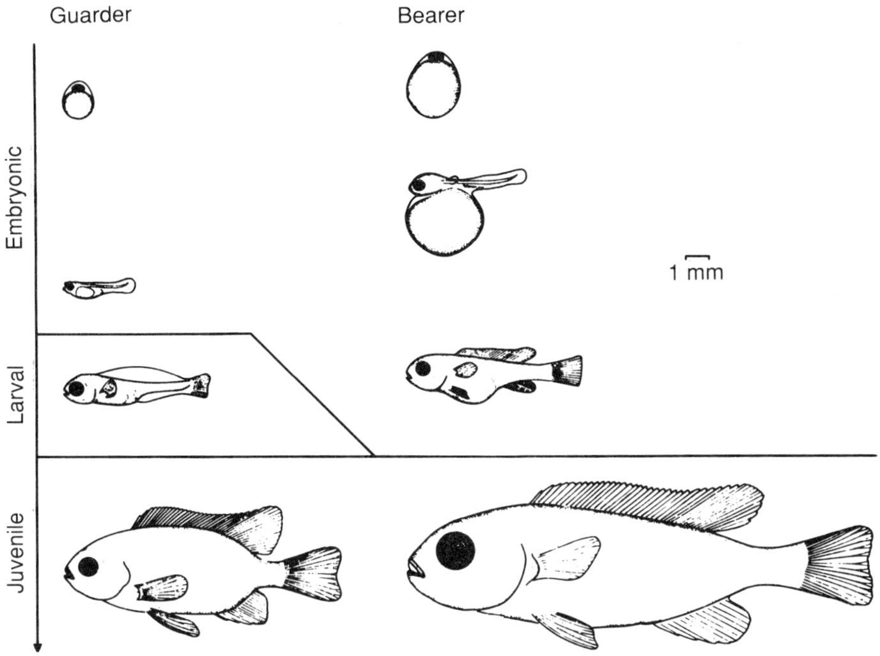

Fig. 10.1 Development in idealized guarder (left) and bearer (right) cichlid species. Vertical arrow indicates the sequence of ontogenetic development from fertilization towards adult. Boundaries between successive periods of development are indicated by horizontal lines; the bearer species does not have a larval period in its ontogeny (reproduced with permission from Noakes and Balon, 1982). Scale bar, 1 mm.

(Figs 10.2, 10.3). During their embryonic, larval and early juvenile periods they are tended and guarded by one or both parents.

The embryos of guarders have a transitional respiratory network of blood vessels on the anal fin fold, while the embryos of bearers have a comparable respiratory network on the surface of the yolk sac (Fishelson, 1966; Hanon, 1975). One species, *Chromidotilapia guntheri*, described as intermediate between guarder and bearer species, has both types of transitional respiratory network (Hanon, 1975). Similarly, Balon (1977) reported that the primary transitional respiratory blood-vessel network in the bearer *Labeotropheus* sp. was on the yolk (although some network was developed on the anal fin fold).

The early behaviour of young of these two kinds of cichlids differs markedly, in relation to the aforementioned differences in the larval and juvenile stages in early development (Figs 10.2, 10.3), as I shall now outline.

Hatching

Hatching is a complex phenomenon, involving at least biochemical, physiological and behavioural aspects (Yamagami, 1988). It is to some extent a physical

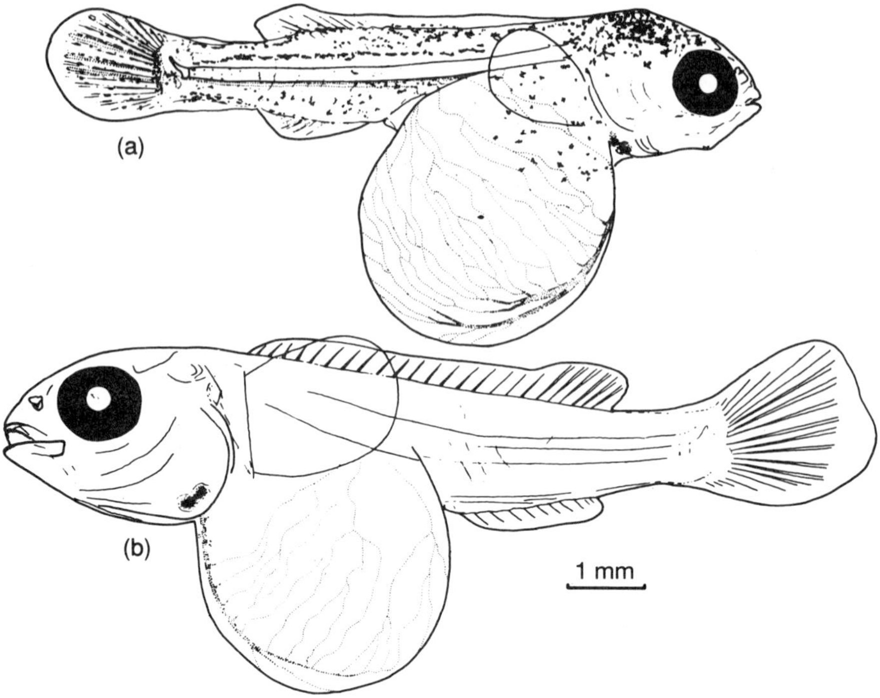

Fig. 10.2 Developing young of a representative bearer, *Labeotropheus* sp. (a) Embryo removed from the female's buccal cavity 8 d after fertilization (mean total time spent in female's buccal cavity is *c.* 28 d). The embryo has hatched from the egg envelope, but still has a large yolk sac. (b) Embryo removed from the mother's buccal cavity 13 d after fertilization. Note complete differentiation of fins and fin rays, and remaining yolk material (reproduced with permission from Balon, 1977). Scale bar, 1 mm.

process, with muscular movements of the embryo rupturing the egg envelope. However, in most if not all species, there is also a chemical breakdown or weakening of the envelope, caused by the proteolytic action of hatching enzymes secreted by specialized glands on the embryo's surface. Surgical release of embryos by cutting and removing the egg envelopes is not necessarily the same as normal hatching. Unless the young fish has reached the appropriate developmental state and competence, its behaviour following surgical release is not typical of that following hatching (Cunningham and Balon, 1986).

Changes in metabolic activity of the embryo during development have also been implicated in the hatching process (Yamagami, 1988). This seems particularly true in those species where there are no specific changes in extrinsic features (e.g. oxygen concentration) that might trigger hatching. Several studies have shown that metabolic activity changes just before hatching. This change might cause increases in spontaneous movement and

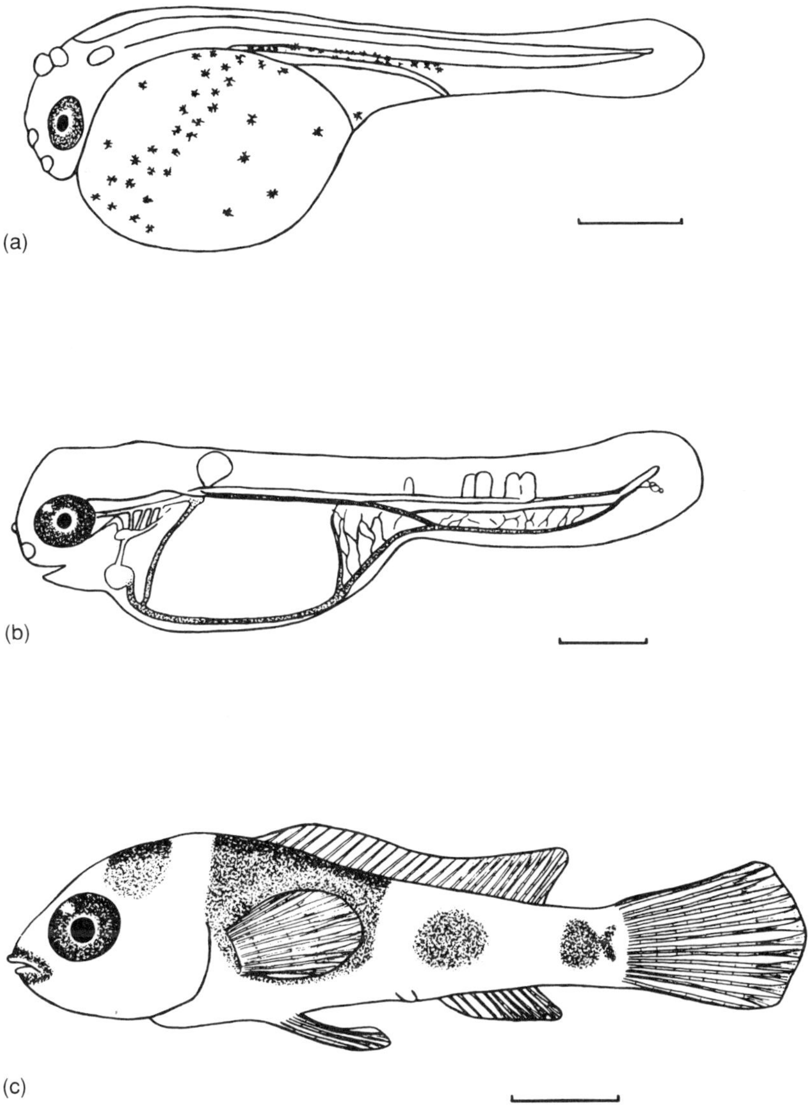

Fig. 10.3 Early development in a representative guarder cichlid species. (a) Embryo at time of hatching from the egg envelope. (b) Young fish, just before transition to exogenous feeding. (c) Fully differentiated juvenile, with median fins and fin supports differentiated, and juvenile morphology and coloration. Scale bars, (a) 0.5 mm, (b) 0.5 mm, (c) 2.0 mm (modified with permission from Jones, 1972).

hence lead to hatching. Alternatively, the increased metabolic demand could decrease the perivitelline oxygen levels, leading to increased embryonic movements and to hatching (Yamagami, 1988).

The precise timing of hatching can be determined by some exogenous cue, so that it takes place at a particular time of day or season. For example, physical agitation can accelerate hatching in various species. Parental guarder cichlids accelerate and synchronize the hatching of their embryos by mouthing and nibbling at them. Thus, orange chromide, *Etroplus maculatus*, embryos kept without parents hatch later, and over a more extended interval, than those tended by parents (Zoran and Ward, 1983).

Hatching can occur at different stages of development within a species, depending on a variety of intrinsic and extrinsic factors, so it cannot be taken as a fixed reference point for either definition or description of 'normal' stages of development (Jones, 1972; Balon, 1985; Yamagami, 1988). Because of this, and even more so because young of different species hatch at markedly different states of development, hatching is not particularly useful for comparisons across species (Noakes, 1978a).

It is tempting to compare the behaviour of young cichlids using their chronological age from initiation of free swimming (guarders) or release from the parent's buccal cavity (bearers). These points are indeed comparable in that they are the times when the young must begin to deal with the external world on their own; in particular, they demarcate the onset of exogenous feeding. Such a simple comparison can be misleading, however, since at the time of free swimming a young guarder is a larva, while at the time of release a young bearer is a juvenile (Fig. 10.1). When using this point of comparison it is hardly surprising, but not very informative, to find that behaviour of these two kinds of young differs. However, the embryonic and larval development of young bearers occurs while they are out of sight, within the parent's buccal cavity. When first released they are juveniles, fully formed in terms of coloration, morphology and behaviour (Noakes and Godin, 1988). Young guarders begin free swimming as incompletely differentiated larvae, with almost every morphological, anatomical and behavioural feature either nonexistent or only incompletely formed (Noakes and Godin, 1988). They do not attain a comparable state until after weeks of parental care. As I will discuss, this fundamental difference is basic to understanding behavioural ontogeny in cichlids.

10.3 FEEDING BEHAVIOUR

The transition from endogenous nutrition (whether from yolk or via specialized 'placental' tissues from the parent) to exogenous feeding is one of the most significant events in the early life history of fishes (Wourms, 1981), and defines the end of the embryonic period (Balon, 1986). The critical importance of first exogenous feeding has repeatedly been stressed (e.g. Blaxter, 1974; Hunter, 1981). A period of mixed feeding (some exogenous food taken before yolk reserves are completely depleted) may occur in some species (e.g. Crane, 1981; Balon, 1985). Developing young may feed on sibling eggs or embryos in some fish species (e.g. Fujita, 1981) (and could be predicted to do so in other bearer species). This behaviour still constitutes endogenous nutrition, as do examples of young fish feeding from the body surfaces of their parents, since the nutrients are provided entirely by the parents (Noakes, 1979).

The teleost alimentary canal develops, by a series of relatively rapid changes coincident with the transitions between embryonic and larval, and then between larval and juvenile, periods (Govani *et al.*, 1986), from a simple tube to a more complex, coiled, elaborate structure. This pattern is in contrast to the more gradual and continuous development of other organ systems (O'Connell, 1981), and corresponds to changes in diet and feeding mode of each of these periods.

When they begin exogenous feeding, precocial young are large, with well-developed sensory systems, locomotor ability, alimentary tract, jaws and feeding structures. In contrast, when altricial young begin exogenous feeding they have less-well-developed alimentary canals, typically consisting of a simple, straight tube (Govani *et al.*, 1986), and they are smaller and less differentiated and developed (Fig. 10.1) (Noakes and Balon, 1982). I shall return to this point later (page 220), since there have been suggestions that young of some bearer species may start feeding on exogenous food in the parent's buccal cavity prior to their release.

These developmental changes in the digestive system and feeding behaviour also include production of digestive enzymes and the digestive process. The rate of food passage through the gut varies directly with food intake, whereas the assimilation efficiency varies inversely with passage time (Govani *et al.*, 1986). Altricial young pass food more quickly through their short, straight guts than do precocial young through their more elaborate, coiled guts. We might expect these developmental features of the digestive system to be reflected in differences in the feeding strategies of young fish. Altricial young would appear to be better suited to feed on contagiously distributed prey in patches of high concentration, whereas precocial young would be better suited to feed on more widely scattered, individual prey (Govani *et al.*, 1986).

Barel (1983) proposed that in terms of feeding behaviour cichlids can be divided into two functional morphotypes, biters and suckers. Biters have shorter, stouter jaw elements and feed mostly by biting or crushing; suckers have longer jaw elements, more protrusible mouths, and feed mostly on fish and invertebrates. Others (e.g. McKaye and KcKaye, 1977; Ribbink *et al.*, 1980; Gottfried, 1986) have shown that these are by no means rigid functional categories, in that individual fish may shift diet opportunistically and switch according to food availability.

Furthermore, there may be considerable overlap, or even switches, between these proposed categories during life. For example, Gottfried (1986) showed that the Midas cichlid, *Cichlasoma citrinellum*, changes from a 'sucker' type of head morphology and feeding behaviour as a juvenile to a 'biter' morphology and feeding behaviour as an adult. Cichlids (and other fish species) commonly pass through such an ontogenetic diet change. Food habits of larvae, juveniles and adults can differ significantly, and so we would predict concurrent changes in trophic morphology. Feeding preferences and specializations can develop (e.g. Ringler, 1933), either as a result of experience (learning), or through maturational changes such as increasing visual acuity (Noakes and Godin, 1988).

Meyer's (1986, 1987, 1988) studies of the development of feeding behaviour in *Cichlasoma managuense*, a guarder species, may be of considerable taxonomic and ethological significance. Young attacked novel prey more readily, and became more efficient in handling prey, with increasing experience. Increasing size (including increasing eye size) probably increased capture success by increasing visual acuity, an effect known in other species (Noakes and Godin, 1988). Meyer also found that size and experience may affect feeding behaviour differently. However, the most intriguing results concerned the effects of diet on morphology. Sibling groups fed on different food types (live *Artemia salina* and a dried-flake food) for a period of 8.5 months differed significantly in shape (particularly in head measurements). These differences were reminiscent of the differences ascribed to 'biter' and 'sucker' morphotypes by Barel (1983). Changing the diet of growing fish tended to produce corresponding changes in head morphology, strongly implying a cause–effect relationship. Meyer (1987) proposed a heterochronic model to account for these effects. The implications of this study are considerable, given the emphasis on trophic morphology in cichlid speciation. The findings and the heterochronic model are also important in relation to studies of ontogenetic processes determination both morphological and ethological characters.

Growth depensation (individual differences in size and growth rates within a group of fish) occurs in many fishes (e.g. Jobling and Reinsnes, 1987). There are suggestions, and some experimental evidence, that this effect can be caused by behavioural dominance among members of the group. Such dominance is commonly observed, and can have physiological consequences (e.g. Noakes and Leatherland, 1977). However, there are also genetic influences on individual differences in growth rate in some fish species. These have been most thoroughly studied in the Salmonidae (Thorpe, 1983, 1987), but the phenomenon has also received attention in cichlids (Francis, 1988; Lagomarsino *et al.*, 1988).

Although growth is an important aspect of intensive culture of cichlids, there have been surprisingly few detailed studies of feeding behaviour and behavioural effects on individual growth. Barlow and Ballin (1976), Barlow and Wallach (1976), and Barlow (1983) reported some factors affecting social dominance in the Midas cichlid, *Cichlasoma citrinellum*. Size is of primary importance, as in most fish species. Larger fish can supplant or intimidate smaller ones, gain access to food, and receive few if any threats or attacks. Body coloration is also important in agonistic interactions in this polymorphic species. More recently, Valerio and Barlow (1986) reported further details of the development of feeding and agonism in the Midas cichlid. Largely as a result of the behavioural mechanisms just discussed, individual growth rates vary considerably in young juveniles. The fastest-growing individuals are soon large enough to cannibalize their smallest conspecifics, and so size differences become yet more accentuated until all the small fish have been cannibalized, leaving a relatively uniform cohort.

10.4 SOCIAL BEHAVIOUR AND THE ROLE OF IMPRINTING

Any social responses of young fish will require that the appropriate sensory and motor systems be sufficiently developed to allow the fish to detect, localize and respond appropriately to the right stimuli (Noakes and Godin, 1988). Developmental age is much more relevant in this respect than chronological age. For example, the approach and withdrawal social responses of young tilapia, *Tilapia mossambica* (Goude *et al.*, 1972) and the timing of dispersal of young jewel cichlids, *Hemichromis bimaculatus*, (Chen *et al.*, 1983) appear to be regulated primarily by developmental processes, rather than by chronological age.

In general, young guarders develop the ability to recognize, and distinguish among, parents, other conspecifics and heterospecifics, using visual and chemical (and possibly other) senses (Noakes and Barlow, 1973a; Barnett, 1977a, b; Coss, 1979). Social interactions with siblings progress from simple physical encounters and direct avoidance, to more elaborate lateral posturing, more complex stimulus–response situations, and eventually to recognizable dominance and sexual interactions (Ohm, 1958; Wyman and Ward, 1973; Valerio and Barlow, 1986). Young bearers, upon release by the parent, behave as independent juveniles, as we would predict. Wyman and Ward (1973) reported that agonistic and other social action patterns tended to increase in both frequency and complexity with age in the orange chromide, *Etroplus maculatus*. Most interestingly, they proposed that a relatively large, complex adult repertoire of social behaviour developed from two initial action patterns in the young. Unfortunately, such detailed ontogenetic information is not available for many species (but see Ohm, 1958); it would be particularly valuable for a group of several closely related species, where it could be used for phylogenetic interpretation as well.

As well as depending upon the state of physiological and anatomical development, early social behaviour can itself affect anatomy and physiology. For example, early social experience can affect features as diverse as interneurones in the optic tectum of jewel cichlids (Coss and Globus, 1978, 1979; Burgess and Coss, 1982; Burgess *et al.*, 1982), growth of the eye and retina in *Haplochromis burtoni* (Fraley and Fernald, 1982), and survival and normal physical development in discus fish, *Symphysodon discus*, and orange chromides (Ward and Barlow, 1967; Minnier and Cole, 1973; Noakes, 1979). Social interactions among juveniles can affect growth rate (page 216). The behavioural consequences of early social experience have attracted perhaps more interest, and are considered next.

Imprinting

It has long been postulated that early experience might have persistent, or delayed, effects on social behaviour. The most commonly proposed mechanism, imprinting (Barlow, 1983; Colgan, 1983), might influence species recognition in general, or mate choice more specifically. The opportunity for imprinting exists particularly in species with prolonged social contact and interaction between parents and their offspring, i.e. guarders. However, the evidence for social imprinting (filial or sexual) is still inconclusive for any fish species, as it is in part contradictory, and equivocal at best (e.g. Seitz, 1940, 1942; Kop and Heuts, 1973; Sjölander and Fernö, 1973; Fernö and Sjölander, 1976; Weber and Weber, 1976; Siepen and Crapon de Caprona, 1986), given methodological and procedural concerns (Barlow, 1983).

In a sense, whether or not imprinting is involved in the effects of early experience on subsequent social behaviour is a question that concerns only the mechanism. The initial question is whether such effects exist. Several studies have suggested that they do in both guarder and bearer species, but others have shown little effect if any. Russock (1986), and Russock and Schein (1977, 1978), investigated the effects of early social experience on filial and subsequent adult social preferences in *Sarotherodon* (= *Oreochromis* = *Tilapia*) *mossambicus*, a bearer species. Young isolated from their parent (the mother in this species) showed peak responsiveness to dummies (maternal mimics) at the ages when young normally reared with the mother show maximal responsiveness to the mother. This responsiveness also declined in a manner similar to that in normal young, until the age at which they usually leave the care of the mother. Russock and Schein concluded that although young had an initial predisposition to respond to the general characteristics of a brooding mother, they would respond to a range of objects. Whatever the object of their early experience, their responsiveness declined with developmental age. This developmental sequence was interpreted in terms of the initial need for young to be attracted by and to remain close to their mother, and then to leave the mother and eventually avoid any large (potentially dangerous) object, such as predatory fish. They (Russock and Schein, 1978) reported that adult social preferences in *S. mossambicus* were not affected by early rearing experience. Sexually mature females and males both preferred to associate with conspecifics, whether they had been reared with mother and siblings, with siblings only, or in social isolation since hatching. However, Russock and Schein did not measure sexual responses, but rather physical proximity, of adults.

However, Kop and Heuts (1973) reported effects of early social experience on some measures of later social preferences in the jewel cichlid, *Hemichromis bimaculatus*, a guarder species. Some of the effects they reported may have resulted from recency, rather than imprinting. More importantly, their sample sizes were absolutely minimal, and the effects seemed to be changed by later experience, so the conclusions must remain tentative.

There is a lot of interest in the effects of early experience on later social preferences, for two different reasons. First, these effects reveal how these fish come to show their usual social preference as a consequence of early experience. Second, early experience also be significant in cases where young are reared by foster parents of their own or other species.

The possibility of fostering is not as far-fetched as might be imagined. There are several examples reported in which such cross-fostering can come about. It has been reported that parents have been found guarding young of other parents or even other species. The mechanism(s) are still a matter of considerable debate (e.g. Lewis, 1980; Kocher and McKaye, 1983; McKaye, 1984; Yanagisawa, 1985; see page 202). The mere occurrence of this phenomenon makes the question of early experience and development of social preferences in cichlids even more crucial.

The prolonged close association of young with each other and parent(s) in both guarder and bearer species provides more-than-ample opportunity for imprinting or other early influences to occur (Hay, 1978). Myrberg (1966, 1975), Noakes and Barlow (1973a, b), McKaye and Barlow (1976b), Barnett (1977b) and others have shown that both the young and parents come to recognize each other, at least to species and perhaps even as individuals, as a result of their close association during parental care. Individual mate recognition was first reported by Noble and Curtis (1939) after detailed studies of social behaviour in the jewel cichlid, a guarder species. At least some of the early chemical and visual cues to which young *Haplochromis* (= *Astatotilapia*) *burtoni*, a bearer species, are exposed can have persistent influences on social responses (Crapon de Caprona, 1982). Some of these effects appear to result from learning, and may be reversed by later experience. However, at least some other effects seem to persist, in a manner that would suggest that imprinting has occurred. For the young, it is during this time that sensory systems and central nervous system are developing, so early experience might have significant and lasting effects (Noakes and Godin, 1988).

10.5 PARENT–YOUNG INTERACTIONS

Guarder species typically form monogamous pairs, and both parents participate in the extended custodial care of their young (Baylis, 1981; Keenleyside, this volume). Pairs in bearer species often are together only for the duration of the actual spawning act, and only one parent, more frequently the female, carries and cares for the developing young until their release. While both kinds of young receive care and protection from their parents, and both face risks from a variety of threats (e.g. Wilhelm, 1980; McKaye, 1981b; Yanagisawa, 1985), parental bearers literally place themselves between the threat and their young. However, parental guarders may also face such potential threats directly, and their investment of time to care for young is also considerable, so any differences in this regard may not be as great as might first appear.

After they have been released by the parent, young bearers typically behave as independent juveniles. The young of some female-bearer African *Oreochromis* species, and of some South American *Geophagus* species (and possibly some other bearer species), are typically taken back into the buccal cavities repeatedly during several days after their first release (e.g. Brestowsky, 1968; Keenleyside, 1979; Yanagisawa, 1985; Kuwamura, 1986a). These young are probably at an earlier stage of development when first released than are the young of other species, which do not return to the parents, but comparative details are lacking. In these latter young that do not return to the parents, all major sensory and motor systems are formed and functional (Noakes and Godin, 1988), and the fish are fully developed and competent in terms of morphology, coloration, locomotion and feeding performance when they are released (e.g. Balon, 1977).

This relatively advanced state of development allows the behaviour of young of these species to be studied. Kuenzer (1975) studied the responses of young *Hemihaplochromis multicolor*, a bearer species, to visual stimuli from artificial dummies to determine which parental features might evoke the young's responses, i.e. approach, make contact and escape. For a few days after their initial release, young tend to remain near the female and return to her buccal cavity when threatened. Increasingly with age, they show avoidance or active escape responses to large (i.e. of adult size or larger) fish. Kuenzer found that patterned dummies were no more effective than uniformly coloured ones, but the size of the dark areas of the pattern and contrast between light and dark areas were important.

Young guarders, by contrast, undergo a prolonged development from a very undifferentiated condition at the time of hatching (Ward and Barlow, 1967; Noakes and Barlow, 1973a; Wyman and Ward, 1973; Ward and Wyman, 1977). The performance of almost every behavioural action is initially either nonexistent or at best rudimentary; sensory and motor systems are then still differentiating and often non-functional (Noakes and Godin, 1988). It is not until these young have grown and become differentiated as juveniles that they leave the custodial care of the parents and behave like the newly-released bearer young.

A series of investigators (e.g. Noble and Curtis, 1939; Baerends and

Baerends-van Roon, 1950; Myrberg, 1964, 1966) have found that parental cichlids actively discriminate against foster young placed with their own young. Noakes and Barlow (1973b) found that the age of the foster young relative to that of the parents' own young seemed to be a critical feature. Midas cichlids, *Cichlasoma citrinellum*, would consistently reject foster young (by eating them) if the foster young were older than their own offspring, but would accept foster (conspecific) offspring that were the same age as, or younger than, their own. Bernstein (1980) reported that parental *C. nigrofasciatum* passed through different parental states, corresponding to the development of their young. This tended to ensure that parents would accept conspecific young close to the developmental stage of their own young.

The behaviour of young cichlids need not be fixed and inflexible during such interactions. Parental discrimination can be altered by the behaviour of the foster young (Shaw and Innes, 1980): in the blue acara, *Aequidens pulcher*, a guarder species, free-swimming offspring 7–17 d after spawning would come to behave like those only 3–5 d after spawning when transferred to foster parents whose own offspring were of the latter age. This suggests that the parental recognition of foster young can depend on behaviour of the young, as well as chemical or 'simple' visual cues (size, coloration, etc.). It also suggests that the young may be capable of learning at least simple responses at this young age, e.g. 'remain motionless near the substrate when approached or pursued by a large fish'.

The young of several cichlids, both guarders and bearers, make deliberate, repeated contacts with their parents' bodies. This behaviour has been a useful measure for investigating the early social responses of the young. Noakes and Barlow (1973a) reported on the development of parent-contacting behaviour in young Midas cichlids, *Cichlasoma citrinellum*. This behaviour is similar, and probably homologous, to responses shown by young of several other cichlid species, both guarders and bearers, towards their parents (Noakes, 1979). More than one action pattern may be involved in some species. Ward and Barlow (1967), Quertermus and Ward (1969), and Wyman and Ward (1973) described two different patterns, micronipping and glancing, in the orange chromide, *Etroplus maculatus*. Micronipping apparently is an ingestive behaviour whereby young remove surface mucus (and perhaps epidermal tissue) from the parents. This action pattern seems to occur widely throughout the Cichlidae (Noakes, 1973, 1979; Ward, 1973). Glancing is a different action pattern, in which the young chafe against the parent's body with a short, scraping movement. Glancing is an important component of developing social behaviour in chromides, and eventually is expressed in more complex juvenile and adult action patterns.

Kuenzer and Peters (1974) reported on the features of parents and young responsible for female brood care in *Nannacara anomala*, a guarder species. As has been reported for some other guarders, adults can show parental brooding behaviour towards unrelated conspecific young or even subjects such as tubificid worms or daphnia: even females that had no prior breeding experience might do so. In general, specific cues from the young (presumably visual, tactile, chemical, etc.) appear to be important not only in the initiation and direction of parental responses, but also in determining the particular responses directed to the young.

Similar findings have been reported with the guarder species *C. nigrofasciatum* (Myrberg, 1964, 1966, 1975). Adult males can be induced to show parental responses to conspecific foster young, but not to heterospecific (*Hemichromis bimaculatus*) young, through repeated presentations of young (Krischik and Weber, 1975). These two species, while both guarders, are native to different continents, *C. nigrofasciatum* to Central America and *H. bimaculatus* to Africa. Consequently, they may be somewhat distantly related (see Chapter 1) and there might be some behavioural or chemical differences between them that make this parental discrimination easier.

10.6 COMPARATIVE STUDIES: THE WAY FORWARD

Two comparative studies of early development and behaviour of guarder and bearer species serve particularly well, both to recapitulate my generalizations about patterns in the ontogeny of behaviour or cichlids.

Fishelson (1966) compared early morphological and behavioural development in three tilapiine species, one guarder species (*Tilapia tholloni*) and two bearer species (*T. nilotica* and *T. macrocephala*). One significant difference among these species was in the respiratory network of blood vessels that develops in embryos. In guarder species this network develops on the anal fin fold, while it develops on the surface of the yolk in embryos of bearer species (page 211). In conjunction with this, Fishelson suggested that early movements of pectoral fins, and whole body undulations ('wriggling'), produced water currents across these respiratory networks (see also Liem, 1981b). Both types of respiratory networks disappear when the branchial circulation is established *c.* 7 d after fertilization species in guarder and *c.* 15 d after fertilization in bearer species.

Other sensory and motor organs and organ systems (e.g. neuromasts, eyes, fins, muscles and axial skeleton) develop in young of these species in generally parallel, synchronized fashion. These systems typically become functional at an earlier age in the guarder species (e.g. 6–7 d after fertilization), corresponding to the earlier free swimming and independent exogenous feeding in these species (Fishelson, 1966). However, the greater amount of yolk, and hence longer period of endogenous nutrition, of bearer species allows them to continue development of these organs and systems at a high rate, so within a few days (e.g. 9–11 d after fertilization) the state of development is more advanced in bearer than guarder species That is to say, the developmental rates are initially higher in guarder species but soon decline, while in bearer species they are initially somewhat lower but remain more or less consistent. The faster initial development of the guarder species allows the young to become functional (e.g. in locomotion and feeding) sooner and to reduce their vulnerability as exposed embryos. The young of bearer species develop within the safety of the parent's buccal cavity, and with a larger supply of yolk can develop more slowly, for a longer time, before having to function independently (Fishelson, 1966).

Brestowsky (1968) also carried out a comparative study of the responses of young to parents in three species of tilapiine cichlids, *Tilapia tholloni* (a guarder species), *T. nilotica* (a bearer species in which the young are carried in the female's buccal cavity), and *T. heudeloti macrocephala* (a bearer species in which young are carried in the male's buccal cavity). His results are particularly instructive.

The normal behavioural development in these species is representative of guarder and bearer (female and male) species. Young *T. tholloni* begin free swimming *c.* 6 d after spawning (hatching occurs *c.* 3 d after spawning). They remain together in a cohesive group, interacting with each other and the parents, for 3–4 weeks after spawning. Young of the female-bearer species are first released from the mother's mouth *c.* 15 d after spawning (hatching occurs *c.* 5 d after spawning). The young stay close to the mother, and are taken back into her buccal cavity, until they are *c.* 21 d of age (since spawning). Finally, young of the male-bearer species hatch *c.* 5 d after spawning, but remain inside the buccal cavity until they are *c.* 23 d of age (since spawning). Once released, they disperse from the male, receive no further protection from him, and do not return to his buccal cavity.

Brestowsky (1968) used an artificial parental dummy, with a cavity to simulate the buccal cavity, and measured the responses of young in terms of approaches, duration of contacts, and time spent within the cavity. He found that there was an initial critical period, beginning when the young became free swimming and lasting for about the next 12 d. During this period only, the young formed a social attachment to the parental dummy; afterwards young that had formed such an attachment remained socially attached to the parental object. Young that had not formed the initial social attachment would not do so during this second (or any other) period.

Details of his experiments in relation to amount of yolk and developmental rates are particularly informative. As is typical for cichlids (Fig. 10.1, Table 10.1), the young of the guarder species (*T. tholloni*) have much smaller amounts of yolk, and become free swimming in a much earlier state of development, than do young of bearer species (especially *T. h. macrocephala*). Perhaps the most interesting finding was the development of behaviour in young *T. h. macrocephala*, the male-bearer species. The yolk sac is largest in this species. Consequently the young spend a much longer time within the father's buccal cavity, and because of the large yolk sac are not capable of free swimming until later, in development when most of the yolk has been absorbed. Accordingly, young *T. h. macrocephala* pass through the initial critical period for forming social attachment while they are still inside the father's buccal cavity, incapable of free swimming. However, when Brestowsky (1968) punctured the yolk sacs of *T. h. macrocephala* embryos, he allowed the young to become free swimming much earlier than normal. When tested, these altered young showed the same initial social attachment and sequence of behavioural development as the other species.

As would be expected, young of the guarder species *T. tholloni* are much less developed at the time of free swimming (Brestowsky, 1968). For the first 3 d of the critical period they are not capable of strong, coordinated locomotion, especially if held and tested singly or in small groups. Not surprisingly, they did not perform as well in tests during the first 3 d of the critical period. The onset of the critical period is clearly determined by the functional competence of sensory and motor systems, but what terminates this period is not so clear.

10.7 PROSPECTS

Several important conclusions have emerged from this review. Despite the current importance of developmental studies in biology, there are surprisingly few programmes using cichlids. Past studies have often been opportunistic or even anecdotal, with a descriptive rather than experimental approach. The ready availability of a great diversity of species, and the ease of raising and observing their young, make cichlids very attractive subjects. Significant advances in our understanding of cichlid taxonomy and systematics can be combined with hypotheses relating ontogeny and life history. Comparative studies of behavioural development in cichlids could be important, not only for understanding developmental processes in these fish, but also for the broader implications of behavioural ontogeny and evolution.

ACKNOWLEDGEMENTS

My studies of ontogeny and social behaviour in fishes have been supported by research funding from several sources, primarily the Natural Sciences and Engineering Research Council (NSERC), Canada (operating grant A6981), the Ontario Ministry of Natural Resources (Renewable Resources Research Grant Program), and the University of Guelph (Faculty Research Leave Award). A number of colleagues, students and research assistants have shared almost every aspect of this research. E. K. Balon, G. W. Barlow and J. R. Baylis provided insightful comments on earlier drafts of this manuscript.

I respectfully dedicate this chapter to Dr Ethelwyn Trewavas, whose work on cichlid fishes has been an inspiration for so many years.

Chapter eleven

Communication

M. H. J. Nelissen

11.1 INTRODUCTION

We all know what the word 'communication' means: subject A tells something to subject B. However, in ethology there is more than one definition of what is meant when we say that animals communicate. The anthropomorphic sense is difficult to use for animals as we are not sure whether they have the same 'intention' as humans when communicating. To avoid this pitfall, different definitions have been proposed for this interesting and comprehensive term.

The conventional definition of communication is the transmission of signals or information between a sender and a receiver. This transmission is a prerequisite for social behaviour. In interspecific communication, the sender and receiver belong to different species.

A more modern definition, inspired by the behavioural ecological way of thinking, looks upon communication as a manipulation of the reactor's behaviour by signals that are produced by the actor. There is still discussion as to whether the sending of signals is an 'honest' act (Zahavi, 1979) or a 'cheating' act (Dawkins and Krebs, 1978) (review, Krebs and Davies, 1987).

In this chapter on communication in cichlid fishes I shall not deal with this discussion or with definitions (Dawkins and Krebs, 1978; Hinde 1981; Caryl, 1982; Krebs and Davies, 1987). I shall only consider the ways in which cichlids send signals to their conspecifics, and how the receiver react to these signals. I shall keep to a few topics in the field of cichlid communication, showing that these animals are a very specialized family of fish. A few studies will illustrate the richness of communication methods that is so typical of the cichlid world. I shall indicate how inventive cichlids are when exchanging information, and why they do it.

11.2 HOW DO CICHLIDS COMMUNICATE?

Many methods of communication – or channels for sending signals – have been described for the animal kingdom. Animals produce visual, acoustic, tactile, chemical electrical and other signals and these are captured by corresponding sense organs (e.g. Busnel, 1963; Sebeok, 1968, 1977; Johnston *et al.*, 1970; Smith, 1977; Hailman, 1977). It is pointless to produce a signal if the 'target' cannot receive or understand it. Therefore there has been a coevolution in the development of signals and the capability to understand them (Wilson, 1975). Most of these communication methods are also used by cichlids. I shall give a few examples of cichlid signal channels.

Visual communication

Signals captured by the eyes are called visual signals. Visual signals can change rapidly and thus transmit more information than is possible in channels with slowly changing signals. More than one reactor can receive the information, and the signals may be long lasting.

Most animals make visual signals by body movement or gesture. So can cichlids, but their communicative capacities are especially rich because they can produce and change colour patterns. Different parts of the body are coloured differently and this pattern changes according to motivation. I shall deal first with colour patterns, then with body movements and gestures.

Colour patterns

Do cichlid fishes see colours? Levine and MacNichol (1982) divided the freshwater fishes into four groups: (1) fishes that confine themselves to the region immediately below the surface and have three visual pigments in the retina, with maximum sensitivity to violet, blue-green and yellow-green; (2) fishes that live in deeper water and also have three visual pigments in the retina, but with maximum sensitivity to longer wavelengths than do the pigments in the first group; (3) fishes active near the surface primarily at dawn or dusk, many of them feeding in deeper water during the day, and having red-sensitive retinal pigments; (4) fishes with only rods and green- and red-sensitive pigments in the retina. According to Levine and MacNichol (1982), cichlid fishes belong to the second group, so they can distinguish colours ranging from green to red. Not all cichlid fishes have been investigated, so the photosensitivity of different species will presumably vary according to habitat.

Colour patterns are a very important channel for communication in cichlids. We may wonder how they developed during cichlid evolution. Baerends (1986, p. 4) proposed the following hypothesis: "They have probably primarily evolved to serve as a protection against predators, mostly for camouflaging and sometimes perhaps for intimidating (e.g. reversed counter-shading). Superimposed on these primary patterns a secondary adaptation to the signalling of motivational states seems to have taken place." This hypothesis was investigated in *Tilapia zillii* by Hulscher-Emeis (1986). The vertical bars in this species appear in situations in which agonistic behaviour occurs to defend a property (e.g. the brood). She argued that the appearance of the vertical bars was associated with a motivational defence factor or tendency, which increased with the value of the property.

Colour patterns are formed by the interaction of thousands of chromatophores, pigment cells that may change the colour of the skin by expanding or contracting under control of the nervous system. Baerends and Baerends-van Roon (1950), whose classic and much-cited work is still relevant, described the following types of chromatophores, each of which is organized into a system.

1. System I. Small melanophores equally distributed over the whole body on the surface of the skin, more dorsally than ventrally. When they expand (in sick or chased fish) they cover the brighter colours produced by other, deeper chromatophores;
2. System II. Larger melanophores that are situated deeper. They may be combined with guanocytes to form melanoiridosomes or they may lie with separate guanocytes on the same level. Many cichlids display vertical bars during aggression; these bars are formed by the melanophores of system II. These chromatophores expand and contract very quickly (in a few seconds), permitting a sudden change in motivation to be communicated rapidly. Whenever the melanophores are contracted, the guanocytes effect a gloss;
3. System III. Large melanophores that are located even deeper and form longitudinal bands usually seen in 'anxious' fish. In system III, contraction and expansion is also very rapid;
4. System IV. This was mentioned for only one species, so does not seem relevant to the whole family;
5. System V. High concentrations of guanocytes may occur in different layers. They contribute to the total colour of the fish (White, yellow, blue, green);
6. System VI. Red erythrophores and yellow and brown allophores occur at different depths. They build up colours such as red, brown, orange and yellow, and take a few hours or days to expand or contract. This system can be used for communicating relatively long-term information (reproductive activity, dominance, etc.).

The colour pattern of a cichlid fish is built up by the contraction of some chromatophores and the expansion of others. As expansion can occur to different degrees and as there are many possible combinations of expanding chromatophores, cichlid fishes may demonstrate a rich collection of colour patterns.

Baerends and Baerends-van Roon (1950) described the colour patterns and their meaning in *Cichlasoma meeki*, *C. biocellatum*, *C. severum*, *C. bimaculatum*, *Aequidens latifrons*, *Hemichromis latifrons*, *Haplochromis multicolor* and *Tilapia natalensis*. They concluded that the greatest change in colour takes place when a fish passes from the non-reproductive into the reproductive condition. This is when a gradual development of the erythrophores and allophores (system VI) can be seen. The rather slow rate of this change suggests that it is controlled by hormonal secretions. Whereas a fish under non-reproductive motivation does not seem able to contract the melanophores of system I, a fish in a reproductive mood usually does. It seems that the potential for change in system I is released by the factors that induce the reproductive mood, but that contraction or expansion then take place under nervous control, since a frightened or defeated fish can expand these melanophores at once and thus mask the underlying erythrophores and allophores. These two last chromatophores never contract in such situations (Baerends and Baerends-van Roon, 1950).

Voss (1977) compared the colour patterns of 20 African cichlid species. His aim was to determine how far the data comparative ethology can be used in the field of systematics and phylogeny to support and supplement conclusions based on morphology. He concluded that colour patterns are extremely useful for differentiating and characterizing not only groups of species or genera, but also individual species. In all studied species, a great similarity could be observed between the colour pattern of free-swimming young fish and adults of both sexes when they were either dominated or frightened. On the level of phylogeny, the comparison of colour patterns and their modalities of appearance enabled Voss to discern different lines of evolution leading towards different specializations: display of invariably coloured structures in *Pelvicachromis*, *Nanochromis* and *Chromidotilapia*; specialization in quick changes of colour patterns in *Tilapia*; simplification of the range of colour patterns in the female and peculiarity of the territorial colour pattern in the male in *Haplochromis* and *Sarotherodon*. On the level of ethology, the existence of two main parental types (biparental and uniparental) was reflected in the range of colour patterns; species displaying biparental behaviour (in which male and female care for offspring) have a range of colour patterns that is complex and consistent in both sexes; species with uniparental behaviour have a reduced range of colour patterns, especially in the guarding individual.

Not only the diversity of colour patterns is interesting. A more important topic from the ethological point of view is the function of these patterns: what information is transmitted via this communication channel? The following example illustrates the range of possibilities. Nelissen (1975) made up the inventory (which is called the ethogram) of *Simochromis diagramma*, a maternal mouthbrooder from Lake Tanganyika, and described the following colour patterns and their association with displays or moods. Males and females did not differ in the form of the patterns (except the sexual pattern), but there were sexual differences in the frequency of occurrence of the patterns (Fig. 11.1). By correlating the observed colour pattern with the circumstances in which it was performed. Nelissen could link the patterns to the animals' mood or state.

In the neutral pattern (Fig. 11.1(a)), nine or ten stripes occur on the flank. This pattern is displayed by animals without any apparent fear, aggressive or sexual motivation. In aquarium tanks it is very rare.

In the dominant pattern (Fig. 11.1(b)), the vertical bars become shorter

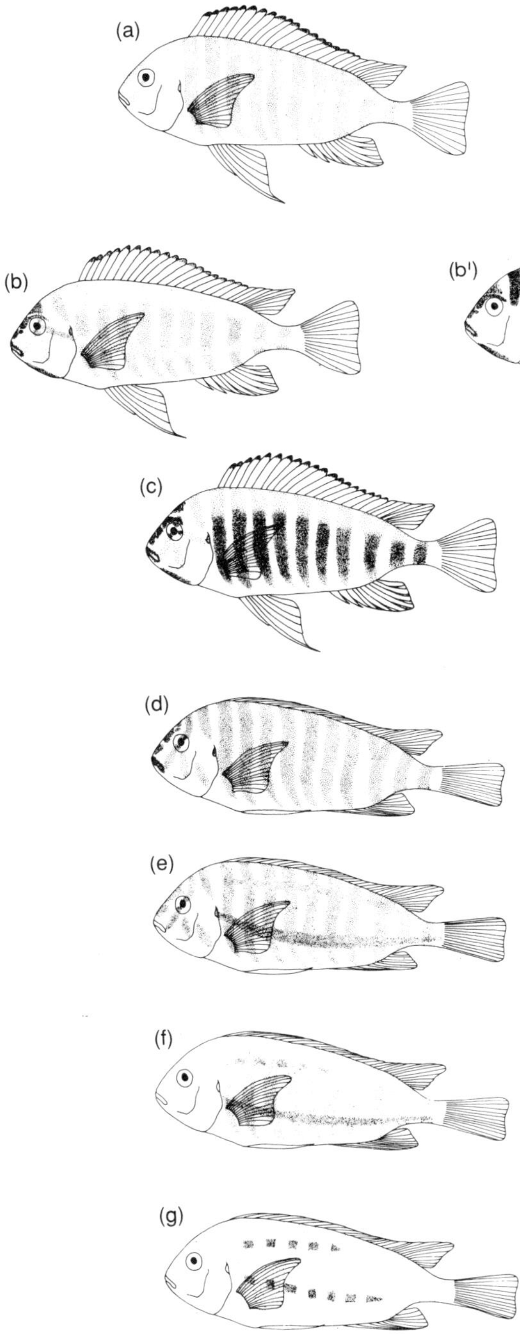

Fig. 11.1 Colour patterns in *Simochromis diagramma.* (a) Neutral pattern. (b) Dominant pattern, (b′) Sexual pattern. (c) Aggressive pattern. (d) to (g) Patterns of inferiority: fear increases from (d) to (g) (reproduced with permission from Nelissen, 1975).

than in the neutral pattern and are interrupted in the middle of the flank by a vague greyish longitudinal line; the head has many black spots. In the male the pectoral fins show a red base. This pattern is shown in dominant fish, which are usually males.

The aggressive pattern (Fig. 11.1(c)) is displayed when the black vertical bars become much darker on the flanks and fade away dorsally above the lateral line; the black markings on the head also become darker. This pattern is observed in dominant or non-dominant aggressive (threatening or attacking) animals.

The pattern of sexually motivated animals (Fig. 11.1(b′)) is similar to that of aggressive fish but there also appears a black occipital stripe between the eyes and the opercular spot. This is the usual pattern in courting males.

In the patterns of inferiority (Fig. 11.1(d) to (g)), with increasing fear the pattern may change progressively in the following way. It starts with a pattern very similar to the neutral pattern, but the head still shows some black spots as in the aggressive pattern (Fig. 11.1(d)); this pattern is displayed by anxious fish and courting females. Two longitudinal stripes are added to the former pattern, and the black markings on the head fade away (Fig. 11.1(e)); this pattern is shown by anxious and chased fish. The upper longitudinal stripe is interrupted at each intersection with the vertical bars, and the stripes and spots on the head are lost (Fig. 11.1(f)); this is seen in fish with increasing fear, for instance when chasing continues. The lower longitudinal stripe is also interrupted at each vertical bar so that a series of black squares is formed (Fig. 11.1(g)); this pattern is rather rare and occurs when fear is extreme.

These are only the basic colour patterns, as more intermediate patterns can be observed when one changes into another. These changes are brief, usually taking only a few seconds, but an intermediate pattern may last longer. The basic patterns described above may therefore represent a somewhat artificial classification.

This example should make clear how much information can be transmitted by a fish, simply by expanding and contracting its chromatophores in a highly ordered way.

Body movements and gestures

Moving the body, or parts of it, or holding it in a particular position is very often a signal and thus belongs to the communicative repertoire of cichlid fishes. The first extensive description of such signals was given by Baerends and Baerends-van Roon (1950). Their nomenclature has been adopted by many others who also added more descriptions of signal movements. It will not be possible here to discuss all these elements, but we can make a selection of those signals that are very common in the cichlid family.

In **lateral display**, the fish makes itself look bigger by erecting the median and the pelvic fins, and often the branchiostegal membrane. A threatening fish that displays this signal holds its body perpendicular to the opponent. In some

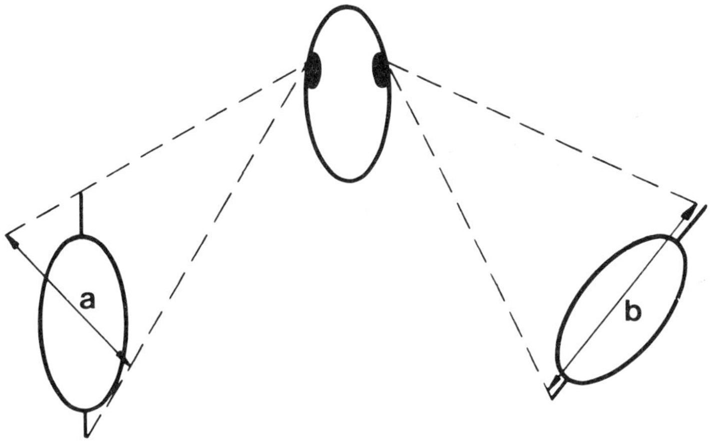

Fig. 11.2 A threatening cichlid fish erects it fins and looks bigger. If the actor's position in the water is lower than that of the reactor, the body may be inclined: a ventral roll permits a maximum body surface to be exhibited (b > a).

cichlids the fish may incline its body if it is not at the same level as the opponent so that it exhibits a maximum body surface (Fig. 11.2).

In **frontal display**, while the threatening fish orientates its head towards the opponent, the operculae and the branchiostegal membrane are erected. Some species may open the mouth widely. Once again, this makes the animal look bigger.

In **aggression inhibition**, a threatened fish makes itself as small as possible by decreasing the body surface seen by the threatening opponent. Mostly this is realized by closing the fins and tilting either in the horizontal or the vertical axis. As with the lateral display, the fish's position in the water relative to that of the threatening animal may influence the degree of tilting.

In **quivering**, the body is curved and quivers with high intensity and low frequency. This signal occurs in a sexual as well as an aggressive context, indicating that one signal may have two types of 'information content'. However, personal observations on *Melanochromis auratus* have revealed that in this species the sexual quivering can be distinguished from the aggressive one by a preceding act: the courting male swims in front of the female for *c.* 20 cm, turns around, swims back and quivers. The turning around is very fast; it is a kind of jerk lasting *c.* 0.2–0.3 s. This swimming to and fro and jerking occurs once or twice before quivering. If quivering is not preceded by one or two jerks, it has an aggressive meaning. This phenomenon is called 'metacommunication' (Wilson, 1975; Lewis and Gower, 1980): the meaning of a signal is changed by another preceding signal. It contributes to signal economy, a topic that deserves more intensive study in cichlids.

In **leading**, when the male of a mouthbrooding species wants to lead a ripe female to the spawning spot, he swims at normal speed from her to the spot he selected, while fanning vigorously with the caudal fin and the soft part of the dorsal and anal fins. Normal speed is maintained by counteracting movements of the pectoral fins.

These few examples illustrate how cichlid fishes may use various combinations of position, orientation and movement of body and fins to communicate visually.

Acoustical communication

From research performed since World War II we know that fishes are not as mute as was often thought. Many sounds produced by fishes have been recorded and analysed (survey, Tavolga, 1977). In cichlid fishes, acoustical communication has been poorly studied. The first description of sound production in cichlids (Bauer, 1963) concerned a deep-pitched crack in courting males of *Tilapia nilotica* which is so loud that it can be heard outside the aquarium tank.

Myrberg *et al.* (1965), in the first detailed analysis of cichlid sounds, studied sounds produced by *Hemichromis bimaculatus*, *Cichlasoma nigrofasciatum* and *Pterophyllum* sp. The first and second species produced sounds of the most intense tone frequency between 300 and 500 Hz. In *Pterophyllum* sp. a sound was noticed with a broad maximum intensity around 3500 Hz and component frequencies over 10 000 Hz.

Acoustic threat behaviour in defence of territories is emphasized in male *Tilapia mossambica* (Rodman, 1966), in which the distribution of energy was uniform in all frequency bands from 50 to 600 Hz.

The Central American cichlid *Cichlasoma centrarchus* may inhibit aggression in a receiver by producing low growing sounds (Schwarz, 1974a, b). Nelissen (1977a, 1978) found a relation between the number of sound types and the number of colour patterns in *Simochromis*, *Tropheus* and *Haplochromis burtoni*: the *Tropheus* species seemed to be specialized in sound production, whereas *Simochromis* displayed fewer sound types but more colour patterns. *H. burtoni* was intermediate. The most common sound in these species was a low-frequency (*c.* 200–900 Hz) br-r-r when fish were quivering aggressively.

Rowland (1978) described the relation of sound production to other behaviour patterns in the jewel fish, *Hemichromis bimaculatus*. Low-frequency thumps and purrs were produced separately or in combination. The thump sounds were associated with a behaviour called ‘nodding’ (several rapidly repeated, low-amplitude, vertical head movements) and probably represented a relatively high level of aggression. Purring was closely associated with the lateral display and tailbeating and occurred during the early stages of courtship and attack. A rarer third sound consisted of a series of low-amplitude pulses concentrated below 100 Hz and was synchronized with the movements of courtship quivering. If these sounds were played back through dummies, the fish displayed a more intense behaviour towards dummies with thumps and purrs than towards a silent dummy or one producing pulses of white noise, indicating that the sounds were indeed used in intraspecific communication.

Knowledge of cichlid acoustics is still relatively sparse, possibly because it is so difficult to conduct critical experiments to determine both the causes and the functions of sound production in fishes.

Chemical communication

In general, chemical communication can be defined as the conveying of information by molecules produced by the actor and captured by chemoreceptors of one or more receivers, where they are translated into nerve impulses. Regnier (1971) called these signals 'semiochemicals'. As in hormonal communication (within one body), the semiochemicals are effective in very minute amounts.

According to the neural endings of the receptors, we can distinguish two major kinds of chemical communication: olfaction and taste. In teleost fishes the nostrils are located dorsally on the head, between the mouth and the eyes. Taste receptors are distributed over the entire body, clustered or singly, but they are concentrated on barbels, filamentous fins, and in the oral and gill regions (Bardach and Villars, 1974).

Water is a good medium for transporting molecules, therefore chemical communication is to be expected in fishes. Indeed, communication by molecules in freshwater fishes was first demonstrated by Wrede (1932) in the European minnow, *Phoxinus phoxinus*, which can detect species-specific substances from individuals in the shoal. In other fishes, the relation between chemical communication and shoaling, homing, alarm calling, fecundity, growth, pair formation and spawning have been studied. In cichlid chemical communication, parental care has been most studied because this social system is so highly developed in the family.

Kühme (1963, 1964) showed that parents of the jewel fish, *Hemichromis bimaculatus*, not only use visual cues to recognize their own young, but also need chemical cues. If the parents could choose among plain water, water with conspecific fry, water with heterospecific fry and water with their own young, they chose the water with their own young. Myrberg (1966, 1975) made similar observations for *Cichlasoma nigrofasciatum*: females could locate and discriminate their own wriggling young from those of other species by chemoreception. The response was weaker in males. When the young became free-swimming fry, vision became the dominant discrimination modality for adults to recognize their own fry.

McKaye and Barlow (1976b) reported that parents recognized young by chemical cues in *C. citrinellum*. Parents lost this ability when their nares were plugged with cotton; by taste alone, they could not discriminate between water from their own young and plain water, proving that they were employing olfaction and not taste. There was also a decrease in the importance of chemical cues at the free-swimming fry stage.

Noakes and Barlow (1973b) demonstrated that taste was also a way of communicating chemically. *Etroplus maculatus* and *E. suratensis* (substrate brooders) use buccal taste to discriminate conspecific fry from strange fry, which they eat. These chemical cues are used together with visual cues. Certainly in mouthbrooders, the importance of taste is to be expected as parents have to recognize their fry both inside and outside the mouth.

Not only do the parents recognize their young, but the reverse also occurs. In the biparental *C. citrinellum* placed in a Y-maze, the free-swimming fry can distinguish chemically between their mother and plain water and between another mother and plain water (Bernett, 1977b). They do not distinguish between the two mothers offered together or between their father and plain water.

Most research on chemical communication in cichlids emphasizes parental care. However, adults also exchange information by chemical cues. Crapon de Caprona (1974, 1980) experimented on isolated male *Haplochromis burtoni* that were confronted with *Tilapia* sp. While the test fish could interact with the *Tilapia*, water from other tanks was introduced: water from a single gravid female, water from a single courting male, or control water. *H. burtoni* approached and courted the *Tilapia* significantly more after the introduction of the water from the gravid female and courting male conspecifics.

Young cichlids can also recognize other young conspecifics through chemical cues: young *Hemichromis bimaculatus* and *H. fasciatus* are attracted to chemical cues of other conspecific young (Kühme, 1964).

Chemical communication can be affected by learning. After rearing young fish either with conspecifics or with a foster species, Crapon de Caprona (1982) found that male *Haplochromis burtoni* learned to recognize the odour of a gravid female of the species with which they had been reared. She also found that females learn the visual aspects of the male rather than the chemical aspects, also across species. Thus, in *H. burtoni* there is a remarkable sexual difference in learning the characteristics of the other sex.

Because there are so many olfactory and taste receptors on the fish's body, future research should reveal more examples of chemical cues in cichlid communication.

Tactile communication

Another often-mentioned channel for conveying information is touch. The behaviour of the reactor is influenced by direct body contact, i.e. by touching another fish. More accurately we could define tactile communication as the exchange of information by somatosensory sensation. The kind of communication is not specifically described in cichlid literature, but some interactions might be considered as tactile communication. The problem, however, is that in aggressive contexts it is hard to say whether body contact is pure communication or whether it is intended to harm the opponent in order to eliminate it.

Some species perform mouth-fighting, in which the opponent's grip each other's mouth and start pushing. For instance, Nelissen (1977b) mentioned mouth-fighting in *Tropheus moorii* when fights escalate. The winner pushes the loser to the bottom. Such fights never end with visible injuries, so it is assumed that the opponents are measuring each other's strength. As soon as the loser determines that the other fish is stronger, he flees, so we may call this communication.

Laterally displaying cichlids may perform some tail-beating: the fish beats the tail sideways to push water at the opponent (it maintains its position by also giving a similar beat to the other side) (Baerends and Baerends-van Roon, 1950). The tail-beating animal does not touch the opponent, but the water does. Presumably this act also communicates the animal's strength. If tail-beating has a communicative function, it can be categorized as tactile, because the reactor experiences a somatosensory sensation instead of a visual, acoustical or chemical one.

When mouthbrooding cichlids are spawning, the male often touches the anal region of the female's flank when she is pushing out the eggs (pers. obs., e.g. in *Simochronis* and *Tropheus* spp.). Such contact may be some kind of encouragement for the female to go on with the egg laying. This action can be compared with that of spawning sticklebacks, in which the male also pushes the female's tail when she enters the nest to lay her eggs. If this assumption is correct, this is also a tactile communication act.

This section revealed the rich world of communicative channels in cichlid fishes. Indeed, cichlids are a very specialized family as far as communication is concerned. It is now useful to consider why they communicate.

11.3 FUNCTIONS OF COMMUNICATION

Why do animals communicate? Wilson (1975, chap. 9) reviews possible answers. Some of these functions also hold for cichlid fishes.

Species recognition

Recognition of conspecifics is a need that is generally to be expected in animals, e.g. to enable reproduction. Certainly in cichlid species that live in multispecific groups, like the Mbuna in Lake Malawi (see page 239), species recognition is of vital importance, not only to find mates but also to distinguish between one's own and heterospecific fry during parental care. Visual as well as chemical cues can be used, possibly successively (see page 233; Myrberg, 1975).

Family recognition

In parental care it is important to distinguish between one's own young and young of other conspecifics. For example, in *Lamprologus brichardi* a new brood is tended not only by the parents, but also by older brothers and sisters (Taborsky and Limberger, 1981; Taborsky, 1984). These helpers are recognized individually by the breeding pair as they are much less frequently attacked than are unrelated young (Hert, 1985; Taborsky *et al.*, 1986).

However, the species-determining cues are not always effective. Mrowka (1987a) reports cases of brood adoption in the mouthbrooding *Pseudocrenilabrus multicolor*; even some broods from substrate-brooding species were adopted. He suggests two explanatory hypotheses: (1) the mother does not recognize her brood; (2) the motivation for mouthbrooding outweighs the motivation to reject the foreign brood. The first hypothesis supposes a lack of sensory equipment or a central filtering mechanism for distinguishing her own or a conspecific brood. This is not supported by literature on recognition mechanisms in cichlids, therefore Mrowka prefers the second possibility.

Individual recognition

When animals have a well-organized social behaviour system, they need to recognize the conspecifics with which they live and maintain the social structure. Nelissen (1985) studied such a social structure in *Melanochromis auratus* from Lake Malawi. In captivity (pers. obs.) and in the field (unpubl. obs. Dr E. Hert), these fish stay together in a stable group. From laboratory observations, Nelissen reported that in such groups a stable linear dominance hierarchy is always established. An alpha animal dominates the whole group, a beta animal dominates all other group members except the alpha, and so on to the omega fish, which is subordinate to all other fish and dominates none. This rank order is maintained with minimum energy expenditure. The group members do not interact aggressively but very frequently exchange signals of dominance (e.g. threatening) or subordination (e.g. avoiding). Such interactions occur primarily between rank neighbours and rarely between group members of very different rank. This structured group life would not be possible if members did not recognize one another individually. It is not yet clear which cues are used for this individual recognition, but the patch pattern on the caudal fin is individually specific, much like fingerprints.

Dominance status

Many cichlids establish dominance relations with conspecifics, in pairs or in groups. The dominance status is often signalled or influenced by colour patterns.

In *Melanochromis auratus* page 236), the dominant animals in a group can be identified by a black colour pattern: the flanks and the head are black, a longitudinal white bar extends from the operculum to the tail, and the back remains white. The subordinate fish have a yellow pattern: they are yellowish with the exception of two longitudinal black bars on the flanks and one on the dorsal fin. If a group member rises in the rank order, the colour pattern may change from yellow to black. In this way the group members immediately know the other's dominance status. This difference is made clear when an individual is separated from the group and offered the choice to 'visit' a yellow or a black conspecific: yellow fish prefer yellow conspecifics; black fish visit the other black ones much more frequently than yellow ones. This topic is still under study.

Dominance relations can also be established in pairs of animals in which rank may also be signalled. For instance, Falter (1987) describes colour patterns that indicate dominance or subordination in pairs of *Oreochromis niloticus*.

Colour patterns not only inform the others about dominance status, they may also influence dominance. In the Midas cichlid, *Cichlasoma citrinellum*, from lakes in Nicaragua, *c*. 8% of the adults are amelanic by mutation: the loss of their melanophores uncovers the deeper xanthophores and makes the animals yellow through orange, called gold (Barlow, 1983). These gold morphs dominate normal fish of equal size, not by being more aggressive but because of the effect of their colour. Barlow (1983; Barlow and Ballin, 1976; Barlow and Wallach, 1976) cautiously hypothesized that the gold morph is a supernormal releaser that inhibits aggression by stimulating fear. The immature golds might be regarded as mimics of mature golds, usually older, larger individuals. "Gold therefore signifies size, which means a superior rival. Small, infrequent golds are then cheaters who, though they are young and weak, mimic the large golds and thereby enjoy a status out of proportion to their size" (Barlow, 1983, p. 245). Indeed, the juveniles learn the association between coloration and dominance.

Aggression

The dominance relations discussed above result from aggressive interactions. Also, when aggression does not lead to a dominance relation, communication may express the animal's motivation. The examples mentioned under 'colour patterns' (page 226) illustrate how cichlids can express their aggression by changing the whole colour pattern.

However, smaller-scale mechanisms are also used by some cichlids. In *Haplochromis burtoni* the attack readiness can be measured by the angular orientation of the black eyebar with respect to the body axis (Heiligenberg *et al.*, 1972): the attack rate increases as the eyebar becomes increasingly parallel to the profile of the forehead, and decreases as the bar approaches an orientation perpendicular to the forehead.

Sexual behaviour

Co-operation between two spawning partners requires a good communication system to ensure synchronization of the male and female acts. Courtship synchronizes sexual arousal. In the courtship phase the activities of the potential partners are controlled by three behavioural systems – the motivations for attack, for escape and for sex (Baerends, 1984). The sex system gradually becomes more dominant over the agonistic ones. In *Chromidotilapia guentheri* this change in motivation is communicated through colour patterns that can be correlated with the occurrence of attack, escape or sex (Baerends *et al.*, 1986). Presumably these signals are necessary for each mate to 'measure' the readiness of the other to proceed with the spawning activities.

Rechten (1980) describes the black belly signal that *Etroplus maculatus* directs at its mate during spawning. She wonders about its meaning, but when such a signal only occurs during spawning it must have a sexual communicative function, perhaps a synchronizing function.

Egg-spots occur on the anal fin in mouthbrooders (Fig. 13.4). As these egg-spots influence the behaviour of other fish, i.e. females, they are real signals. Wickler (1962b) considered these spots as an example of auto-mimicry and described how they enhance the fertilization of the eggs. Immediately after the female has laid eggs, she takes them into her mouth, and while she attempts to 'pick up' the yellowish egg-spots on the male's anal fin she inhales his sperm and thus fertilizes the eggs in her mouth. In some species the male does not have spots on the anal fin, but a genital tassel, which has the same function (Wickler, 1965b).

Hert (1986) eliminated the anal fin spots by freeze-branding the skin surface with dry ice. She compared the reproductive success of males with and without egg-spots. Contrary to common belief the spots were not necessary for fertilization, although they were attractive to females in *Haplochromis elegans*. Males with egg-spots fertilized more clutches of eggs than males without. In choice experiments, females always chose males with spots (Hert, 1987). They also preferred males with more spots and dominant males, but did not use the spots for species recognition. Hert concluded that the spots serve as intraspecific sexual advertisement.

Assessment

The mechanisms for assessing the capacities of conspecifics are important enough to be dealt with separately, even if the assessment concerns aggression or sexual motivation.

Especially in fighting, assessment is very important. Most fights end with one contestant giving up before injuries have been inflicted. Parker (1974) has argued that during a fight the individual should gain information about the opponent's fighting ability relative to its own. This information should be used to decide whether to continue, to escalate, or to give up. How do cichlids assess the opponent's abilities? Animals with very different body lengths generally will not start fighting. Body length may be an assessment signal. If a fight has started, this signal may also be used (Turner and Huntingford, 1986). Lateral display, which mostly occurs at the beginning of a fight, might indicate size. Enquist and Jakobsson (1986) discuss the problem in interpreting such signals by the opponents: body size is only a valuable signal for assessment if it can be compared with the animal's own body size, but how does an animal know its own size? This problem might be solved by tail-beating, in which the fish pushes a water stream, usually against the opponent's head (see page 234), ('Tactile communication'). The reactor can measure how water stream affects its position and the actor can observe the effect. Another way to measure relative strength is mouth-fighting, in which the fighting fish grasp each other's jaws and start pushing. This direct measure of relative strength yields even more information than tail-beating (Jakobsson *et al.*, 1979).

The sum of all behaviours in a fighting contest may give information that can be used to decide whether to continue or stop. Barlow *et al.* (1986) found a difference between long-term and short-term experiments with *Cichlasoma citrinellum* as to what determines the outcome of a fight. In their long-term experiment, which they compared with intraterritorial disputes in the field, prowess (determined by weight) determined the winner, while in their short-term experiment, which they compared with establishing a territory in the field, daring to escalate determined the winner of the fights.

Interspecific communication

Communication is not restricted to the members of the same species. Signals are often understood by individuals of another species (e.g. in parental care of a mixed-species brood).

In cichlids interspecific communication is especially important for species that live together in multispecific groups, e.g. the Mbuna in Lake Malawi. Very little research has been done in this field, but unpublished observations by Andries and Nelissen reveal that among the Mbuna, fish of various species interact to establish interspecific dominance hierarchies.

11.4 DO CICHLIDS REALLY BENEFIT FROM COMMUNICATION?

A general discussion of the benefits of communication is clearly beyond the scope of this chapter. However, it should be mentioned that the question has been raised as to whether or not cichlid fishes benefit from giving and receiving information. Turner and Huntingford (1986) studied fighting behaviour in *Oreochromis mossambicus* and found significant differences between eventual winners and losers in the frequencies of several behavioural measures, even at the beginning of the contests. This is contradictory to the predictions of game theory, in which it is believed that fighting animals should not provide their opponents with any information about how long or how fiercely they intend to fight, and consequently the behaviour of the eventual winner and loser are not expected to differ (Maynard Smith, 1982). Turner and Huntingford suggest that there are circumstances where there may be drawbacks, as well as benefits, resulting from the concealing of intentions.

Chapter twelve

Sexual selection and the evolution of the cichlid fishes of Lake Malawi, Africa

K. R. McKaye

12.1 INTRODUCTION

Sexual selection by 'female choice' is a topic which has been rediscovered and is generating considerable interest in evolutionary biology. Darwin's (1871) examination of the evolution of seemingly maladaptive traits in *The Descent of Man and Selection in Relation to Sex*, and Fisher's (1930) theory of 'runaway sexual selection' and its role in generating new species have stimulated numerous more recent studies on a wide variety of plants and animals (Kirkpatrick, 1987). Cichlid fishes in particular provide extraordinary opportunities to study the process of mate choice and its importance to evolutionary theory. Understanding the role sexual selection has played in the evolution of mating behaviour and cichlid morphology will certainly assist in explaining both the amazing speciation of cichlids and how so many closely related species coexist.

The sexual selection literature is expanding rapidly. Major reviews of the theory describe examples from many animal and even plant groups. The reader is referred to Bateson (1983), Arnold (1985) and Kirkpatrick (1987) for excellent discussions of mate choice, genetics and sexual selection by female choice in particular. Dominey (1984), in the first paper to discuss the importance of sexual selection to cichlid speciation, provides the framework for much of the following discussion. In this chapter, I concur with his basic premise that runaway sexual selection may provide the explanation for the explosive speciation of the cichlid fishes of Lake Malawi, Africa. Major papers in the field will be discussed, recent data concerning mate choice by Lake Malawi cichlids will be presented, and suggestions for future studies will be made.

12.2 OVERVIEW OF SEXUAL SELECTION THEORY

Sexual selection has been proposed as a major factor in the proliferation of polygamous species (Lande, 1981) such as cichlid fishes (Dominey, 1984). Where males provide no resources or parental care and females have numerous males among which to choose, 'extravagant' male secondary characteristics could result solely from sexual selection (Darwin, 1871; Fisher, 1930; Lande, 1981). Theoretically, positive 'runaway' feedback provides a mechanism that could accelerate the joint evolution of female mate choice and male secondary sexual attributes. Fisher (1930) hypothesized that 'an evolution of sexual preference due to this cause would establish an effective isolation between two differentiated parts of a species, even when geographical and other factors were least favorable to such separation.'

Furthermore, Fisher (1930) postulated that the following conditions would result under his model of runaway sexual selection: (1) the development of the male character and female choice for such a character would advance together until checked by severe counter-selection; (2) the rate of development of this character would be proportional to the development already attained; (3) the selected character would increase with time exponentially as a result of this relationship; (4) this 'runaway' process could begin from small differences soon achieving major proportions.

Lande (1981) expanded upon Fisher's (1930) original thesis and concluded that male traits most likely to undergo this runaway process are 'those under weak natural selection'. Characters such as size, colour or behavioural patterns subject to courtship have the potential for changing quickly. Divergence in such characters as driven by sexual selection could originate rapidly throughout the range of a species due to local population differences in female choice (Lande, 1982). In particular, Arnold (1983) argued that lek-breeding species should exhibit 'extraordinary geographic variation in male attributes'.

Do cichlids qualify as lekking species for application of these models? Bradbury (1981) suggested four criteria for a rigorous definition of a lek-mating system: (1) males engage in no care of the offspring; (2) males must be territorial and clustered; (3) no resources within the territories should influence female choice, (4) unrestricted mate choice by females should be evident. Many (if not most) of the cichlids of Lake Malawi fit this restricted definition of a lekking species (McKaye, 1983, 1984). Therefore, the multi-species assemblages of African cichlids are an ideal vertebrate group in which to evaluate this Fisherian 'runaway' hypothesis and its potential role in speciation (Dominey, 1984).

12.3 FIELD STUDIES OF CICHLID MATE CHOICE

The few studies that have examined cichlid mate selection *in situ* have concentrated primarily upon species which engage in biparental care of the young (see Chapters 8 and 9). In these species, both parents defend broods from predators and there is a strong positive correlation in size between the males and females (Perrone, 1978a; Neil, 1984; McKaye, 1986). Natural selection arguments are made concerning adaptive value of large size in defending the brood, taking a territory, and producing more eggs. These could clearly account for such mate choice. Assortative mating by colour morphs of *Cichlasoma citrinellum* (Barlow, 1983; McKaye and Barlow, 1976a; McKaye, 1980) can also be explained by variation in selective pressures. For *C. citrinellum* the aggressive advantage associated with being gold is balanced by increased vulnerability to predators associated with being more conspicuous (Barlow, 1983). Such pre-mating choice might eventually lead to complete reproductive isolation between the two colour types (McKaye, 1980), resulting in turn in the formation of two distinct species (McKaye and Stauffer, 1986). Though sexual selection might play a role in these mating systems, natural selection seems to be the predominant process driving them (McKaye, 1986).

For the African cichlids, Fryer and Iles (1972) hypothesized that the critical parameters of mating success were variation among males in either intensity of male coloration or 'nest' form. Though not explicitly stated, their hypothesis implies that sexual selection might be operating. How then could such differences in colour or the male's display site evolve through sexual selection, and what is the evidence that sexual selection might occur?

Coloration is important in distinguishing among Lake Malawi cichlid species (Holzberg, 1978; Marsh *et al.*, 1981; McKaye *et al.*, 1982, 1984; Ribbink *et al.*, 1983a). The conclusion that breeding-colour differences among males are clearly associated with reproductive isolation is based solely upon rock-dwelling cichlids. If female choice is important, the ultimate cause of such choice is unlikely to be a single factor applying to all species. The evolution of female choice might result from: (1) 'non-adaptive' runaway sexual selection; (2) a mechanism allowing females to choose mates that are relatively resistant to parasites (Hamilton and Zuk, 1982); or (3) a competitive selective advantage such as the aggressive edge gold morphs of *Cichlasoma citrinellum* have over the normal grey form (Barlow, 1983).

Although a model of sexual selection for male coloration is attractive, how would sexual selection explain the diverse community of sand-dwelling cichlids in Lake Malawi? These are not as colourful or variable in their colour patterns as the rock-dwelling cichlids (McKaye, 1984). At first glance this observation would appear to falsify any general theory of sexual selection in explaining the evolution of this species-rich group, yet the key might lie in their display sites as a character upon which female choice may operate (McKaye,

1983; Dominey, 1984; McKaye, 1984; McKaye and Stauffer, 1988; McKaye *et al.*, 1990). Is there any evidence of differential female mate choice which would fit Lande's (1980, 1981, 1982) models and support Dominey's (1984) hypothesis?

12.4 LANDE'S MODEL OF SEXUAL SELECTION UPON POLYGAMOUS SPECIES

Many models of sexual selection now exist and deal with 'good gene' hypotheses, the 'handicap' hypothesis, and others (Zahavi, 1975; Karlin and Raper, 1979; Janetos, 1980; O'Donald, 1980; Nur and Hasson, 1984; Maynard Smith, 1985; Heisler, 1985; Watt *et al.*, 1986). It is not my intention to discuss them here, nor to review the sexual selection literature for other groups such as birds (however, see Andersson (1982) and Moller (1988) for examples of experimental tests of sexual selection hypotheses for the development of extreme male characters), but rather to focus on Lande's (1981) model and apply it to cichlid fishes. This decision does not mean that these other variants of sexual selection theory do not apply to cichlids. But since field data concerning cichlid mate choice are scarce, I prefer to focus on the most general model relevant to cichlids. For a detailed treatment of the theory see Lande (1980, 1981, 1982) and Arnold (1983, 1985).

Lande's polygenic model considers genetic change of a continuous character in a population. Such a character could be the colour or form of a bower (nest) in which variation could be produced both by several genes and by the environment. Diet might be important to colour (Barlow, 1983), or in the case of bower form, substrate type or erosion patterns could affect its ultimate expression.

For any sexual selection model, including Lande's, to be realistic and of heuristic value to the study of cichlids, several conditions must be met. There must be some variation in the male character subject to female choice (Fig. 12.1), there must be variation in female mating preferences (Fig. 12.2), and the trait must be heritable (Fig. 12.3).

As a consequence of Lande's models, there will be a genetic coupling of male and female characters resulting eventually in positive assortative mating. With this genetic coupling, females that prefer males with the largest bower or brightest colours will also have sons and brothers that construct large bowers or are the most colourful (Fig. 12.2). The genetic regression (genic covariance divided by genic variance) will define the slope of this genetic coupling and subsequent assortative mating tendency. The greater the genic covariance relative to the genic variance, the steeper the slope and the faster the resulting evolutionary trajectory.

Under Lande's models, two outcomes can occur which have important implications for cichlid speciation. If the line of equilibrium (determined by the

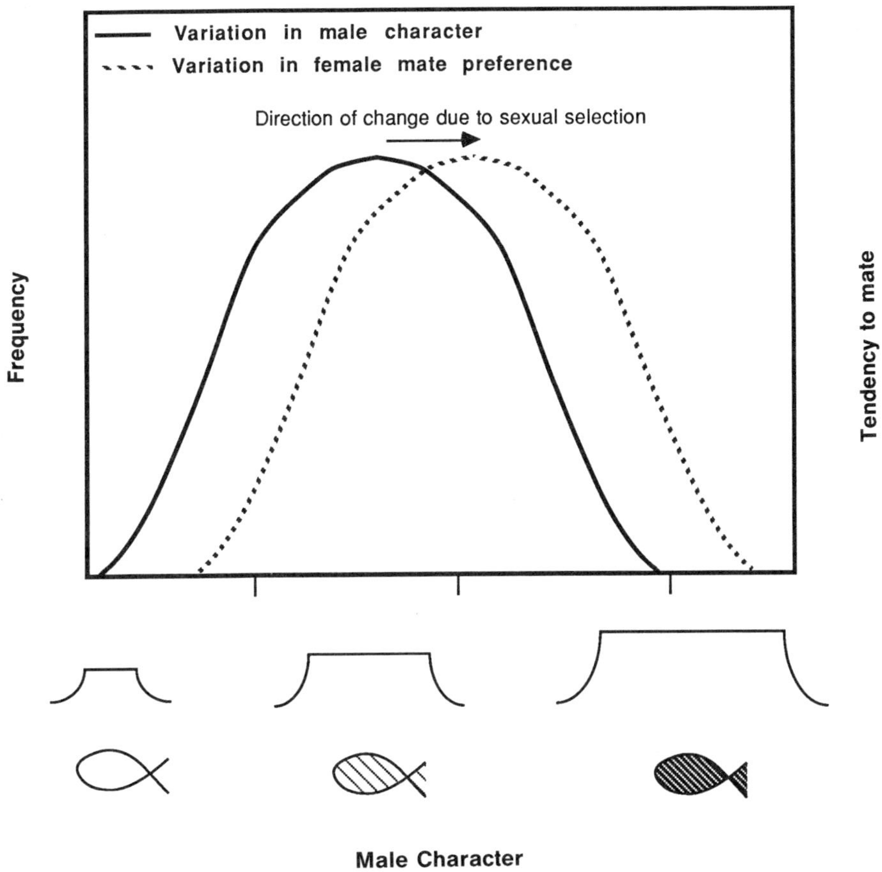

Fig. 12.1 A hypothetical character such as male colour (fish outlines: darker hatching signifies brighter coloration) or size of bower (plateau outlines) is normally distributed (solid curve) around a mean. Female mating tendency (broken curve) in this case favours more brightly coloured males or larger bower size. Tendency through time would move the male character in the population to the right (arrow) as males with such characters would have more breeding opportunities.

stereotypy of female mating preference and the strength of natural selection on the male character) is stable, populations will stop evolving when they reach the line (Fig. 12.3). Nevertheless, numerous possibilities are available for populations to finish at different points depending on where they started. However, if the slope of the genetic regression is greater than that of the line of equilibrium, populations will evolve away from the line of what have now become unstable equilibria (Fig. 12.4).

If natural selection against the trait is relatively weak compared with the stereotypy of female choice, then the number of final end points for a trait is

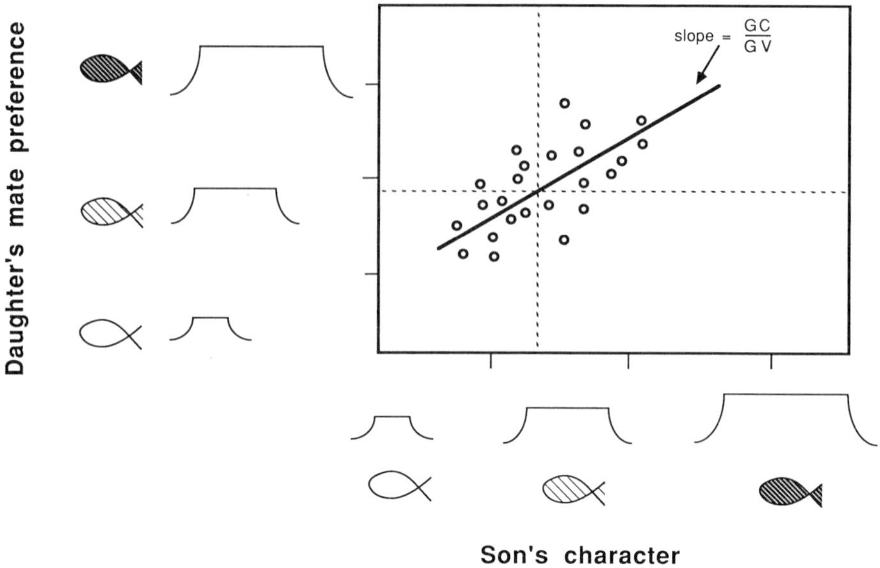

Fig. 12.2 Each point represents the averages of brothers and sisters produced by a single father with multiple mates. The genic covariance GC is the covariance for the two breeding values, female mate preference and male character. The genic variance GV is the variance in the male character. GC/GV, the genetic regression, is the least squares fit of sister's preference on brother's character. The steeper the slope, the greater the linkage between female preference and male's character state (modified with permission from Arnold, 1983).

almost unlimited, and no adaptive explanations for the existence of the characters are required. What is the field evidence from Lake Malawi?

12.5 COLORATION AND BOWER FORM; MALE CHARACTERS UNDERGOING SEXUAL SELECTION?

Ribbink *et al.* (1983a) reported numerous colour variants of reproductively isolated rock-dwelling cichlids (Mbuna) throughout Lake Malawi. Every Mbuna community at isolated rock outcroppings has high diversity with different colourful species for which no adaptive pattern has been ascertained. Many of these brightly coloured forms, which were once thought to be morphs of a single species on the basis of morphological and meristic evidence, have now been shown by behavioural (Holzberg, 1978; Schröder, 1980; Marsh *et al.*, 1981) and genetic data (McKaye *et al.*, 1982, 1984) to be sibling species. The initial realization that these colour morphs were different species came from field behavioural studies of mate choice (Holzberg, 1978) which determined that female mate choice was correlated with male coloration.

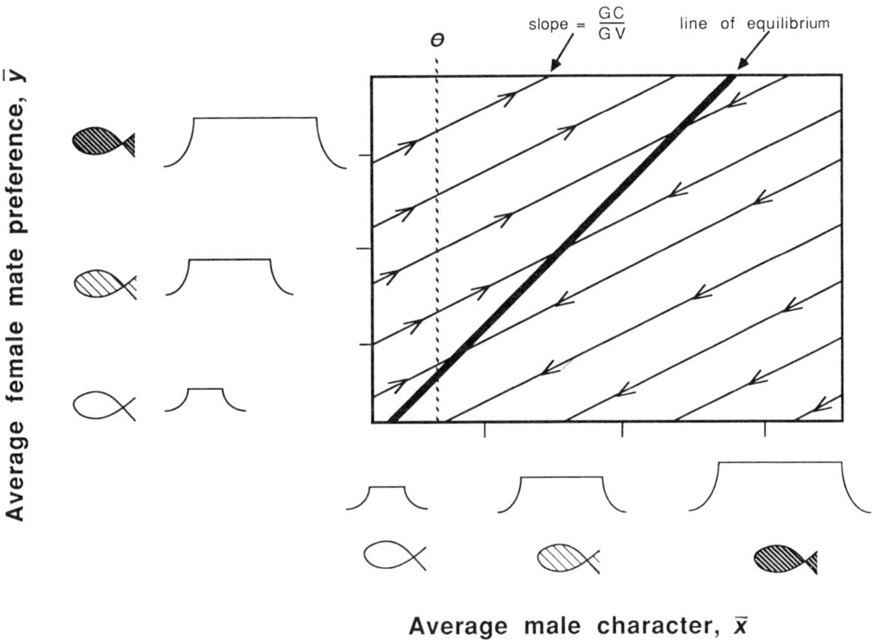

Fig. 12.3 The genetic regression GC/GV defines the trajectory along which the population will evolve (arrow heads indicate direction) until it hits the line of equilibrium, which is determined by the stereotypy of female mating preference and the strength of natural selection on the male character (Lande, 1981). The dashed line, θ, indicates the optimal male character for function which, in the case of bower size, might be the balance between the energetic costs of sand movement and the benefits derived from reducing egg predation before the female gets the eggs in her mouth (McKaye, 1984). In the case of colour it might be the most cryptic colour to camouflage the fish from predators (modified with permission from Arnold, 1983).

Similar behavioural studies of sand-dwelling cichlids also indicate that many more species exist than were previously recognized. These species can be distinguished by the form of their display sites (McKaye, 1983, 1984; McKaye and Stauffer, 1988). Fryer and Iles (1972), Keenleyside (1979), McKaye (1983, 1984) and McKaye and Stauffer (1988) have used the term 'nest' to describe these display sites. However, McKaye *et al.* (in press) propose the term '**bower**' as defined in the ornithological literature (Collias and Collias, 1983; Borgia, 1985; Borgia *et al.*, 1985). They are solely display sites, constructed independently of the need to care for eggs and young (McKaye, 1984). The female deposits eggs in the bower and then immediately picks them up in her mouth, where they are fertilized and brooded. The males engage in no parental care as found in other fishes with 'nests' (Keenleyside, 1979). McKaye (1983)

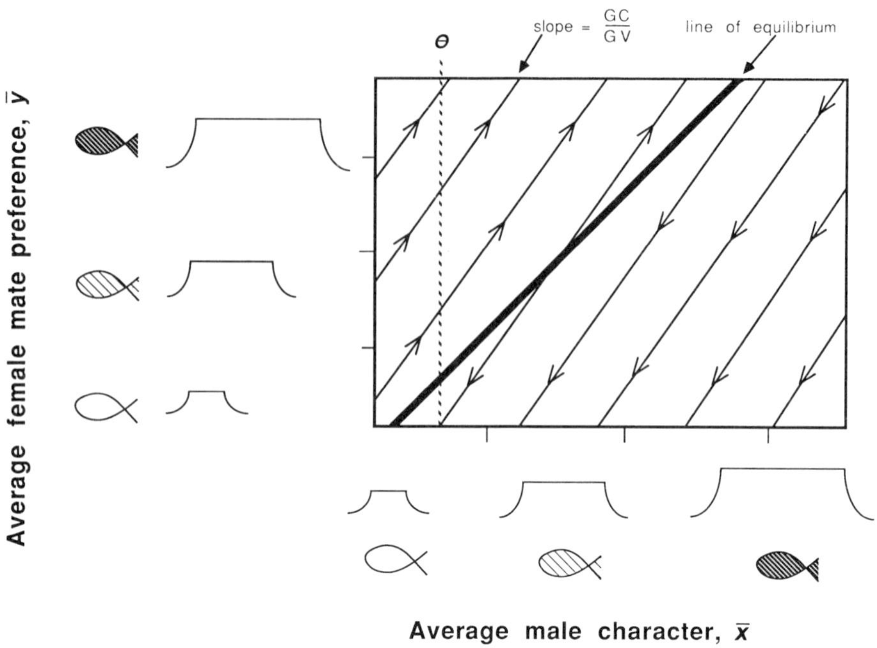

Fig. 12.4 The runaway process occurs when the genetic regression GC/GV is greater than the slope of the line of equilibrium. Populations will evolve away from the line of equilibrium in either direction (arrow heads). As a result a male's bower could be either smaller or larger than that at the starting point, or male coloration could become brighter or duller (modified with permission from Arnold, 1983).

hypothesized that bower size itself could be the critical component of bower form in determining female choice and relative male success.

12.6 MORPHOLOGICAL AND GENETIC EVIDENCE OF MONOPHYLY OF BOWER FORM

Since 1980 my colleagues and I have studied the reproductive ecology of approximately 50 haplochromine fishes in southern Lake Malawi. We have observed a wide variety of bower forms among the cichlid species (McKaye, 1983, 1984; McKaye and Stauffer, 1988). These bower forms can be divided broadly into ten types (Fig. 12.5). They range in size from giant craters 3 m in diameter (McKaye and Stauffer, 1988) to small depressions in the sand, and to 'sand castles' (photograph: Bass, 1988: 152–3) with base diameters of 1 m (McKaye, 1984). Given the wide variability in bower shape, with craters, mounds and 'bumps' in the sand, it appears that no single bower form or size is the best structure for the laying of eggs.

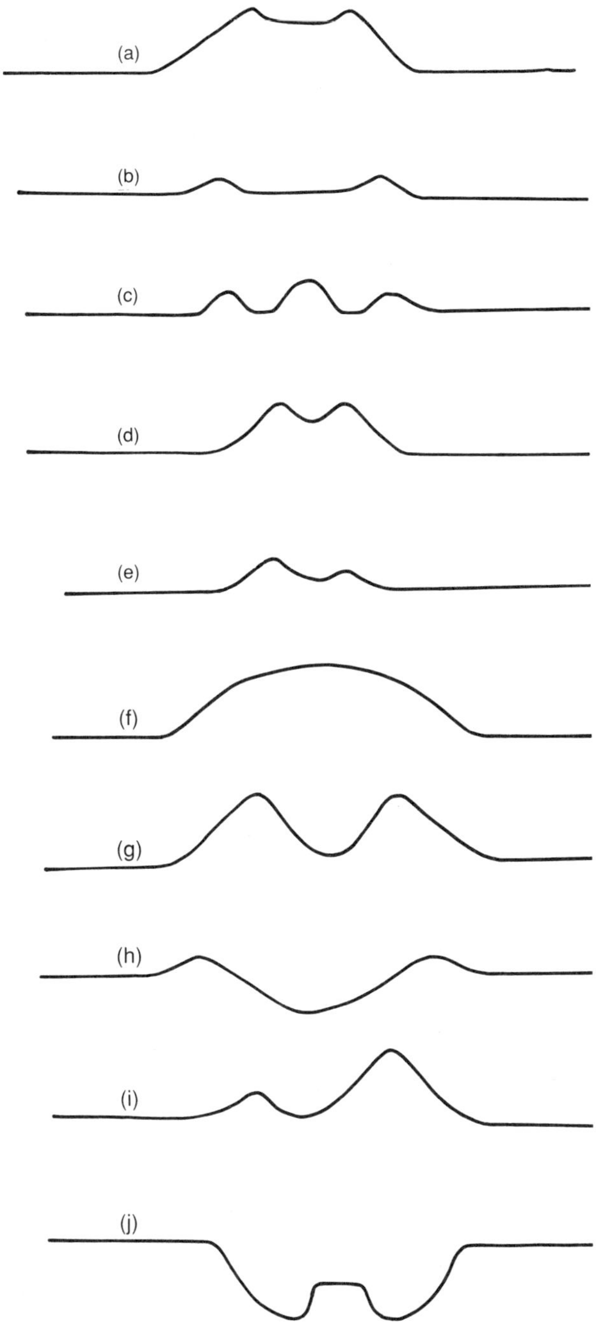

Fig. 12.5 Profiles of the ten major bower forms in Lake Malawi. (a) *Lethrinops lituris* group (see text). (b) *Lethrinops furcifer, L. furcicauda*. (c) *Lethrinops aurita*. (d) *Cyrtocara eucinostomus, C. argyrosoma* (see text) (McKaye, 1983). (e) *Cyrtocara kirkii, C. labidodon*. (f) *Cyrtocara marginatus*. (g) *Trematocranus microstoma, Cyrtocara placodon*. (h) *Cyrtocara kiwinge*. (i) *Cyrtocara atritaeniatus*. (j) *Oreochromis squamipinnis, O. saka* (McKaye and Stauffer, 1988).

If selective processes and responses to local ecological conditions are not driving the evolution of bower form, then rapidly evolving species with similar bowers would be expected to have arisen from a common ancestor. Our initial work supports this conclusion as the 'endemic tilapias', *Oreochromis*, of Lake Malawi have crater bowers (Fig. 12.5(j)) with a central platform which is unique unto them (McKaye and Stauffer, 1988).

Also, species of the *Lethrinops lituris* group build a mound which is distinct from the other nine Lake Malawi cichlid bower types (Fig. 12.5(a)). Their bowers consist of mounds with flat breeding platforms (see photograph in Bass, 1988, pp. 152–3). Species of the *L. lituris* group, which will be included in a new genus, share morphological characters unique among cichlid fishes in Lake Malawi. These include a keel on the pharyngeal bone, and low gill raker counts (5–8; see also Trewavas, 1931). This independent morphological evidence of apomorphic characters, which are congruent with bower form, further supports the hypothesis of common ancestry among cichlids with similar bowers.

With the *L. lituris* group, bower forms of five populations were measured and examined by principal component analysis (McKaye *et al.*, 1990). We also used starch-gel electrophoresis to examine genetic variation of enzyme loci from breeding males and females of the same five populations. The phylogenies derived from both the similarity of bower form and genetic distances were identical ($P < 0.01$: 105 different phenograms were possible (Sneath and Sokal, 1973)).

If bower form was governed by local ecological conditions leading to adaptive functional responses in bower building or variance between populations which was non-heritable, then no such congruence would be expected. Results from this study, therefore, provide preliminary evidence that bower form in these cichlid populations (species?) is correlated with genetic distance. Such differences probably arose from non-functional differences in female choice.

12.7 EVIDENCE OF FEMALE CHOICE AMONG LEKKING CICHLIDS IN LAKE MALAWI

What is the evidence for female choice among these cichlids? We are studying this process in several species of the *L. lituris* group and in *Cyrtocara eucinostomus* and *C. argyrosoma*, focusing first upon species with major differences in bower size. We have completed the analysis of data from the two *Cyrtocara* species (McKaye *et al.*, 1990).

In these two species and all species of the *L. lituris* group that we have studied, males establish themselves on the lek before they build their bowers and mate. The females move over the arena in a non-random fashion and lay eggs with several males (Fig. 12.6). The entire egg-laying process, from the

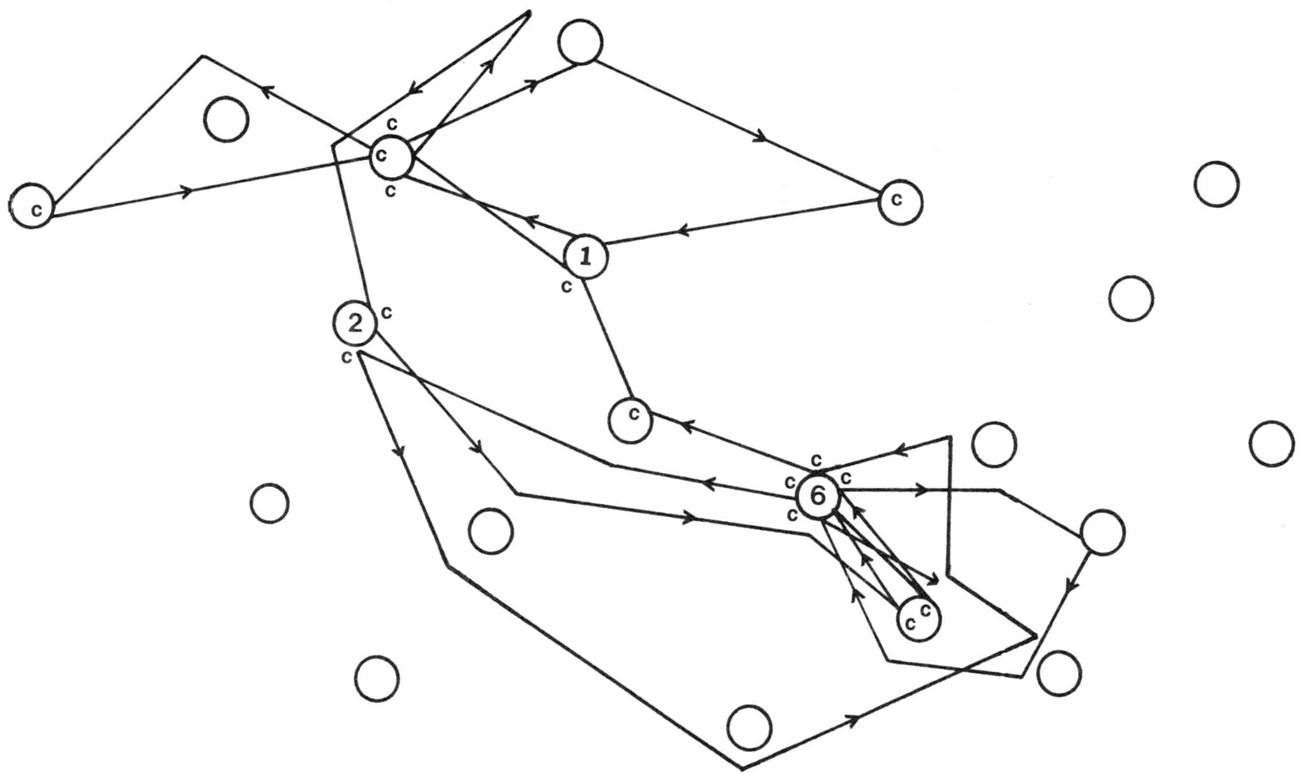

Fig. 12.6 Movements of a single female *Cyrtocara argyrosoma* over an arena while laying eggs. Each circle represents a bower. Every turn indicates that a male courted her; c, circling in the bower with the male; 1, 2 or 6 denotes number of eggs laid during that bout. The SCUBA diver ran out of air before this female completed laying all her eggs.

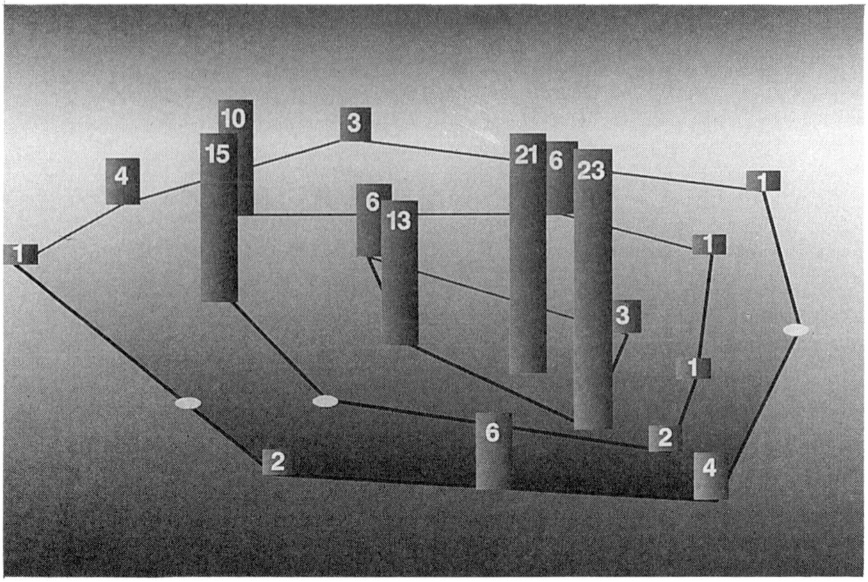

Fig. 12.7 *Cyrtocara argyrosoma* arena. Height of histogram and number indicates number of times during observation periods that females circled with individual males in each bower. Note that females enter bowers in the centre of the arena more than those on the periphery. Lines connecting bowers are derived from successively connecting outer bowers. This technique most accurately reflects central tendency in the arena. For example, a female arriving at the arena will pass an equal number of outer nests from any direction to reach the 'central' male.

time a female lays the first egg until she leaves with the eggs in her mouth, takes anywhere from 25 to 65 minutes.

Research on the breeding behaviour of *C. argyrosoma* and *C. eucinostomus* (McKaye *et al.*, 1990) has demonstrated that the process of female mate choice is complex. For example, *C. argyrosoma* females selectively chose males from the centre of the lek (Figs 12.7, 12.8), while *C. eucinostomus* females chose males that built the largest bowers (Fig. 12.9, McKaye *et al.*, 1990). The males of these two species have similar bower forms (Fig. 12.5(d)), but the sizes of bowers and arenas are different. *Cyrtocara argyrosoma* males form arenas ranging in capacity from 20 to 50 males, whereas *C. eucinostomus* males will display in a huge lek that may have over 50 000 males at the height of the breeding season (McKaye, 1983, 1984).

For either of these species, the size of the male was not correlated with female choice. Often, a small male held a large bower and vice versa. As might be expected from the female selection process, *C. argyrosoma* males have small

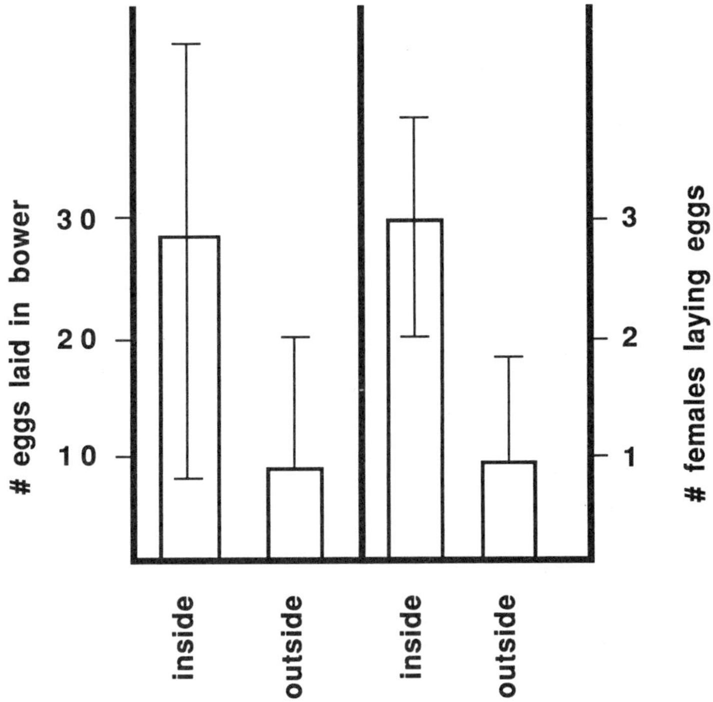

Fig. 12.8 *Cyrtocara argyrosoma*: mean number of eggs laid with, and of females that visited, the males in the central five bowers (inside) versus the 16 outer bowers (outside) of the arena shown in Fig. 12.7. Error bars indicate 2 S.D. ($P < 0.01$ for both measures, Kruskal-Wallis test).

bowers (*c.* 4.0 cm in height) and *C. eucinostomus* males build bowers with an average height of more than 15 cm (McKaye, 1983; McKaye *et al.*, in prep.).

When bowers of *C. eucinostomus* were experimentally removed, they were rapidly rebuilt, and returned to control sizes after two weeks (Fig. 12.10). However, within 1 d, the average top diameter of the new bower, or 'courtship platform', was not significantly different from the mean control diameter of 29 cm. Although platform shape and size are specific for all species we have examined, height and base diameter can vary owing to local conditions and the age of the arena.

Our results suggest that within *Cyrtocara eucinostomus* a specific character, bower size, can influence the mate preference of females, and that males will

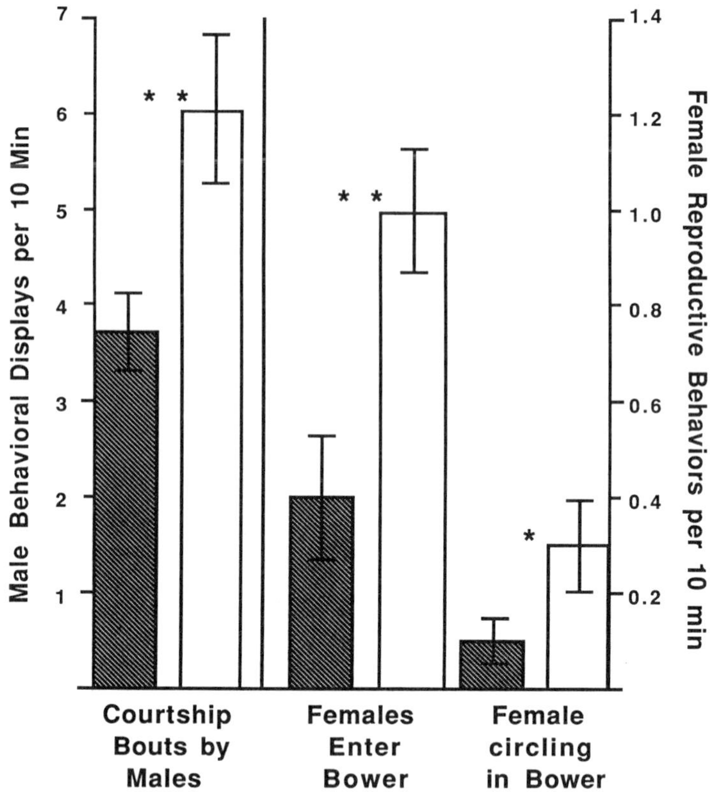

Fig. 12.9 *Cyrtocara eucinostomus*: behaviour of males and females on paired bowers, expressed as mean (± 1 S.E.) over 10 minutes. Hatched columns, 'short' bower (mean height 10.4 cm, $N = 43$); open columns, 'tall' bower (mean height 24.9 cm, $N = 43$). All differences in behaviour between tall and short bowers were significant (Wilcoxon test, *, $P < 0.05$; **, $P < 0.01$).

evolve behaviour which causes them to try to increase the size of their bower. Yet in some other Malawi lekking cichlids the bower is nothing more than a shallow depression in the sand. No bower form seems ideally adapted for reducing egg predation between laying and the female taking the eggs into the mouth, since at least ten major different bower forms have evolved in Lake Malawi (see page 249). Hence we conclude that bower form evolves primarily in response to female choice.

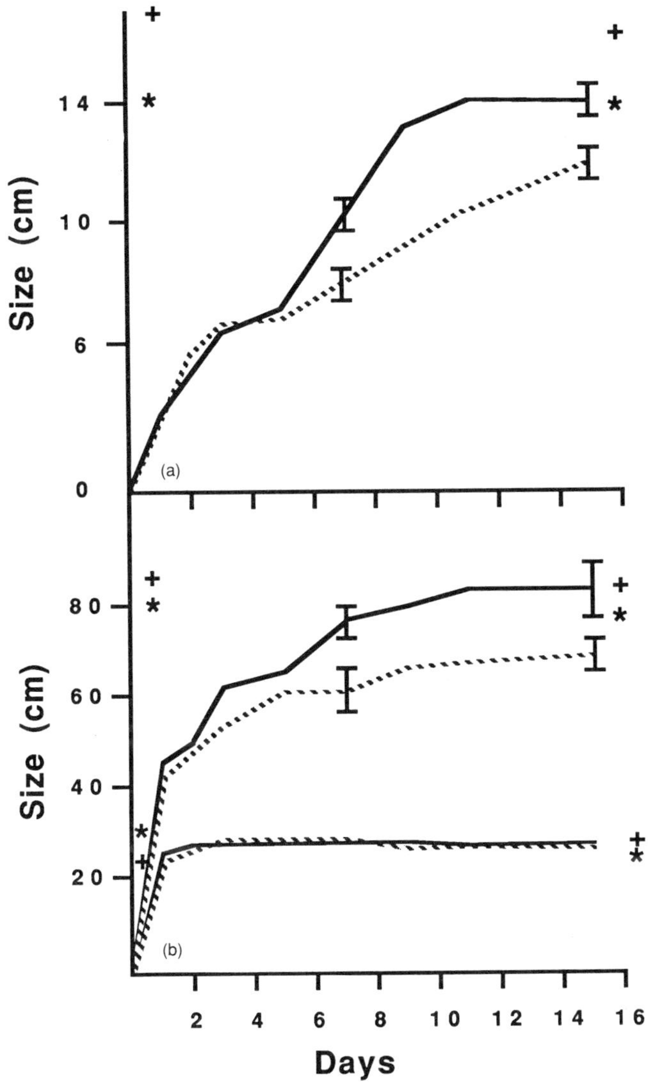

Fig. 12.10 *Cyrtocara eucinostomus*: growth of bowers following experimental removal of existing bowers. Solid curves, shallow bowers; broken curves, deep bowers; +, mean shallow control, *, mean deep control. Control means were determined at beginning and end of experiment. Vertical bars indicate 95% confidence interval. (a) Height of bower. (b) Diameter of bower at base (upper curves) and top (lower curves).

These initial studies have demonstrated only that the necessary first conditions for sexual selection are met: (1) variation in a male character and (2) female use of this trait as a criterion in mating. Demonstration of the heritability of this trait will be required as further support for the sexual selection hypothesis of cichlid speciation. Nevertheless, underwater exploration within Lake Malawi supports Arnold's (1983) assertion that lek-breeding species should exhibit "extraordinary geographic variation in male attributes". On the east side of the Nankumba Peninsula we have recently discovered two cichlid taxa which 'key out' morphologically as *C. eucinostomus*. One of these entities breeds at a smaller size and has bowers less than 1 cm in height in 15 m of water, and the other has bowers greater than 35 cm in height in 4.5 m water. The intriguing question of whether or not these differences could have arisen through differences in mate preference by females, and caused the splitting of these forms into new species, is now the subject of further investigation.

If variance in female preference for bower size can be demonstrated and if this variance is heritable (see Majerus *et al.*, 1982, 1986; O'Donald and Majerus, 1985, for such evidence in ladybirds), then Dominey's (1984) view that 'runaway' sexual selection played a major role in the explosive evolution of several hundred sand-dwelling cichlids will be strongly supported.

Future studies of female choice will require field observations followed by rigorous laboratory experiments which realistically duplicate the field conditions. Individual female choice patterns must be observed, and then the patterns of choice by daughters and the character of her sons must be determined. Large holding facilities with clear water are required for such studies. Since females will lay eggs with more than half a dozen males, it is important to observe the courtship behaviour directly. The presence of sneaky males (see page 180) and multiple matings could confuse results obtained under extremely confined conditions or limited visibility. To pursue this fascinating research, heritability studies will probably have to be linked to reproductive studies of cichlids in aquaculture ponds.

This hypothesis of cichlid speciation might initially appear to be an example of 'non-Darwinian' evolution. However, that is not so. It was Darwin (1871) who first discussed sexual selection of 'maladaptive' traits in male fishes in the lead chapter of the second volume of his two-volume work, *The Descent of Man and Selection in Relation to Sex*.

> "In many species the male alone is ornamented with bright colours; or these are much brighter in the male than the female. The male, also, is sometimes provided with appendages which appear to be of no more use to him for the ordinary purposes of life than are the tail-feathers to the peacock... There is reason to suspect that many tropical fishes differ sexually in colour and structure...."

Darwin (1871) was also made aware of the bright colour of cichlid males during the breeding season by Professors Günther of the British Museum and

Agassiz from Harvard University's Museum of Comparative Zoology. It is interesting to speculate how evolutionary theory might have developed if Darwin had been able to dive in Lake Malawi and collect and observe these amazing fishes instead of Galapagos Islands birds. Following Gould's study of his bird collection in 1837, Darwin became a proponent of geographic speciation driven ultimately by natural selection (Mayr, 1982). Would an extensive collection of brightly coloured Malawi cichlids have resulted also in a theory of speciation by sexual selection?

As Einstein wrote to Heisenberg, "It is theory which decides what we can observe" (Heisenberg, 1971), and the classical Darwinian theory of allopatric speciation driven by natural selection has been the predominant view of cichlid evolution (Fryer and Iles, 1972). Now we have an exciting alternative theory to test. The necessary research will have to involve long-term multi-disciplinary studies integrating behaviour, morphology and genetics. No one approach by itself will provide sufficient data to explain the intriguing biological phenomenon now occurring in the evolutionary laboratories of the Great Lakes of Africa. I am sure that no single theory will explain the origin of these species.

ACKNOWLEDGEMENTS

I am grateful to G. W. Barlow, J. Hoogland, J. Howard, R. Morgan, and J. R. Stauffer Jr for their comments on the manuscript and to B. McKaye and S. Rohwer for producing the figures. I would especially like to thank C. Hollingworth for his detailed efforts at copy-editing. The Malawi Fisheries Department provided the initial facilities which made this work possible and I am extremely grateful for all of the help members of this department have given me. The Department of Animal Sciences, Bunda College of Agriculture and the Department of Biology, Chancellor College, University of Malawi, provided valuable logistic support in the latter portions of this study, which was made possible by a cooperative agreement between the Universities of Malawi and Maryland. The work was supported by National Science Foundation grants BSR 82-14603 and BNS 86006836. Contribution 2017-AEL, University of Maryland, Center for Environmental and Estuarine Studies.

Chapter thirteen

The haplochromine cichlids in Lake Victoria: an assessment of biological and fisheries interests

C. D. N. Barel, W. Ligtvoet, T. Goldschmidt, F. Witte and P. C. Goudswaard

13.1 INTRODUCTION

Until recently the fish fauna of the world's largest tropical lake, the East African Lake Victoria, was dominated by haplochromine cichlids. Among vertebrates they were exceptional or even unique in various aspects: with *c.* 300 endemic species, the haplochromine flock of Lake Victoria was only rivalled by that of Lake Malawi, which is estimated to harbour 300–500+ species (Ribbink *et al.*, 1983a; Fryer, pers. comm.), but there is no other endemic group of cichlids or other vertebrates with such a wide range of niches realized with so little anatomical diversity.

At the onset of the 1980s, the Nile perch, *Lates ?niloticus*,* which had been introduced into Lake Victoria some 20 y previously for fishery purposes, suddenly increased vastly in number. It virtually obliterated the haplochromine flock lakewide within 7 y. Most probably the majority of haplochromine species from Lake Victoria are disappearing or have already become extinct. The fish fauna of the Lake is now dominated by the Nile perch, which sustains a greatly increased fishery. To establish whether such mass

*Preliminary taxonomic investigations of *Lates* specimens from Lake Victoria failed to identify these fishes unambiguously as *L. niloticus*. There remains the possibility that the *Lates* introduced into Lake Victoria belonged to other species, e.g. *L. macrophthalmus* or *L. longispinis*, or that it is a hybrid.

destruction of species is balanced by the benefits of the new fisheries requires a complex evaluation (Norton, 1986; Wilson, 1988) which is beyond the scope of this book. The best contribution we can make to that evaluation is to summarize the biological and fishery significance of the original haplochromine species flock and of the present fish fauna of the lake. On this consideration we based the structure of this chapter.

13.2 THE ORIGINAL HAPLOCHROMINE SPECIES FLOCK

Particular attributes of the haplochromines of Lake Victoria

The proliferation of species and adaptive radiation of the African Great Lakes cichlids make the species flocks from these lakes outstanding subjects for comparative and evolutionary biology.

As well as similarities such as speciosity, ecological diversity and repetition of ecotypes (Fryer and Iles, 1972), there are differences between the flocks. Within the total range of anatomical diversity exhibited by the African Great Lakes cichlids, the range of the Lake Victoria haplochromines is narrow and centred on a so-called generalized [=modal] morphotype (Greenwood, 1974a, 1981a), anatomically described by Anker (1978, 1984, 1986, 1987, 1989), Barel *et al.* (1976, 1977) and Ismail (1979). How can so many morphologically similar species coexist? Concerning this question, the haplochromine cichlids of Lake Victoria provide a great challenge to comparative vertebrate biology. For instance, the 45 km^2 of our research area in the Mwanza Gulf of Lake Victoria (Fig. 13.9) harboured *c.* 200 haplochromine species. The freshwater fishes of Europe comprise about the same number of species, and they belong to 29 families (Maitland, 1977). In the survey of research into the coexistence of the Lake Victoria haplochromines we shall demonstrate that so far there is no need for strikingly new theories explaining their coexistence. Rather, this species flock reveals how subtle can be the ecological differentiations of vertebrates at an incipient stage of speciation.

Survey of the haplochromines before the Nile perch impact

With a surface area of *c.* 69 000 km^2 (Greenwood, 1974a), Lake Victoria (Fig. 13.9) is the world's second largest lake. Physiographically the lake is by no means uniform (Greenwood, 1974a; Beadle, 1981; van Oijen *et al.*, 1981), and the distribution of the haplochromine species often agrees with the physiographic variation (e.g. Greenwood, 1974a, 1981a; van Oijen *et al.*, 1981; Witte, 1981, 1984a; Goldschmidt *et al.*, 1990).

Greenwood (1974a) estimated a total of 150–70 haplochromine species for Lake Victoria, but little was then known of pelagic species and of haplochromines from the deep (20–90 m) and mid-water (6–20 m) habitats (Fig. 13.9). The surveys by the Haplochromis Ecology Survey Team (HEST) in southern Lake Victoria included these habitats as well as the shallow waters. The resulting collections contain numerous new species, which raised the estimate to *c.* 300 species. As only a small section of the lake, namely the northern and southern parts, has been inventoried, any species that are geographically restricted to uninventoried areas are not counted. Another major bias in our estimate may result from the criteria for species distinction (page 268). However, owing to the serious reduction in the species flock these estimates can hardly be made more accurate.

Prior to the upsurge of the Nile perch in Lake Victoria, its original species flock of mainly endemic haplochromine Cichlidae surpassed in species number, biomass and ecological diversity all other fish families represented in the Lake (Greenwood, 1974a; Lowe-McConnell, 1987; pers. obs.). Each habitat contained its own community of haplochromine species. The narrow restriction to a particular habitat may be illustrated by the species distribution within the small ($< 0.5\,km^2$) Butimba Bay, which includes distinct rock, mud and sand habitats (Fig. 13.9) with entirely different haplochromine communities (Witte, 1984a).

Provided that our criteria for habitat similarity are relevant to haplochromine ecology, it appears that geographically separated habitats of the same type do not necessarily contain the same species. Although many species are found in the same habitat, both in the northern and southern parts of the lake (van Oijen *et al.*, 1981), others are known only from small areas (Witte and Witte-Maas, 1987; Goldschmidt and Witte, 1990; Sevenster, pers. comm.).

Besides speciosity and habitat restriction, trophic diversity is another well-known hallmark of the African Great Lakes cichlids. The Lake Victoria haplochromines tapped their food from virtually all resources, but each species included only a limited number of food categories in its diet. Based on these diets, the following trophic groups have been distinguished: epilithic and epiphytic algivores, detritivores, phytoplanktivores, zooplanktivores, insectivores, piscivores (including a scale eater and predators on cichlid eggs and larvae), molluscivores, prawn eaters, two species feeding on fish parasites, a macrophytophagous species and a crab-eating species (Greenwood, 1974a and many other publications, reprinted in Greenwood, 1981a; Hoogerhoud *et al.*, 1983; van Oijen *et al.*, 1981; van Oijen, 1982; Witte, 1981, 1987; Witte and Witte-Maas, 1981; Goldschmidt *et al.*, 1990). Except for the last three, all other trophic groups are represented by more than two species: there are *c.* 120 species of piscivores and 10–20 species in most of the remaining trophic groups (Witte and van Oijen, 1990).

Biological research on the original haplochromine fauna

History of the research

Until World War II, research on the haplochromines from Lake Victoria mainly concerned species descriptions based on museum collections. The first description (Hilgendorf, 1888) also introduced the taxon *Haplochromis*. Most papers from this period are by Boulenger, Regan and Trewavas, all taxonomists at the British Museum (Natural History).

Although the description of species remained a major part of the research, interest in the Lake Victoria haplochromines broadened after World War II. During the 1950s and 1960s most investigations were undertaken by scientists linked with EAFFRO (East African Freshwater Fisheries Research Organization), operating from Jinja (Uganda) at the lake's northern end (Fig. 13.9). The majority of publications from that time are by Greenwood and still concern species descriptions, but Fryer, Greenwood and Welcomme, the main EAFFRO representatives working on haplochromines, also dealt with haplochromine phylogeny, conservation, ecology and fisheries. During the 1970s Greenwood extended and synthesized his studies on the East African haplochromines, including those of Lake Victoria.

Fisheries research in the beginning of the 1970s revealed that the Lake Victoria haplochromines constituted a huge stock which was only partly exploited (page 274).

Since 1975, biologists from Leiden University (The Netherlands) have investigated the Lake Victoria haplochromines; their field team (HEST)* has operated since 1977 in southern Lake Victoria (Fig. 13.9). The group's research, and that of biologists associated with it, is wide ranging, e.g. ecology, fisheries, biological control of bilharzia, taxonomy and functional morphology.

The comprehensive bibliography on Lake Victoria by Crul *et al.* (1989) contains over 200 references on haplochromines. Although no longer complete, and partly outdated, Greenwood (1974a) is still the best general introduction to the haplochromine species flock of Lake Victoria.

Research on coexistence

Taxonomic and ecological knowledge of the Lake Victoria haplochromines increased considerably during the 1960s and 1970s (Greenwood, 1974a, van Oijen *et al.*, 1981; Witte, 1981). It revealed a community of some 300 closely related species that exploited virtually all resources in the lake. These observations led to two complementary questions concerning the coexistence of these species: (1) how do these species subdivide the available resources? (2) how can such a wide array of resources be exploited with so little anatomical diversity?

Among species using the same general category of food, there is often a further, trophic differentiation, which can be described qualitatively (e.g. prey

*HEST reports, quoted in this chapter, are available from F. Witte.

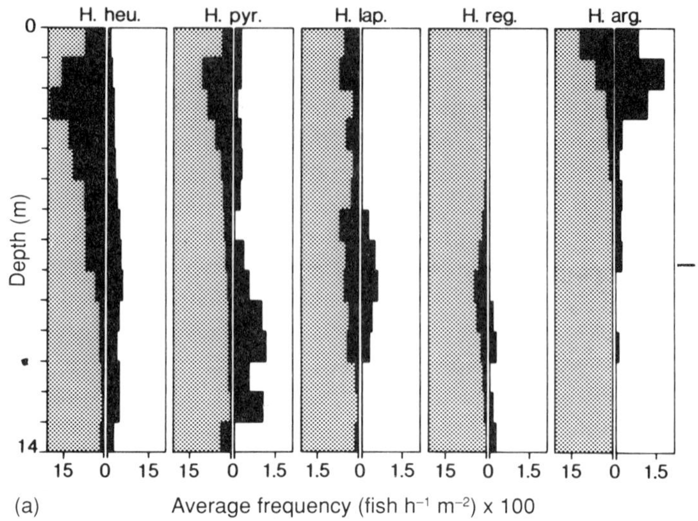

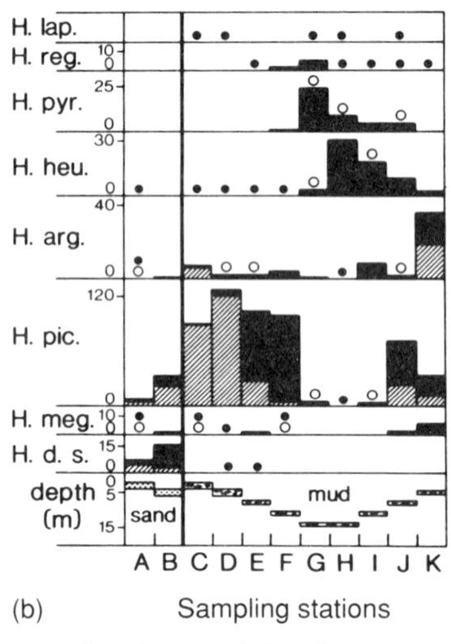

Fig. 13.1 (a) and (b) Horizontal and vertical distribution patterns of zooplanktivorous haplochromine species in the Mwanza Gulf. (a) Vertical distributions, derived from gill-net catches of station G (Fig. 13.9). Black bars denote number of fish; stippled background, noctural samples (22.00 to 05.00); white background, diurnal samples 10.00 to 17.00). Note differing horizontal scales. (b) Horizontal distributions, average numbers of fish in a 2.5 kg sample taken from trawl catches from a transect over the Mwanza Gulf (see Fig. 13.9). Black columns, adults; hatched columns, juveniles; ●, < 1 adult; ○, < 1 juvenile. Note differing vertical scales *H.* arg., *Haplochromis* 'argens'; *H.* d.s., *H.* 'double stripe'; *H. heu.*, *H. heusinkveldi*; *H. lap.*, *H. laparogramma*; *H. meg.*, *H. megalops*; *H. pic.*, *H. piceatus*; *H. pyr.*, *H. pyrrhocephalus*; *H.* reg., *H.* 'reginus'. (Adapted with permission from Goldschmidt *et al.*, 1990).

taxa selected by piscivores (Fig. 13.5; van Oijen, 1982) or quantitatively (e.g. zooplanktivores eating the same items but in different proportions; Witte, 1984a, 1987; Goldschmidt *et al.*, 1990). Species having similar diets can be segregated in their horizontal or vertical distributions, e.g. insectivores (Hoogerhoud *et al.*, 1983), molluscivores (Witte, 1981, 1984a, 1987) and zooplanktivores (Fig. 13.1; Goldschmidt *et al.*, 1990; Goldschmidt and Witte, 1990). Such segregations may exist temporarily under extreme conditions (e.g. very low prey densities, Goldschmidt, 1989c) or periodically, i.e. seasonally, monthly or daily (Fig. 13.1; Witte, 1984a; Goldschmidt *et al.*, 1990). Niche segregation concerns not only the trophic and spatial axes: among a group of sympatric zooplanktivores, the taxonomically most similar species differed most when breeding characteristics were taken into account (Goldschmidt and Witte, 1990).

The results summarized above indicate that each species has a unique combination of ecological attributes, which are generally known to contribute to species separation. The precise mechanisms by which these attributes were instrumental in niche segregation in Lake Victoria mostly remain unknown, but the following examples of anatomical and behavioural differentiation provide clues.

In contrast to previous opinions that adaptive radiation mainly concerns the feeding apparatus (Greenwood, 1974a, 1984; Sage *et al.*, 1984), more recent research suggests that all anatomical systems are involved, e.g. locomotor apparatus (Barel, 1983), gills (Fig. 13.2; Galis and Barel, 1980; Hoogerhoud *et al.*, 1983), eyes (Fig. 13.3; van der Meer and Anker, 1984; van der Meer, 1989), ovaries (Goldschmidt and Goudswaard, 1989; Goldschmidt and

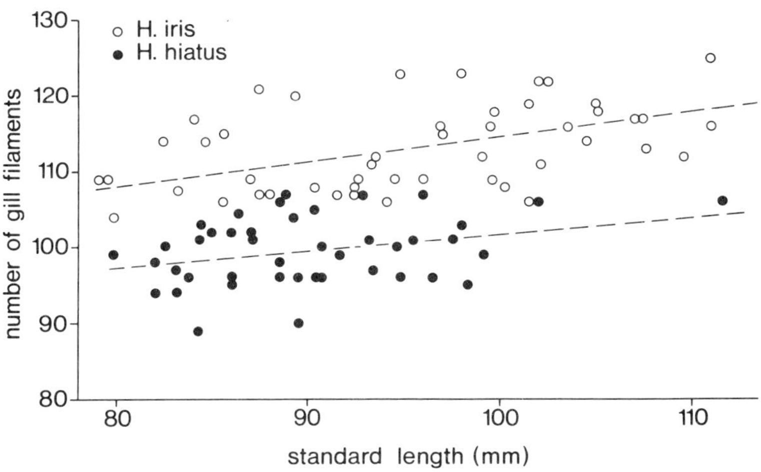

Fig. 13.2 Ecomorphological differences in gills. *Haplochromis hiatus* (●) and *H. iris* (○) are two closely related parapatric species. The oxygen concentration in the habitat of *H. iris* was at times considerably lower than that in the habitat of *H. hiatus*. (Reproduced with permission from Hoogerhoud *et al.*, 1983.)

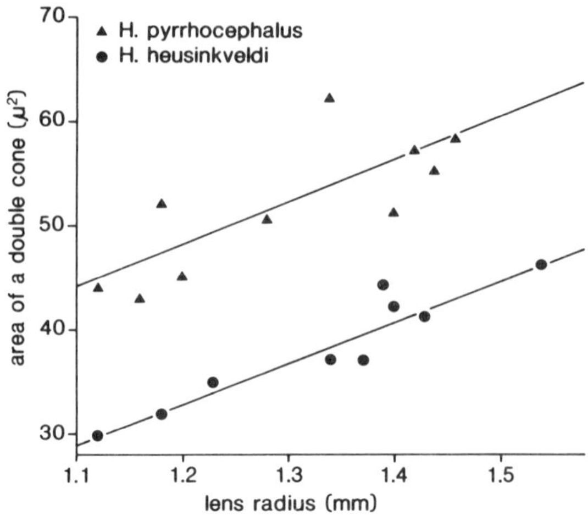

Fig. 13.3 Structural differences in the retinas of two microsympatric species, *Haplochromis pyrrhocephalus* (▲) and *H. heusinkveldi* (●) (see Fig. 13.5). Ecologically they differed mainly in the proportions of food items in their diets, which were otherwise highly similar. Among eyes of the same size (i.e. same lens radius), a larger cone area is an indication of greater sensitivity and reduced acuity. (Reproduced with permission of H. J. van der Meer.)

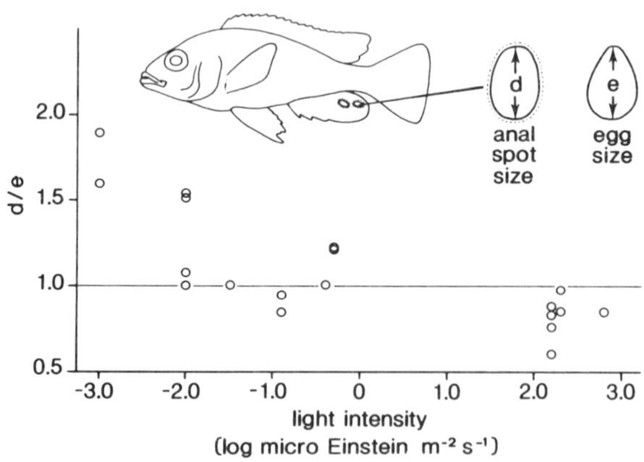

Fig. 13.4 Ecomorphology of anal spots. The size of these spots may be a compromise between two demands: mimicking an egg and conspicuousness. In brighter habitats the spots were smaller than the eggs, in darker habitats they tended to be larger, and in habitats with intermediate light intensities there was a good match between egg size and spot size. Each dot represents one species. (Reproduced with permission from Goldschmidt, 1989b and de Beer, 1989.)

Witte, 1990) and secondary reproductive characters (Fig. 13.4; Goldschmidt, 1989b).

Besides morphological differentiation, species-specific and intraspecific differences in foraging behaviour (cf. Slootweg, 1987; Galis and de Jong, 1988) can also contribute to segregation. Behavioural differences may result in the tapping of different fractions of the same resource population. The piscivores (Fig. 13.5) and the molluscivores provide a case in point.

Among the paedophages, i.e. piscivores feeding on cichlid broods (Greenwood, 1974a), *Haplochromis barbarae* steals eggs from a spawning female in the short period between laying the eggs and taking them into her mouth (Witte-Maas, 1981). In contrast, *H.* 'rostrodon' obtains its prey by ramming mouthbrooding females and engulfing their snouts (Wilhelm, 1980; but see page 162). If the number of eggs spawned exceeds the carrying capacity of the buccal cavity (Welcomme, 1967b), such 'overproduced' eggs would represent a fraction of the resource not exploited by the engulfing paedophages. The anatomical differences among paedophagous cichlids correspond with the

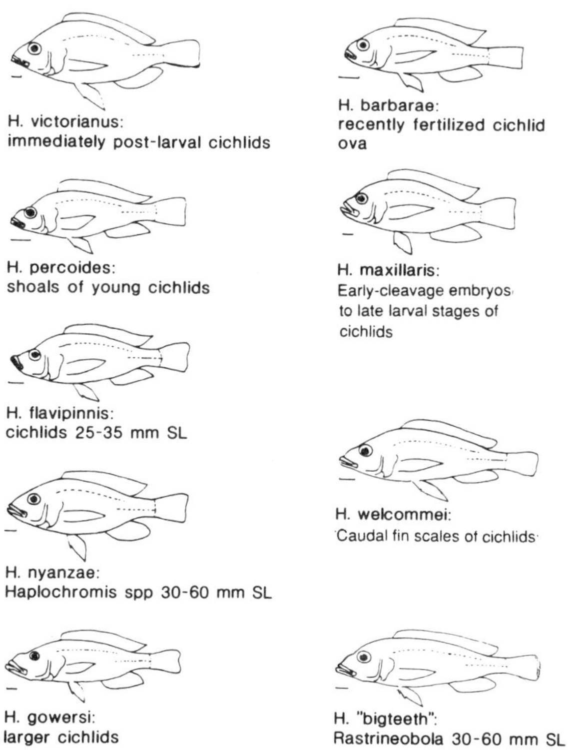

Fig. 13.5 Intratrophic differentiation of haplochromine piscivores. (Adapted from Barel *et al.* (1977) and Greenwood (1981a); reproduced with permission from van Oijen, 1982.)

behavioural differences: the oral features of *H. barbarae* are similar to those of bottom-feeding insectivores, whereas the suction apparatus of *H.* 'rostrodon' resembles that of an ambush-hunting piscivore (Barel *et al.*, 1976, 1977; van Oijen in: Dullemeijer and Barel, 1977; Barel, 1983). The anatomical differentiation of the 20+ species of paedophages suggests further differentiation of their feeding behaviour (Wilhelm and Barel, unpublished data). These species were among the first to disappear (Witte and Goudswaard, 1985) and direct observation of their behaviour is therefore impossible.

Following Greenwood (e.g. 1974a), the molluscivorous species can be divided into two groups: oral shellers and pharyngeal crushers. The crushers possess hypertrophied pharyngeal jaws and muscles, whereas the shellers have short, stocky oral jaws powered by thick adductor muscles (Barel, 1983). Laboratory observations (Slootweg, 1987) demonstrated that shellers can feed on larger snails than crushers. With its oral jaws a sheller grasps a large snail by its foot and then, by vigorously shaking its head, rips the snail from its shell. A crusher always cracks the mollusc between its pharyngeal jaws; swallowing is preceded by a time-consuming process during which shell fragments are spat out (Hoogerhoud, 1986a, 1987).

Both feeding strategies have advantages and disadvantages. Oral shelling requires an accurate, visually guided predation technique, which is liable to failure: e.g. under aquarium conditions the snail often retracts its foot in time (Slootweg, 1987). Probably the necessary visual cues can be seen in the sandy habitat to which most oral-shelling species are restricted (Witte, 1981, 1984a), but not in mud bottoms, the other main snail habitat. Pharyngeal crushers rarely fail once a prey is encountered, and their feeding technique allows them to live over various bottom types. Many crushing species do occur over muddy substrates (Witte, 1981, 1984a). The drawbacks of the mollusc-crushing technique include restriction to smaller prey and time-consuming prey-processing. A biophysical explanation of these constraints is given by Hoogerhoud (1986a, 1987).

Further trophic differentiation occurs within both the shelling and crushing species groups (Hoogerhoud, pers. comm.), but the underlying mechanisms have scarcely been investigated.

The foregoing illustrates how an analysis of trophic behaviour and morphology can be instrumental in explaining niche segregation. More complex morphological constraints on ecological segregation have been investigated by Barel and co-workers (Barel *et al.*, 1989), who argue that the number of resources which a species may exploit is limited by the compatibility of the anatomical requirements for coping with these resources. For instance, constructional morphology (*sensu* Dullemeijer and Barel, 1977) demonstrates that the size of the buccal cavity is related not only to food but also to the locomotor apparatus (Hoogerhoud and Barel, 1978; Barel 1983), the eyes (Barel, 1983, 1984), the gill apparatus (Galis and Barel, 1980) and to the number and size of the eggs produced in the ovary (Goldschmidt and Barel, in prep.). This complex interrelationship implies that the various ways of feeding, swimming, breathing and breeding known for haplochromines in general, cannot be combined haphazardly. Combinations incompatible on anatomical grounds predict the trade-offs among the above functions, each of which is a co-determinant of fitness. As Sibly and Calow (1986) conclude, exposing such trade-offs has proven to be surprisingly difficult. The Lake Victoria haplochromines apparently would have been suitable objects for an attempt to solve this problem with an a priori (deductive) approach.

So far 'only' the observed niche segregation has been related to the anatomical and behavioural capacities of the species. However, there remains the important question of the extent to which the species is restricted to that niche, in other words how far it could widen or change its niche once the ecological circumstances changed. For understanding the evolutionary processes which shaped the cichlid communities these questions are interesting, because they help to identify which selection pressures are active.

Natural experiments provided by the Lake Victoria ecosystem gave some insight into the role that competition played in niche segregation (Goldschmidt, 1989c).

Phenotypic plasticity is another clue to potential selection pressures. Again the haplochromine species that may take molluscs in their diet provided suitable objects (other examples of phenotypic plasticity in cichlids: Witte, 1984b; van der Meer and Anker, 1986; Meyer, 1987; Witte *et al.*, 1990). *Astatoreochromis alluaudi* from several lakes had less hypertrophied pharyngeals (Fig. 13.6.). As far as information is available, these fishes had no molluscs in their diet and fed on insects, probably a more profitable prey than molluscs (Hoogerhoud, 1986a, b). By raising *A. alluaudi* on either snail or snail-less diets, Hoogerhoud (1986a) demonstrated that the pharyngeals of these fish developed in the same way as those of wild-caught fish with and without snails in their diet. Similar indications have been obtained for other molluscivorous species of Lake Victoria (van Overbeeke, pers. comm.). From his observations, Hoogerhoud derived the intriguing hypothesis that *A. alluaudi* (or other potential pharyngeal crushers) in competition with an anatomically better-adapted insectivorous species would be forced in early ontogeny to feed on less profitable items (snails), and that the hypertrophy of its pharyngeals, induced by molluscivory, would make it progressively less efficient at feeding on the more profitable insects. Lake Victoria contained many morphoclines (viz., series of species gradually differing in their ecologically relevant structures, e.g. Fig. 13.7) which would have been ideal subjects for experimental research on the significance of competition and its implications for speciation. In all probability, only a few have survived the Nile perch impact.

Research on taxonomy

Among the African Great Lakes cichlids, the anatomically little-diversified haplochromines of Lake Victoria probably offer the most complex taxonomic

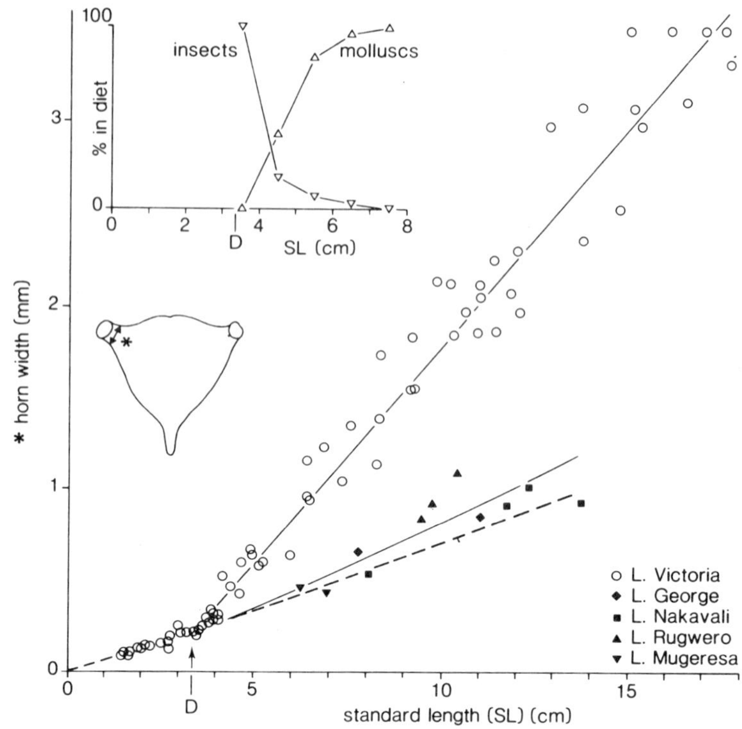

Fig. 13.6 Variability in hypertrophy of the lower pharyngeal jaw of *Astatoreochromis alluaudi*. Horn width is a good measure of the overall degree of hypertrophy. Broken line, regression through horn widths prior to D; solid symbols, specimens which, as far as could be determined, had not fed on molluscs; D, size at which molluscs enter the diet. *Inset*: Diet composition of *A. alluaudi* from Lake Victoria. (Adapted with permission from Hoogerhoud, 1986a and Witte *et al.*, 1990.)

riddle, both at the generic and at the species level. The narrow clustering around the modal form, the occurrence of all intergrades within morphoclines of 'species' (Greenwood, 1974a, 1981a), and biochemical similarity (Sage *et al.*, 1984; Verheyen *et al.*, 1985; Rompaey *et al.*, 1988; Meyer *et al.*, 1990) all suggest that the Lake Victoria haplochromines were at an incipient stage of radiation. As we experienced only too often with these fishes, species distinction at such an evolutionary grade is not only difficult practically, it is a fundamental problem for which current taxonomic procedures offer no solution (cf. Heywood, 1988; Patterson, 1988).

In contrast to the attention paid to supraspecific classification, remarkably little has been published on the technical and fundamental problems of haplochromine species distinction. Describing these species was for many years a speciality of museum curators like Hilgendorf, Boulenger, Pellegrin, Regan, Trewavas, Poll and Greenwood. Only recently, when ecological and

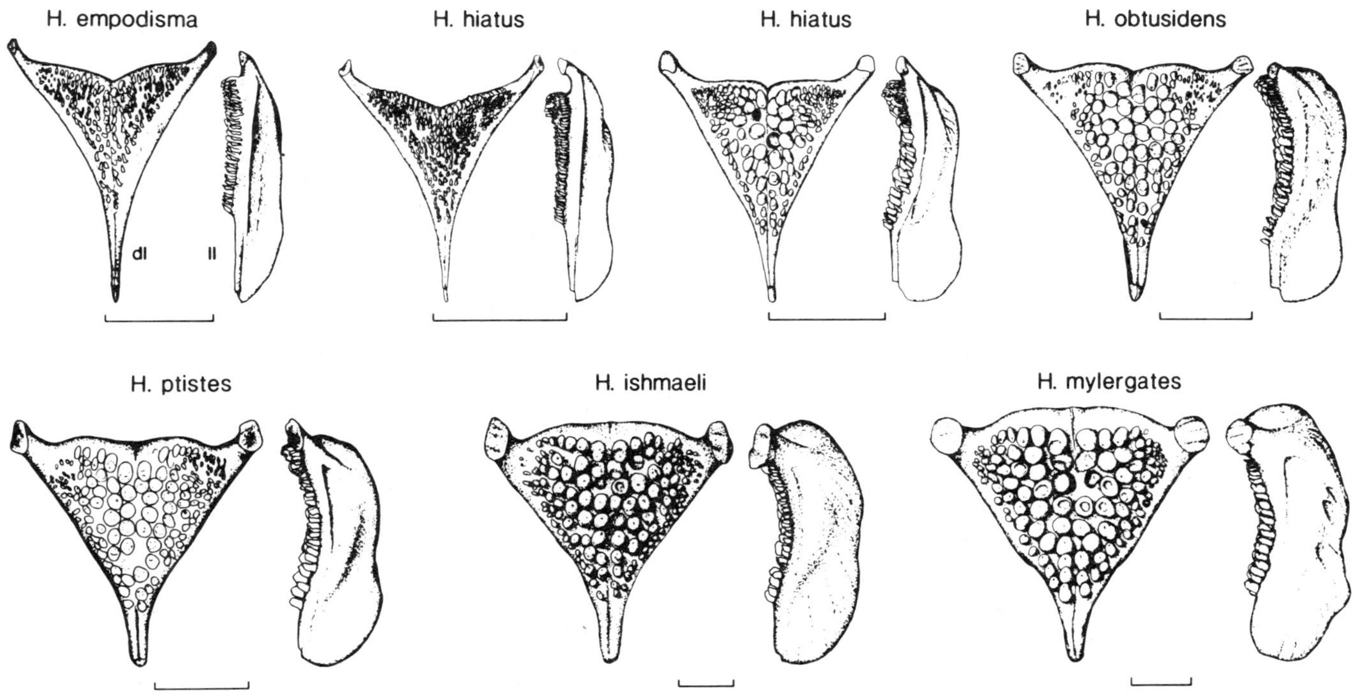

Fig. 13.7 Morphocline of lower pharyngeal jaws of haplochromines from Lake Victoria. (Reproduced with permission from Hoogerhoud, 1984.) Scale bar, 5 mm.

fishery research confronted biologists other than museum systematists with species identification, the description of species, especially of the difficult haplochromines of Lake Victoria, required a more public discussion (e.g. Barel *et al.*, 1977). In general, biologists trained in distinguishing haplochromines taxonomically produce similar classifications. However, these mainly intuitive classifications lead to a practical and a fundamental problem, namely the objective description of the forms distinguished and the question whether these forms are biological species. Except for the study of Crapon de Caprona and Fritzsch (1984) and anecdotal aquarium observations of Witte (1984a), there is little direct evidence that the so-called 'species' of haplochromine cichlids from Lake Victoria fulfil the criteria of a biological species as set by Mayr (e.g. 1964). The best indirect evidence comes from ecological and morphological attributes from which genetic isolation can be inferred. Among morphologically similar forms, coloration of sexually active males is such a character. However, as Witte and Witte-Maas (1987) experienced in their attempt to distinguish species among forms that morphologically were only slightly different, there remain major problems.

If these species are so difficult to delimit, what biological meaning do the forms distinguished have, or are they only classification artefacts? Our research indicates that sympatric and parapatric forms, at least, behave as different species. They differ in male breeding coloration and often too in other characteristics such as the location of spawning areas and the timing of the breeding season (Goldschmidt and Witte, 1990; Hoogerhoud *et al.*, 1983). However, species delimitation and description becomes problematic when, in addition to a considered form, there are other, allopatric forms which differ only slightly in characters that indicate genetic isolation. Alas, this problem, well known to taxonomists, is staggeringly complex and only too common for forms of Lake Victoria haplochromines (e.g. Witte and Witte-Maas, 1987; Goldschmidt and de Visser, 1990). The question with such forms is not only which to include in one species, but also what subspecific status to give to the included forms. These forms provide biologically meaningful clusters of ecological and morphological data, of which the coherence must not be lost by lumping them within one species. The administrative techniques of taxonomy do not provide a formal procedure for labelling these biologically increasingly-relevant units. Witte and Witte-Maas (1987) suggested a practical procedure for coping with the foregoing problems but several fundamental questions remain unanswered.

As yet, taxonomic analysis of the Lake Victoria haplochromines has not solved the problems of species distinction nor that of supraspecific relationships. The latter is not treated here, but is discussed extensively elsewhere (Greenwood, 1980, 1981a; Barel, 1984; Hoogerhoud, 1984; Witte and Witte-Maas, 1987; van Rompaey *et al.*, 1988). It is possible (1) to distinguish different forms, which in areas of potential interaction behave as different species, and (2) to distinguish a limited number of supraspecific groups which might be different evolutionary lines. However, the precise taxonomic status of these forms and groups is not clear and their objective delimitation is problematic. The problems offered by the Lake Victoria haplochromines are probably typical of incipient radiation.

It is a sad conclusion that the largest extant example of such a radiation among vertebrates was destroyed during a period when its potential significance to evolutionary and comparative biology was revealed and widely acknowledged. What is left of the original species flock of Lake Victoria haplochromines consists of extensive collections of field notes, colour slides and preserved specimens. The incomplete taxonomic synthesis of these data will seriously impede the accessibility of the Lake Victoria haplochromines for studies on comparative biology.

13.3 THE IMPACT OF THE NILE PERCH ON THE ORIGINAL HAPLOCHROMINE FLOCK

Transition period

After its introduction into northern Lake Victoria around 1960 (Coulter *et al.*, 1986), the Nile perch population remained for two decades at low levels, i.e. a catch rate of less than 1.5% by weight of the total catch (Kudhongania and Cordone, 1974). At the end of the 1970s a sudden increase of the Nile perch stock was noted in the shallow Wynam Gulf in northern Lake Victoria (Fig. 13.9). Here trawl catches of Nile perch constituted in 1981 more than 90% by weight of the catch. In the period 1981–83 the Nile perch contributed not more than 7% of the total Ugandan catch, of which the haplochromines still formed 86% (Okaronon *et al.*, 1985). After spreading along the shores in a more or less clockwise direction, the Nile perch began to increase strongly in southern Lake Victoria a few years later. Trawl catches in various Tanzanian waters showed tremendous expansion of the Nile perch stock since 1984 (Fig. 13.8), with catch rates rising within 3 y from almost nil to 90–99% of the catch in 1987.

With the increase of the Nile perch, there was a simultaneous decline of most other fish stocks in the lake, especially of the haplochromine stock (Fig. 13.8; Arunga, 1981; Müller and Benda, 1981; Okemwa, 1981, 1984; Okaronon *et al.*, 1985; Hughes, 1986; Goudswaard, 1988; Goudswaard and Ligtvoet, 1988). The rapidly declining body size (dwarfism) and falling catches of the inshore haplochromines in heavily fished areas indicated the vulnerability of this stock to intensive artisanal (Marten, 1979) and industrial fishing (Witte and Goudswaard, 1985, 1987; Goudswaard and Ligtvoet, 1988). However, the following data demonstrate that the lakewide decline of the haplochromine stock must be attributed mainly to the impact of the Nile perch. In the Emin Pasha Gulf (Fig. 13.9), where neither an intensive artisanal nor a trawl fishery was conducted, the haplochromine stock also declined strongly while the Nile perch stock increased. Haplochromines are now virtually absent from the

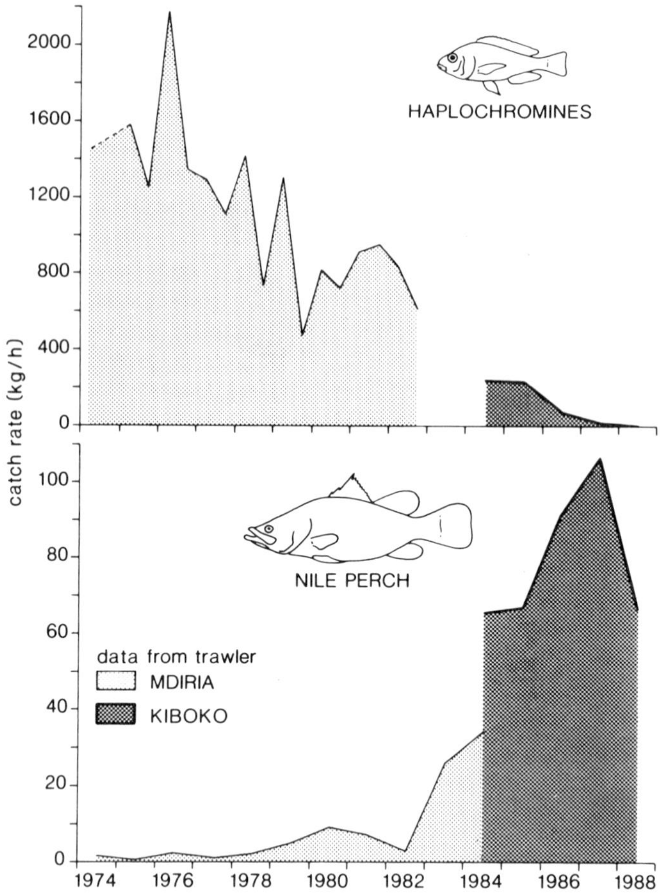

Fig. 13.8 The decline of the haplochromines (a) and concomitant upsurge of the Nile perch (b) in the Mwanza Gulf areas fished by two trawlers; the fishing capacities of the two trawlers are different. (Adapted from Goudswaard and Ligtvoet, 1988.) Light stippling, data from *Mdiria*; dark stippling, data from *Kiboko*; the *Kiboko* became operational in 1984).

deeper waters (25–50 m) near Mwanza (Goudswaard and Ligtvoet, 1988), where at the end of the 1960s the catch rates of haplochromines were considerable (Kudhongania and Cordone, 1974) and about as high as those in shallower water. Since fishing hardly occurred in the deeper waters, whereas the Nile perch in the trawl surveys did increase from zero to 100 $kg\,h^{-1}$ in these areas (Goudswaard and Ligtvoet, 1988), the disappearance of the haplochromines from the deeper part of the lake is likely to result from predation by the Nile perch. Originally the predation level of fish on haplochromines was set mainly by piscivorous haplochromines and by the bottom-dwelling catfishes *Clarias* and *Bagrus*, of which the latter was the most important (Chilvers and Gee, 1974).

With the upsurge of the Nile perch, the level of predation by fish on haplochromines increased markedly in the lake. Until the haplochromines became scarce, the Nile perch preyed predominantly on them (Gee, 1969a, b; Hamblyn, 1966; Okedi, 1971; Ogari, 1985, 1988; Ligtvoet and Mkumbo, 1990). Once the haplochromines were depleted, the Nile perch changed to the shrimp *Caridina nilotica*, the cyprinid *Rastrineobola argentea* and to cannibalism (Ogari and Dodzie, 1988; Ligtvoet and Mkumbo, 1990).

Current status

By the end of 1987 the haplochromine stock in the areas investigated had virtually disappeared. Numerous species vanished, including those which once constituted the bulk of the haplochromine catches. Some pelagic, zooplanktivorous haplochromine species have survived, but their abundance has dwindled dramatically. Haplochromine species from rock habitats and from shallow inshore waters seem less affected by Nile perch predation. Later data from trawl surveys in the Kenyan and Ugandan waters are not available, but in general the developments of the haplochromine and Nile perch stocks appear similar to those of the better-surveyed, Tanzanian areas (Hughes, 1986, but see Anonymous, 1988)

In September 1986, haplochromines still formed the most important food source of the Nile perch in the Mwanza area, although their stock was already much reduced. However, in June and September 1987, the benthic prawn *Caridina nilotica*, the cyprinid fish *Rastrineobola argentea* and juvenile Nile perch had become the main food items of adult Nile perch (Ligtvoet and Mkumbo, 1990). Compared with haplochromines the Nile perch is extremely flexible in its diet: large (up to 25–30% of body length) and small prey (2 cm *Caridina*) are readily taken.

Future status

Considering the enormous impact of the Nile perch upon the haplochromine stock in Lake Victoria, and in view of the situation in Lakes Albert and Turkana, where haplochromines and Nile perch have long coexisted, it is unlikely that the original haplochromine flock will recover. In Lakes Albert and Turkana, where haplochromines constitute no more than *c.* 20% of the species (Lowe-McConnell, 1987), they form a minor part of the total fish population and are ecologically far less diverse than in Lake Victoria (Trewavas, 1938; Greenwood, 1974b). In Lake Kyoga, where the Nile perch was introduced around 1955, the effects resemble those in Lake Victoria. The formerly abundant haplochromine species are now rare, with relict populations restricted to inshore areas or to areas with aquatic macrophytes (Ogutu-Ohwayo, 1985). Over a period of less than 15 years the following changes in the diet of the Nile perch in Lake Kyoga were observed (Ogutu-Ohwayo, 1990). Soon after its introduction into Lake Kyoga haplochromines and small mormyrids were the main prey of the Nile perch. A few years later the diet shifted to include tilapiines, and still later *Rastrineobola argenta*, which previously had not been recorded as prey but then became even more important than cichlids.

Developments in the Nile perch stock will depend largely on developments in the stocks of their main prey. Despite predation on them by the Nile perch, the stocks of *Rastrineobola* and *Caridina* are increasing (Wanink, 1988, pers. comm.). The ecological mechanism is not yet fully understood, but competitive release due to the disappearance of the detritivorous haplochromines (competitors of the prawns) and of zooplanktivorous haplochromines (competitors of *Rastrineobola*) may play a role. These developments may indicate that the current population level of Nile perch can be maintained.

The above expectations are speculative and should be looked upon with great caution. The ecosystem is still changing and there is little experience with this developing system. Anecdotal evidence suggests an increase in fish kills (Ochumba, 1988; pers. obs.), in populations of molluscs and lake fly (chaoborids) and eutrophication. For the benefit both of fisheries and of academic biology, great efforts should be made to follow this 'natural' experiment with care. The former, complex ecosystem has proved sufficiently resilient to withstand perturbations caused by increased fishing pressure. Although some target species were reduced to low population levels (Marten, 1979), the general structure of the ecosystem was not affected. However, the large perturbation which the Nile perch imposed upon the system apparently exceeded its resistance: numerous species covering all trophic levels were depleted or destroyed, leading to a structurally different and – in terms of species number – more simple fish community (Ligtvoet and Witte, in press).

13.4 FISHERIES IN LAKE VICTORIA AND THEIR SOCIOECONOMIC EVALUATION

Fisheries before the Nile perch upsurge

The fisheries in Lake Victoria have been documented only from the onset of this century. The main pattern is that after more valuable target species became over-fished, efforts were directed to other fish groups, which subsequently also declined.

During the period 1905 to 1916, when gill nets were introduced into different parts of Lake Victoria, the tilapiine cichlids *Oreochromis esculentus* and *O. variabilis* were the main target species. Catches decreased from 25–100 tilapias per net in 1916 to 4–7 in 1928 (Graham, 1929). During the following decades the catch dropped further and the fish became smaller.

In addition to the tilapiine fishery, there existed local fisheries for *Protopterus aethiopius*, *Bagrus docmac* and *Clarias gariepinus*. In the late 1950s the highly valued *Labeo victorianus*, caught by blocking rivers during its seasonal spawning migration into them (Fryer, 1973), was reduced to very low population levels.

At the beginning of the 1950s *Tilapia zillii T. melanopleura*, *Oreochromis niloticus* and *O. leucostictus* were introduced into the lake (Welcomme, 1967a), but this did not improve the catches of tilapias (Lowe-McConnell, 1987). By 1970 the tilapiine catches had declined to less than 0.5 fish per net (Fryer, 1972; Kudhongania and Cordone, 1974).

Since the late 1960s, gill nets have been used to harvest smaller fish, such as haplochromines, *Synodontis* spp. and *Schilbe mystus* (Marten, 1979). In the 1970s, beach seine fishing, which captures haplochromines and large numbers of spawning and juvenile tilapiines, spread around the lake (Marten, 1979). Meanwhile light fishing on the cyprinid *Rastrineobola argentea* developed in both Kenyan and Tanzanian waters (Marten, 1979; Okedi, 1981).

Because of their small size and bony texture, the haplochromines were among the last groups to be exploited. As the catches of the larger, non-haplochromine species continued to decline, beach seine and gill net fishermen started to concentrate on haplochromines. They became an important protein source in a wide belt around Lake Victoria in the early 1970s and were eaten fresh, dried or smoked.

A lakewide survey at the end of the 1960s estimated the total demersal fish biomass of the lake to consist of at least 80% haplochromines (Kudhongania and Cordone, 1974). These findings led to the development of a small-scale trawl fishery in the Tanzanian part of the lake. A fish meal plant became operational in Mwanza in 1976. It should have converted about $60\,t\,d^{-1}$ of wet fish into fodder (Bon and Ibrahim, 1975), but it never reached this level. Simultaneously a trawl fishery for direct and local consumption developed. By the end of the 1970s about ten trawlers were working in the Mwanza area. Although this trawl fishery was meant to be restricted to areas with a depth exceeding 20 m, in practice it mainly operated between 6 and 22 m.

The highest yields of haplochromines were recorded in 1977 (CIFA, 1982). However, within a decade after the onset of intensified fishing, the haplochromines in the Mwanza and Nyanza Gulfs showed signs of over-fishing (Marten, 1979; Witte, 1981; Witte and Goudswaard, 1985).

Fisheries during and after the Nile perch upsurge

The changes in the fish community brought about by the Nile perch resulted in structural changes in the fishery. In Kenya, the commercial catch of Nile perch in 1979–82 rose from virtually zero to almost 60% of the total catch (Okemwa, 1984; Ssentongo and Welcomme, 1985). Meanwhile in Uganda, the contribution of Nile perch in the commercial catches rose from *c.* 13% to almost 70% (Okaronon *et al.*, 1985). Fish-landing data show the onset of the Nile perch fishery in Tanzanian waters from 1982 onwards; within 3 y its production encompassed nearly 40% of the total catch (Ligtvoet *et al.*, 1988). Although the figures produced by the three riparian states may be disputed (CIFA, 1988), their general trend, revealing a fourfold increase over the last 5 y of the total fish landed, is not unrealistic. Currently the Nile perch fishery is the most important in the three riparian states. The increase in fisheries is due not only to the growing Nile perch catches, but also to increasing catches of *Oreochromis niloticus* and *Rastrineobola* (CIFA, 1988). In general it seems that the lake's multi-species fishery changed to a fishery with three main species of commercial importance.

Socioeconomics of the Lake Victoria fisheries

Comparing the higher yield of the present fisheries with that of former times, one may question whether such an increase could have been obtained by a carefully controlled intensification of the fishery on haplochromines. We (W.L., P.C.G.) consider this option non-viable. Because of the narrow and highly specific habitat ranges and low fecundity of these cichlids (Goldschmidt, 1989a; Goldschmidt *et al.*, 1990; Goldschmidt and Witte, 1990), the haplochromine stock has proved extremely sensitive to fishing pressure (cf. Marten, 1979; Witte and Goudswaard, 1985; Goudswaard and Ligtvoet, 1988). Furthermore, haplochromines were not very popular table fishes: the planned commercial fishery for haplochromines would mainly have produced fish meal.

Except in Uganda, where the Nile perch was already known from Lakes Kyoga and Albert, the people around Lake Victoria were reluctant to eat the new fish, which wiped out the traditional food fishes, destroyed their fishing gear and was disliked for its oily nature (Barel *et al.*, 1985; Coulter *et al.*, 1986; Ribbink, 1987). However, the fishing methods, feeding customs and fish-processing techniques were soon adapted. The Nile perch is widely appreciated as a 'money maker' and a good, reasonably priced table fish. In Tanzania, this formerly despised species is now nicknamed 'mkombozi' (saviour) (Ligtvoet, 1988), reflecting a changed and unambiguous attitude.

The expansion of the Nile perch stock, and the following strong increase in total yield, triggered a lively business. Despite the growing supply of the Nile perch, ex-vessel prices remained relatively high, indicating strong demand. The combination of high yield and high prices led to high and increasing revenues from the fisheries. Reynolds and Greboval (1988), assessing the gross economic benefits from the transformation of the Lake Victoria system, consider the effect of the Nile perch to be exceedingly positive.

Although accurate data are lacking, the number of fishermen seems to have increased significantly (CIFA, 1988; Reynolds and Greboval, 1988). In the Mwanza area of the lake, over the period 1980–88, the percentage of fishermen aiming for Nile perch rose from less than 5% to 80%. Most fishermen say they switched from other fisheries to the Nile perch fishery, but since 1984 an increasing number of persons entered the fishery from other occupations (Ligtvoet *et al.*, 1988). With a conservative assumption that the accessory services and post-harvest sectors employ around three persons per fishermen, Reynolds and Greboval (1988) estimated that direct employment of the entire fisheries provided many thousands of extra jobs after the establishment of the Nile perch.

13.5 CONCLUDING REMARKS

There is little hope that the haplochromine species flock from Lake Victoria will recover from the devastating impact of the Nile perch. In the Mwanza area, the only area of the lake monitored before, during and after the Nile perch started to boom, numerous haplochromine species have disappeared. Most of these species could formerly be collected from every catch, but shortly after 1984 (when the Nile perch became dominant) they vanished and so far they have not been caught again (Goudswaard, pers. obs.; Witte *et al.*, in prep.). In this way at least 60% of the haplochromine species from habitats now occupied by Nile perch are estimated to be extinct. Only in a few habitats do the original communities of haplochromine species seem less, or not at all, affected by the Nile perch. Some of these habitats, such as very shallow waters and rocky areas, are probably less accessible to Nile perch.

In view of the poor cichlid faunas of other lakes with Nile perch, there is no reason to expect a major recovery of Lake Victoria's now vestigial haplochromine species flock. Rumours suggesting otherwise declare that haplochromine species survived (e.g. Seegers, 1987; Anonymous, 1988), but fail to provide crucial evidence: for instance, a comparison of the haplochromine species before and concomitant with Nile perch domination. The unwarranted suggestions from such publications that 'things are not so serious' are typical misinterpretations of the decline of biota which are taxonomically little known.

The vastly increased Nile perch stock and simultaneous increases in the *Oreochromis niloticus* and *Rastrineobola* stocks have greatly improved the fisheries although the ecosystem is severely perturbed. As long as there is no catastrophic collapse of these stocks, the prospects for these fisheries seem good. However, the wood required for artisanal processing of the Nile perch aggravates seriously environmental problems, namely deforestation around the lake (Barel *et al.*, 1985; Reynolds and Greboval, 1988).

Among biologists the decline of the Lake Victoria species flock has caused great concern and indignation (Barel, 1986). The flock offered an exceptionally fertile research object for studying incipient radiation of vertebrates. Similarity in the ecological requirements of many Lake Victoria haplochromine species suggested to Greenwood (1974a, 1981a, 1984) that basic ecological principles were violated by the Lake Victoria haplochromines. However, later research triggered by these suggestions revealed that the ecological and ecomorphological similarity was more apparent than real. The scale by which to measure the interspecific differentiation needed to be far finer than any hitherto applied. The results demonstrated consistent differences in diet, breeding, microhabitat, behaviour and anatomical structures, all of which are potentially relevant to ecological isolation. The next step, a critical test of the actual mechanisms by which these differences contributed to keeping species separated, is now impracticable for most species. Most of the investigated species have disappeared and their habitats have undergone major changes. However, as an ardent opponent of Nile perch introduction into Lake Victoria pointed out objectively, there are also interesting aspects to academic biology in the current events: a large-scale field experiment on the evolutionary effect of adding a top predator (Fryer, pers. comm.). We should not miss the chance to study the effect on the surviving haplochromine species.

If our estimate of the number of haplochromine species recently destroyed by the Nile perch in Lake Victoria is correct, it sets a deplorable record of vertebrate extinction rates caused by human intervention. The evaluation of this mass destruction of species should not be limited to balancing academic against regional socioeconomic interests. From a global standpoint, yet another ecosystem has been irreversibly affected. The intricate implications of evaluating such impacts by Man on nature are beyond the scope of this book, but well exposed in Norton (1986) and Wilson (1988). For a survey of the effects of other fish introductions, the reader is referred to Welcomme (1984) and Coulter *et al.* (1986).

The murky waters of Lake Victoria rule out direct observations on the fish. This constraint has hampered ecological research considerably and has precluded the vistas which contributed so much to the popularity of the cichlids from Lakes Malawi and Tanganyika. Scientific interest in the Lake Victoria haplochromines was further frustrated by the taxonomic intricacies and anatomical similarity of these fishes. Compared with the interest in the other Great Lakes cichlids, the haplochromine species flock of Lake Victoria was treated until quite recently like an unwanted stepchild in a fairy-tale. It is not a happy ending when Cinderella, dressing for the ball, dies.

ACKNOWLEDGEMENTS

Drs G. Ch. Anker, G. Fryer, C. Hollingworth, R. J. C. Hoogerhoud, M. J. P. van Oijen and J. de Visser are acknowledged for their critical reading of the manuscript and suggestions for improvements. Mr M. Geerts kindly provided information from his extensive collection on popular and scientific cichlid literature. Mr M. Brittijn prepared the figures and Mrs M. K. I. Vogelsang-Liem typed the various drafts. The field research of the Haplochromis Ecology Survey Team is financed by the Netherlands Foundation for the Advancement of Tropical Research (WOTRO) and by the Section for Research and Technology of the Netherlands Ministry for Development Cooperation.

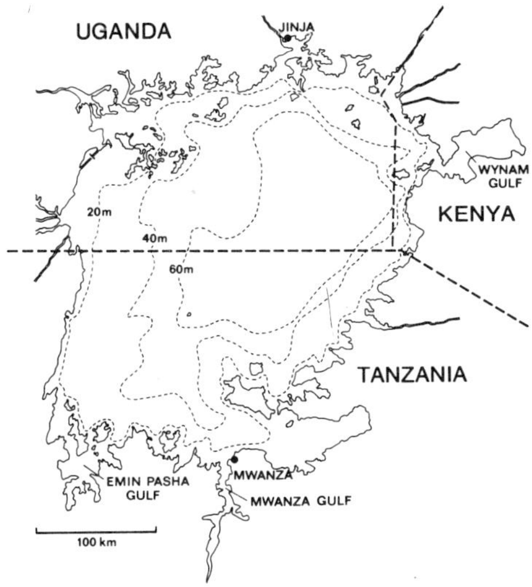

(a)

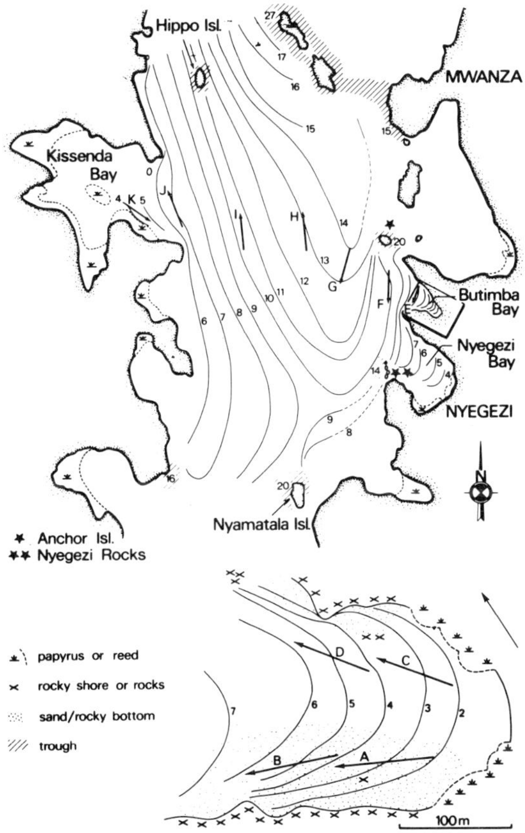

(b)

Fig. 13.9 (a) General map of Lake Victoria (adapted from Green Wood, 1974a). (b) The northern section of the Mwanza Gulf, the main research area of HEST. Details of Butimba Bay. Arrows A–K indicate the main sample stations of HEST. ((a) and (b) adapted from Witte, 1981.)

Chapter fourteen

Cichlids in aquaculture

R. S. V. Pullin

14.1 INTRODUCTION

It is relatively easy to define a cichlid, but surprisingly difficult to define aquaculture within the wide range of systems from which such fish are produced – from capture of unmanaged wild populations to intensive aquaculture with all life-history stages under close management. Operations such as fish ranching, culture-based fisheries and fisheries enhancement through brushparks and other devices, have long been considered a grey area between capture and culture. This has posed such problems for the compilation of meaningful statistics on aquaculture that an Aquaculture Steering Committee of the Fisheries Department of the Food and Agriculture Organization of the United Nations (FAO) formulated in December 1987 the following new definition of aquaculture, which is adopted here:

> "Aquaculture is the *farming* of aquatic organisms, including fish, molluscs, crustaceans and aquatic plants. Farming implies some form of intervention in the rearing process to enhance production, such as regular stocking, feeding, protection from predators, etc. Farming also implies individual or corporate ownership of the *stock* being cultivated. For statistical purposes, aquatic organisms which are harvested by an individual or corporate body which has owned them throughout their rearing period contribute to *aquaculture*, while aquatic organisms which are exploitable by the public as a common property resource, with or without appropriate licences, are the harvest of *fisheries*."

To this definition, the committee added the following explanatory notes and designations:

> "The ownership of the structure in which or around which organisms are caught e.g. fish aggregating devices (FAD's) or brushparks does not confer ownership of the organisms themselves – thus their production is 'fisheries'. The juvenile output of hatcheries is aquaculture but the

> production of market-sized organisms resulting from the growth of those juveniles contributes to fisheries production if (as in inland fisheries enhancement) they are available to the public to harvest. If the juveniles produced by hatcheries remain privately owned (by an individual or a corporate body) until harvested the result is aquaculture production."

The vast majority of cichlid aquaculture is inland, in fresh and, to a lesser extent, brackish water. There is a growing realization that inland aquaculture is more akin to agriculture than to fisheries and will become increasingly integrated into mixed farming systems (Edwards *et al.*, 1988).

Tilapias, together with carps, have been recommended by ICLARM (1988) as appropriate species for the Consultative Group on International Agricultural Research (a global research system which has hitherto considered only terrestrial food production) to use to initiate a new venture in international research for small-scale integrated farming. This venture would be a major step towards bringing aquaculture, and the strategic research needed to support its development, into the mainstream of agriculture.

This review is limited to aquaculture of cichlids for human food and does not consider the ornamental cichlid trade. It is therefore largely concerned with the most commonly cultured cichlids, the tilapias, for which the nomenclature used here follows Trewavas (1982, 1983). Trewavas (1983) summarizes the features that distinguish the tribe Tilapiini from other cichlids. Her nomenclature is now followed by the foremost international aquaculture journals (*Aquaculture* and *Aquaculture and Fisheries Management*) and by an increasing number of scientists and publishers, the main exceptions being some North American journals which still use the monogeneric nomenclature (*Tilapia*) and whose contributors often cite a checklist (Robins *et al.*, 1980) as the reason. It is particularly important for aquaculturists that cichlid nomenclature be standardized as far as possible, not only to facilitate communication between farmers and researchers but also for computerized information retrieval (Pullin, 1982; see comments on FAO statistics, page 287).

The following section explains why some cichlids are good candidates for aquaculture. It is followed by regional summaries of cichlid aquaculture which incorporate discussions on the potential of cichlids.

14.2 WHY CULTURE CICHLIDS?

General considerations

There are many cichlids with excellent attributes for aquaculture. Chief among these advantages is that the larger cichlids are a good source of 'white fish' and fish products. In most Third World countries they are marketed whole – fresh or dried – but they can also be processed into a wide variety of value-added products; for processing, their lack of fine intramuscular bones confers on them an advantage over the other most widely cultured family, the Cyprinidae.

The size of fish preferred by consumers varies greatly from country to country. In the Philippines, for example, diners prefer individual fish, hence tilapias of *c.* 100 g are used. In Chinese cuisine, however, the family usually prefers to share a large fish of 500 g or more. Production of value-added tilapia products, such as fillets, usually requires fish of several hundred grams. Tilapias, particularly *Oreochromis niloticus*, can be grown to suit all these purpose.

O. niloticus was the original tilapia to be cultured in Africa – in ancient Egypt (Chimits, 1957a) – and its increasing use in tropical aquaculture worldwide suggests that tilapia could become an international white fish commodity (Pullin, 1984). Other cultured cichlids such as the pearlspot, *Etroplus suratensis*, in South Asia and some Latin American species are also highly prized as food fishes, but generally have slower growth and reach smaller sizes than the cultured tilapias, e.g. most *Cichlasoma* species attain less than 50 g.

Feeds, or fertilizers to produce natural feeds in ponds, are usually the major component of variable costs in aquacultural production. This applies to both extensive/semi-intensive Third World aquaculture and to intensive commercial aquaculture. Intensification of tilapia culture is discussed by Balarin (1984). Intensive aquaculture is very similar in concept to feedlot livestock production and its feeding costs are usually 60% or more of total variable costs. The feeding habits of cichlids are paramount to this kind of aquaculture. The other main requirement for aquaculture is seed, i.e. the necessity to breed adequate numbers of high-quality fry and fingerlings.

Before feeding and breeding are discussed in relation to aquaculture, it must be emphasized that the amount of published information on the biology of cultured cichlids is staggering. This applies especially to the tilapias. The major sources of reference and bibliographies are Chimits (1955, 1957b), Thys van den Audenaerde (1968b), Fryer and Iles (1972), Balarin and Hatton (1979), Balarin and Haller (1982), Pullin and Lowe-McConnell (1982), Fishelson and Yaron (1983), Schoenen (1982, 1984, 1985) and Pullin *et al.* (1988). In addition, there is an increasing literature in Asian languages: e.g. Lai and Huang's (1981) bibliography of publications from Taiwan. A collection of 180 papers on tilapias in Russian is available from A. Ivojlov, Biological Institute of Leningrad University.

The excellent attributes for aquaculture of tilapias, dubbed 'aquatic chickens' by ICLARM, are summarized by Pullin (1985). Unfortunately, the first species spread throughout the world for aquaculture (*O. mossambicus*) has poorer growth and breeding characteristics for that purpose than most other cultured species. An index of growth performance ($\phi' = \log_{10}K + 2\log_{10}L_{\infty}$), where K and L_{∞} are parameters of the von Bertalanffy growth equation;

Moreau *et al.* (1986) allows the growth potential of different species to be compared. Pauly *et al.* (1988) have used this index to compare tilapias in aquaculture and open waters. Such comparisons confirm the high suitability of some *Oreochromis* species, particularly *O. niloticus*, for aquaculture.

The feeding characteristics of cultured cichlids

Undoubtedly the most important feeding characteristic of cichlids in aquaculture is that most species feed low in the food web. This means that they can be fed on cheaper feeds than can carnivorous fishes such as catfish and groupers. Feeding fish on vegetation (terrestrial plants or aquatic macrophytes) or on natural feed produced in fish-ponds (bacteria, plankton and detritus) is by far the most cost-effective basis for freshwater aquaculture in most Third World countries. For semi-intensive aquaculture, a polyculture of carps (a macrophytophagous species and one or more water-column/benthic microphagous species, the feeding niche(s) of which are enriched by faeces from the relatively inefficient digestive processes of the macrophytophagous fish) has for centuries been the main strategy in Chinese aquaculture. However, there are no cichlid species that can match the cyprinids, particularly the voracious grass carp, *Ctenopharyngodon idella*, as macrophyte feeders.

Among the tilapias, *Tilapia rendalli* and *T. zillii* are the species for which aquatic macrophytes and inundated terrestrial plants are major food items. The former species is preferred for aquaculture (Pullin, 1986); Caulton (1982) reviews the energetics of its herbivorous habit. Among the non-tilapiine cichlids, *Etroplus suratensis*, a South Asian species, is a reasonably effective macrophyte feeder and a good prospect for aquaculture (Jayaprakas and Phil, 1980; Costa, 1983; de Silva and Perera, 1983; Samarakoon, 1985; Chandrasoma, 1986) but grows slower than *T. rendalli* (Macintosh, 1987). There is a view in Sri Lanka that *E. suratensis* will usually displace *T. rendalli* in open waters, not by feeding competition but by competition for breeding sites. Like the tilapias, *E. suratensis* is not wholly a macrophyte feeder and is somewhat detritivorous. Therefore, although polyculture of macrophytophagous and microphagous tilapias holds promise (Pullin, 1986), polyculture of herbivorous cyprinids and tilapias may be a better strategy (Edwards *et al.*, 1988).

There is a large choice of microphagous cichlid species, especially among the tilapias. However, recent experience has narrowed the choice of candidates for polyculture to a few omnivorous, opportunistic species, such as *O. niloticus*.

All species listed in Table 14.1 will accept feed pellets in intensive aquaculture systems and will feed effectively on natural food produced by fertilizing fish-ponds with inorganic fertilizers or organic byproducts from agriculture, such as livestock excreta. *O. niloticus* (Fig. 14.1) is particularly versatile as it has the fine gill rakers of a planktivorous filter feeder, good

Table 14.1 Tilapias of current or potential importance in aquaculture

Species	Evaluation
The most widely cultured species and hybrids	
Oreochromis niloticus	Fast growth, especially in the tropics; versatile feeder.
O. aureus	Fast growth; versatile feeder; tolerant of cold but difficult to seine in ponds, so best grown in cages or used as a male parental stock for hybridization (see next entry).
Oreochromis monosex male hybrids, principally *O. niloticus* × *O. aureus*	Fast growth, especially on pelleted feeds.
Other cultured species of current or potential importance	
Tilapia rendalli	Macrophyte feeder; potential for polyculture with microphagous tilapias; excellent keeping qualities.
Oreochromis spilurus spilurus	Fast growth; tolerant of salinity; good grazer on epiphytic algae.
Oreochromis andersonii	Reasonable growth and tolerance of cold.
Sarotherodon melanotheron	Tolerant of salinity; good growth in experimental separate-sex culture.
Red tilapias – a group of single species (principally *O. niloticus* and *O. mossambicus*) or more often hybrids of these and/or *O. aureus* and/or *O. urolepis hornorum*	Mixed information, but generally good growth and sometimes preferred to normal-coloured fish in markets: some hybrids do well in seawater cage culture (e.g. *O. mossambicus*/*O. urolepis hornorum* hybrids).

*Sources: Pullin (1983), Lowe-McConnell (1988).

pharyngeal and jaw teeth for browsing, grazing and detritivory, and the ability to take particulate food, such as cereal brans, and larger food items. It will also eat aquatic macrophytes such as *Lemna* and *Azolla* – it is truly an omnivore.

An excellent review of tilapia feeds and feeding is given by Jauncey and Ross (1982). Tilapias also figure prominently in Moriarty and Pullin (1987), wherein Bowen (1987) indicates the relevance of detritivory in *O. mossambicus* to manipulating detrital foodchains in fish-ponds.

Apart from the herbivorous/omnivorous tilapias, there are a few other species whose feeding habits suit them well to particular aquaculture systems. For example, the piscivores *Cichla ocellaris*, *Cichla temensis*, *Cichlasoma managuense* and African species such as *Hemichromis fasciatus* and *Serranochromis* spp., have potential for control of recruitment in mixed-sex tilapia populations cultured in ponds. Browsers on epiphytes and aufwuchs, such as *Sarotherodon melanotheron* and *O. spilurus*, are also useful.

Fig. 14.1 Nile tilapia, *Oreochromis niloticus*, at the Institut des Savanes, Bouaké, Côte d'Ivoire; one of the pioneering research stations in African aquaculture. The current strain is descended from a mixture of Nile and Volta strains (detailed history, Pullin, 1988).

Many tilapias and the Latin American cichlid *Astronotus ocellatus* feed on aquatic invertebrates. There have been proposals to use cichlids to control aquatic disease vectors and intermediate hosts, particularly mosquito larvae and snails that release bilharzia cercariae. Such threats to public health constrain the development of aquaculture in some parts of Africa. It is unlikely, however, that cultured cichlids can appreciably control vectors and intermediate hosts, except perhaps mosquito larvae wherever fish-ponds are the principal breeding grounds for mosquitos. Even there, the tendency of most cichlids towards omnivory may leave some larvae uneaten. Where other water sources are abundant (i.e. throughout the humid tropics), cultured cichlids are unlikely to dent the mosquito population significantly. Snail control by culture of mollusc-eating cichlids such as *Haplochromis* spp. and *Cyrtocara* spp. has been proposed, but because the cercarial output from each snail is so high, it must be doubted whether a significant reduction in the risk of disease to fish-farm workers could be achieved. The solution to these terrible diseases is better health and sanitation through tackling their root cause – poverty.

14.3 BREEDING CULTURED CICHLIDS

The cultured cichlids, particularly the tilapias, are easily bred in captivity without complex hatchery equipment or hormonal induction of spawning.

Indeed they can be bred year-round in the tropics in a wide range of pond, hapa (net cage) and tank hatchery systems. However, this important asset for aquaculture is accompanied by liabilities that have forced culturists to consider a wide range of methods for controlling tilapia reproduction. The basic problem is that a mixed-sex, freely-breeding tilapia population, kept in a confined space such as a fish-pond, matures early and breeds prolifically, thereby overpopulating the pond with undersized fish. The extensive literature on this problem is full of terms like 'stunting' and 'a hypothetical switch between growth and reproduction'. This phenomenon is still not fully understood (discussions, Lowe-McConnell, 1982; Noakes and Balon, 1982). Clearly it allows wild fish to populate new, favourable environments very rapidly, e.g. when dried-up lakes become refilled and recolonized from residual stocks in the feeder streams. A major factor seems to be space, e.g. *O. mossambicus* may mature at a size of 5 g in aquaria, but at hundreds of grams in lakes.

Whatever the basis for this property, it is clearly unsatisfactory for culturists striving to grow fish to acceptable market size in confined spaces. Moreover, it has given exotic tilapias, principally *O. mossambicus*, a bad reputation in many countries because escapees have spread rapidly into natural waters and aquacultural facilities and have become problematical competitors with other wild and cultured fishes.

Most culturists therefore prevent breeding in production systems while maximizing reproductive output in separate hatchery systems. Cultured tilapias have relatively low fecundity compared with most other cultured finfish. They breed more or less year-round in the tropics and have a short generation time of *c.* 4–6 months, but have fecundities of the order of a few hundred to a few thousand swim-up fry per individual spawning, for the mouthbrooders and substrate spawners respectively. Hatchery operators are continually trying to improve broodstock management, whether in hapas, ponds or tanks, so as to maximize reproductive output. The key variables are sex ratios, age/size of broodstock, turnover procedures and fry-collection methods. Smith *et al.* (1985) and Guerrero (1987) provide good examples of a range of tilapia hatchery methods in a Third World country.

Much of the tilapia culture practised in Third World countries is still mixed-sex culture, i.e. the problems of stunting and overpopulation of production ponds are either tolerated or at best addressed by selective harvesting of larger fish – but note that serious genetic consequences may arise from culling the best growers and retaining the early-maturing runts for breeding. However, in all progressive tilapia culture systems the goal is to grow an all-male population in order to prevent breeding and to capitalize on the faster growth of tilapia males over females (*c.* 20% for the cultured tilapias). Guerrero (1982) reviews the wide range of methods available. Only three have seen significant use: (1) manual sexing, which although not very efficient, as half the seed are discarded and errors allow some females to enter the production stock and breed, can be improved by stocking a predatory fish to consume any recruits (Hopkins *et al.*, 1982); (2) production of monosex male F_1 hybrid progeny from certain interspecific crosses; (3) treatment of first-feeding fry with androgenic steroids to reverse the sex of the females.

For many years (review, Wohlfarth and Hulata, 1983), the main research thrust of breeders in the most progressive countries culturing tilapias (Israel and Taiwan) was to perfect the technology for producing all-male F_1 hybrid fry from interspecific crosses such as *O. niloticus* × *O. aureus* and *O. niloticus* × *O. mossambicus*. Such interspecific crosses give very high percentages of male progeny (*c.* 95–8%). Crosses between *O. urolepis hornorum* as the male parent and *O. niloticus* or *O. mossambicus* females give 100% male progeny. These crosses seem exciting prospects for culturists, but have serious flaws. First, if the proportion of male progeny is less than 95–8% (as it can be, from poorly managed parental stocks), then significant breeding will still occur in the production stock – the same problem as with manual sexing. Second, the strict management needed to keep separate two parental lines and to prevent hybridization and backcrossing of progeny to parents (which destroys the all-male progeny effect) is extremely difficult, particularly in Third World situations where technology and expertise are lacking. Therefore in general the attempts to transfer this technology to Third World countries have not prospered. Lovshin (1982) provides a useful review of hybridization work in Brazil and elsewhere.

Culture of F_1 interspecific hybrids as production stock will probably contract as far as the mainstream of tropical tilapia culture is concerned. However, they will probably continue to be used in subtropical countries where cold tolerance is an important commercial trait, e.g. Israel and Taiwan. It was revealed in workshop discussions in 1987 (Pullin, 1988, pp. 62–3) that the Israeli tilapia industry cultures all-male *O. niloticus* × *O. aureus* F_1 hybrids rather than sex-reversed *O. niloticus* because of the highly valuable cold tolerance of *O. aureus* that is present in the hybrids. *O. aureus* is not cultured alone in ponds because it is exceedingly difficult to harvest by seining.

Compared with the limited scope of interspecific hybridization to produce all-male progeny, hormonal sex reversal offers excellent prospects. The problem was for many years the difficulty in scaling this technique up from experimental to farm use. However, this is now well-established (Buddle, 1984) and the technology is practicable in the Third World (Macintosh *et al.*, 1985; Guerrero, 1987). Moreover, it is safe, as the hormone used (normally 17α-methyltestosterone) is rapidly eliminated from the growing fry (Johnstone *et al.*, 1983). It also gives the treated fry a growth advantage – the well-known anabolic steroid effect. This effect has not yet been well quantified on farms, but is real enough to have convinced most Taiwanese hatchery operators to treat all their fry, single species and *O. niloticus* × *O. aureus* hybrids (pers. obs.).

Macintosh *et al.* (1988) discuss the effects of steroid treatment on growth.

Therefore, although the breeding of cichlids for aquaculture presents problems, solutions are now established.

One cannot leave the topic of breeding without considering genetics. Throughout most of the history of aquaculture, culturists have paid scant attention to genetic improvement and to avoidance of genetic deterioration of their stocks. However, this attitude is now changing rapidly. Aquaculturists are still far from applying the techniques commonly used in livestock breeding, such as pedigreeing, but the advantages of applying genetics, particularly quantitative genetics and selection, in aquaculture are beginning to show (Gjedrem, 1985). For the cichlids, little genetical research has been performed other than that reviewed for the tilapias by Wohlfarth and Hulata (1983) and more recent studies, many of which are discussed in Pullin (1988). The problems and prospects for genetic improvement of tilapias are discussed by Pullin and Capili (1988) and Tave (1988). A consensus is emerging that research into cichlid genetics should move away from all-male hybridization technology towards a systematic effort to apply quantitative genetics in breeding programmes. That effort requires thorough documentation, evaluation and, eventually, utilization of the wild genetic resources. For the tilapias, this approach has been thoroughly discussed (Pullin, 1988) and new founder stocks for breeding programmes are being collected. The future for tilapia breeders is therefore exceedingly bright because of demand for improved breeds and improved hatchery technology for monosex fry production.

14.4 WORLD PRODUCTION OF CICHLIDS IN AQUACULTURE

Cichlid aquaculture, particularly tilapia culture, is in a dynamic growth phase. Rapid developments are often the result of a shift to different and more appropriate species; for example a change from *O. mossambicus* to *O. niloticus* in the Philippines.

Accurate statistics are extremely difficult to collect in many Third World countries where cichlids are cultured. Moreover, statistics reported to FAO vary in the detail to which the fish are identified. Some statistics refer to individual species, usually tilapias, whereas others refer to cichlids not elsewhere identified. Furthermore, some countries report blanket statistics for culture of freshwater fishes which, particularly in Africa, often include many cichlids. The appended data (page 298) constitute at best a rough guide to cultured cichlid production in 1984–86.

Another complicating factor is that FAO statistics hitherto have also listed 'tilapias not elsewhere included' and give the Latin nomenclature of this entry as '*Sarotherodon, Tilapia* spp.' while having a separate listing for 'Tilapias oreochromis' described as *Oreochromis* spp. If one follows Trewavas (1982,

1983) and uses all three generic names, then the only *Sarotherodon* species cultured to any significant extent are *S. melanotheron* and *S. galilaeus*. As neither of these has been spread for culture much beyond Africa, most if not all of the FAO statistics described as '*Sarotherodon*' are really listing *Oreochromis* spp. The confusion arises from the transitional period when Trewavas proposed a bigeneric split (*Sarotherodon* and *Tilapia*). Clearly a move towards accurate collection of statistics classified by species and hybrids, following a consistent nomenclature, is essential. The collection and processing of cichlid aquacultural statistics will improve rapidly, owing to the new FAO definition of aquaculture (page 279) and an increasing effort to verify information.

Cichlid aquaculture has progressed much more in Asia than in Africa and other regions (see page 298–307). There is now great scope for cooperation among regions, particularly between Asia and Africa, to assist aquacultural development from Asian experience and to spur tilapia genetic improvement world-wide from African genetic resources.

Africa

The leading countries in cichlid aquaculture are probably Côte d'Ivoire, Kenya, Nigeria, Zaïre, Zambia and Zimbabwe, but the picture is changing rapidly as countries with renewed interests in aquaculture prepare for expansion. Clearly the future lies in the use of good species (*O. niloticus* or native equivalents) in farming systems appropriate for Africa, i.e. systems that can transcend climatic and socioeconomic constraints. The adoption of integrated crop–livestock – fish farming principles, as practised in Asia (Edwards *et al.*, 1988), may hold the key to future expansion (Fig. 14.2). The tilapias will be the most important species (Huisman, 1986; Powles, 1987; Balarin, 1988). Moreover, from 1984, FAO has been producing a series of national reviews for aquacultural development. The series has various authors, principally J. D. Balarin, and is designated *FAO Fisheries Circular* no. 770. Reviews have been completed on Bénin, Cameroon, Central African Republic, Egypt, Ethiopia, Kenya, Liberia, Malaŵi, Sierra Leone, Tanzania, Uganda, Zambia and Zimbabwe. The potential of tilapias for aquacultural development is considerable, except in the high-altitude, cool areas.

A large and increasingly heated debate concerns which aquaculture systems make sense in Africa: small-scale ponds or cages for the many needy farmers in the rural areas, medium-scale commercial enterprises for what have been termed 'emergent farmers' (usually relatively wealthy professional people on the urban fringes), or large-scale commercial enterprises similar to estate farming, often managed by large companies which feed the fish to their workers. The debate on suitability not withstanding, there is probably scope for all three systems to coexist and to increase protein supply and income in urban and rural areas.

Fig. 14.2 A typical small holding pond on a farm in Malawi. The species raised are *Oreochromis shiranus* and *Tilapia rendalli*. Farmers cannot yet afford sufficient conventional feeds and fertilizers to produce high fish yields. Research is in progress to optimize the use of on-farm resources such as grass, compost and wood ash.

Asia

The scale and diversity of cultured tilapia production in Asia is impressive (Fig. 14.3). The leading producers are China (Peoples' Republic and Taiwan), Indonesia, the Philippines, Taiwan and Thailand, but many other countries could expand their production rapidly as the culture of *O. niloticus* becomes more widespread. The climatic constraints to tilapia culture in Asia apply chiefly to East and Northeast Asia. The tilapia culture industries of the more northern provinces of China, Korea and Japan have little scope for expansion.

Attitudes to tilapia may be changing in South Asian countries that traditionally have preferred carps and (except for Sri Lanka) have benefited little from the historical introductions of *O. mossambicus*. Sri Lanka has important reservoir fisheries for *O. mossambicus*. *O. massambicus* was introduced to Dhaka from Bangkok in 1954 and was distributed all over what is now Bangladesh and Pakistan. Naik (1973) concluded that it had some advantages for culture where carps could not survive, but many disadvantages, chiefly small size and slow growth, and therefore poor marketability. He recommended *Etroplus suratensis* as a better species for development. Indian experience with *O. mossambicus* was similar. However, we now know that *O. mossambicus* is a much poorer species for culture than *O. niloticus*. Moreover, the founder stocks of *O. mossambicus* spread amongst Asian and Pacific nations all derive from a tiny founder population (two females and three males discovered in Java, Indonesia in 1938) (Schuster, 1952). Their genetic variation is probably extremely low compared with that of African stocks. introductions of *O. niloticus* (and *O. aureus* in Taiwan, China) have allowed major industries to develop.

The International Center for Living Aquatic Resources Management (ICLARM), Manila, commenced in mid-1988 a project for the Asian Development Bank and the United Nations Development Programme to develop better strains of *O. niloticus* for culture. The strategy is to collect new founder stocks from Africa and to apply strong selection for fast growth in specific farming systems (ponds, cages, rice fields, etc.). It is a measure of the growing interest in tilapia culture in Asia that the following countries, when asked through this project whether they needed improved *O. niloticus* breeds for aquacultural development, all responded with enthusiasm: Bangladesh, India, Indonesia, Malaysia, Pakistan, Taiwan, China and Vietnam. With such increasing interest, tilapia culture is set for very large expansion in tropical and subtropical Asia. The prospects for expansion of *Etroplus suratensis* culture in South Asia are less clear: much more research is needed on its cost-effective culture.

Caribbean

In the Caribbean, tilapia culture is at a fairly low level of production but is becoming more widespread. It is proceeding along two levels of development – freshwater pond culture and saltwater tank and cage culture. In addition to the appended statistics (page 304–5), there is experimental or pilot-scale tilapia culture in Dominica, Montserrat and Saint Lucia. Subsistence tilapia culture also has a long history in Haiti.

The leading nations are undoubtedly Jamaica and Cuba. The former boasts more than 200 tilapia farms and has a national association of tilapia farmers. The history of tilapia farming in Jamaica is reviewed by Cooke and Moo-Young (1983). Production grew from 2.2 t in 1977 to 69.6 t in 1982. The growth of the large-scale commercial sector is illustrated by Karlan (1985) in an account of the development of a 'superintensive' farm run by Jamaica Aqualapia Ltd. The company's target for the last three quarters of 1986 was 800 t. For Cuba, the statistics are difficult to interpret. There is large production of tilapias from Cuban reservoirs: Henderson (1982) projected an annual yield of 12 000 t. Although this productivity would fall outside the definition of aquaculture, it clearly depends to some extent on hatchery activities.

(a)

(b)

Fig. 14.3 Two styles of Asian tilapia culture. (a) to (c) Taiwan: (a) intensive cage culture of *Oreochromis* hybrids in the Tseng Wen reservoir; (b) intensive tank culture of red *Oreochromis* hybrids at the Fu Hai farm, Pingtung County; (c) moving tilapia broodstock. (d) to (f) Philippines: (d) feeding rice bran to tilapia broodstock ponds in a small village hatchery; (e) an alternative to pond breeding – a hatchery with broodstock held in 'hapas' made of mosquito netting; (f) grow out in net cages in the eutrophic lake Laguna de Bay.

(c)

(d)

(e)

(f)

Fig. 14.4 Nile tilapia, *Oreochromis niloticus*, in experimental sewage-fish culture near Lima, Peru.

Freshwater culture of *O. niloticus* and saltwater culture of salt-tolerant hybrids, particularly red tilapias, are ready for rapid expansion in the Caribbean.

Latin America

Aquacultural production is generally low in Latin America, where meat is often preferred to fish. Nevertheless, the culture of cichlids is well established in a few countries and some non-tilapiine cichlids are cultured. *Cichla ocellaris*, *Cichla temensis* and *Astronotus ocellatus* have high potential. Most *Cichlasoma* species are too small to use in aquaculture, but *Cichlasoma urophthalmus* (the

Table 14.2 Attributes of indigenous Latin American cichlids*

Species	*Optimum temp.* (°C)	*Market size* (*g*)	*Production time* (*months*)	*Growth rate* ($\%d^{-1}$)	*Production* ($t\,ha^{-1}\,y^{-1}$)	*Culture method*†
Cichla sp.	25–30	300–500	24	1.0	0.5–1.0	P, MC
Astronotus sp.	25–30	>500	18	0.9	–	P, MC or PC

*Source: Saint-Paul (1987).
†P, pond; MC, monoculture; PC, polyculture.

(b)

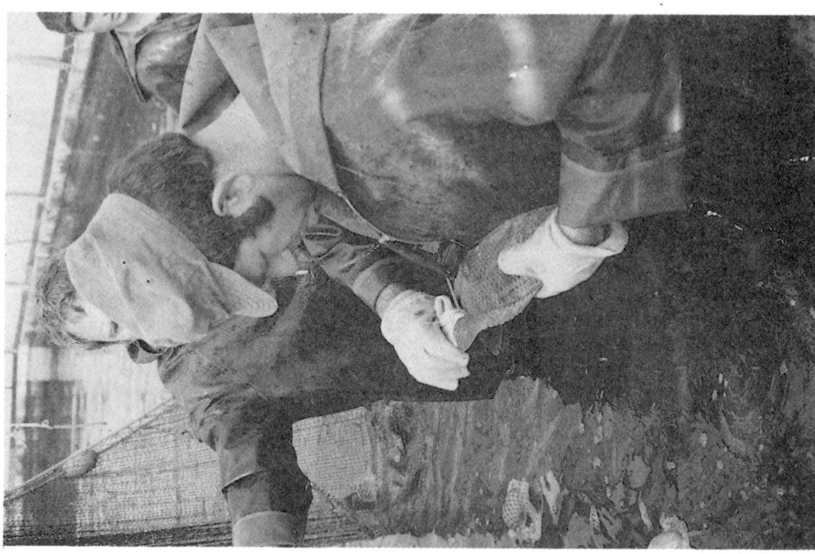

(a)

(c)

(d)

Fig. 14.5 Tilapia culture technology in Israel. (a) Tagging selected broodfish; parental stocks of *Oreochromis niloticus* and *O. aureus* are held separately before breeding to produce F1 hybrid production stock. (b) A tilapia egg incubator at an intensive hatchery practising hormonal sex-reversal of fry. There are various designs of incubators. Eggs are kept in gentle motion by upwelling water or aviation. (c) Harvesting by escalator from a production pond. (d) Overwintering structures to protect broodstock and holding ponds from the cold. (e) Harvested hybrid tilapias (still living) entering a fully automated plant for killing, gutting and production of value-added product such as fillets.

(e)

Mexican mojarra) can grow to 600 g and is said to taste better than tilapias. Basic research on *C. urophthalmus* is investigating its physiology and behaviour in relation to aquaculture (Martinez-Palacios and Ross, 1986). Table 14.2 summarizes data on cultured *Cichla* sp. and *Astronotus* sp., additional summary information is reviewed by Bonetto and Castello (1985) and Saint-Paul (1986).

The brightest prospects for Latin American cichlid culture are probably the tilapias, particularly *O. niloticus* (Fig. 14.4) and *T. rendalli*. The need for expansion of all-male fry production is well recognized. Dunseth and Bayne (1976) describe the combined culture of *O. aureus* with the cichlid predator *Cichlasoma managuense* to control recruitment. The countries with good prospects for expansion of cichlid aquaculture are mainly those in which it can be integrated with other existing agricultural systems, e.g. crop–livestock–tilapia systems in Costa Rica and Panama and the use of residues such as coffee pulp (Bayne *et al.*, 1976) in tilapia feeds.

Mediterranean/Middle East

Except in Egypt and the Levant, cichlids (tilapias) are not native to Mediterranean countries and the Middle East. The climatic constraints are obvious; cold winters or large day-night temperature changes. Egypt and Israel are the leading producers. Israel has long been a leader in tilapia culture technology (Fig. 14.5). It is difficult to foresee an expansion of tilapia culture in countries bordering the Mediterranean Sea. The costs of over-wintering broodfish and the short growing seasons are too restrictive except for very capital-intensive systems. However, capital-intensive systems may be appropriate to oil-rich states eager to invest in food security programmes. Therefore, Kuwait (Hopkins *et al.*, 1986a, b), Saudi Arabia, Bahrain and the United Arab Emirates (which are also experimenting with tilapia culture) may foster expansion of tilapia culture in pumped brackish groundwater or in sea cages.

Other countries

In the other countries listed (page 307), there is clearly very limited scope for expansion of cichlid aquaculture. The USA has many other inexpensive high-protein foods, the USSR is generally too cold, and aquaculture at power stations is everywhere bound to be limited in scale. The Pacific nations and Hawaii may have better prospects. Despite the relatively low populations of the Pacific region and the abundance of marine fish, large tilapias are well accepted in the markets. The culture of *O. niloticus* could expand in the region, with Fiji most likely to assume leadership.

14.5 CONCLUSIONS

Cichlids, particularly tilapias, have an ancient history, a major role and a bright future in aquaculture. *Oreochromis niloticus* is emerging as the species of choice for tropical farmers, but many other species will remain locally important. It is difficult to make a realistic estimate of the current total production of cultured cichlids, but it must be *c.* 250 000 t y^{-1}. This may well double by 1999 as more farmers in Third World countries see the benefit of incorporating aquaculture into mixed farming systems.

ACKNOWLEDGEMENTS

I am greatly indebted to Michael Robinson and his colleagues in the Fisheries Statistics Division of the Food and Agriculture Organization of the United Nations, Rome, for their help and advice in presenting cichlid-aquaculture statistics. I also thank colleagues and friends in ICLARM* and elsewhere for numerous helpful suggestions which made the difficult task of summarizing cichlid aquaculture a pleasure to undertake.

*This is ICLARM Contribution No. 498.

14.6 CICHLIDS IN AQUACULTURE: PRODUCTION APPENDIX

This appendix summarizes the production of cichlids from aquaculture during 1984–86. The data are taken from the most recent information available to the Food and Agriculture Organization of the United Nations (FAO) unless another reference is cited, and are based on the new FAO definition of aquaculture. All remarks have been added by the present author. Missing entries do not necessarily indicate zero production; many reflect the difficulties of obtaining reliable information, or the necessity of revising previous estimates in the light of FAO's new definition of aquaculture.

Country	*Species cultured*	*Production (t)*			*Remarks*
Africa		1984	1985	1986	
Angola	*Tilapia sparrmanii*	–	2	–	Not a prime species for culture; culture of other species is anticipated.
Bénin	Various tilapias	15	11	15	This probably includes *Oreochromis*, *Sarotherodon* and *Tilapia* spp. Exotic species, including *O. niloticus* and *O. spilurus*, have been introduced recently.
Burkina Faso	*Oreochromis niloticus*	157	150	150	Further development of aquaculture will probably concentrate on this species.
Burundi	Various tilapias	–	–	–	Burundi produced 2 t of freshwater fishes from aquaculture in 1986. This probably included cichlids (tilapias).
Cameroon	Various tilapias	–	–	–	Production is $< 200\,t\,y^{-1}$ (Satia, 1987). The best prospect for expansion is *O. niloticus* culture.
Central African Republic	*O. niloticus*	288	200	200	There are good prospects for expansion of *O. niloticus* culture.

Congo	*O. niloticus*	29	–	88	There are good prospects for expansion of *O. niloticus* culture. Probably the 1985 entry for various tilapias is mainly *O. niloticus*.
	Various tilapias	–	21	–	
Côte d'Ivoire	*O. niloticus*	373	550	660	There are good prospects for expansion of *O. niloticus* culture. There is also a large research effort to culture in brackish-water lagoons *S. melanotheron*, salt-tolerant exotic *Oreochromis* species and hybrids between *O. niloticus* and exotic *Oreochromis* species.
	Various tilapias	–	–	7	
Egypt (see Mediterranean/Middle East)					
Ethiopia	*O. niloticus*, *T. zillii*	–	–	–	These species have been tried in pond culture but there is no significant production. Much of the country is too high and cold for cichlid aquaculture.
Gabon	Various tilapias	28	–	–	There is scope for expansion of aquaculture using brackish- and freshwater tilapias.
Ghana	Various tilapias	390	321	331	Ghana cultures *O. niloticus*, *S. galilaeus*, *S. melanotheron* and *T. zillii*.
Guinea	Various tilapias	–	–	–	Guinea produced 5 t of freshwater fishes from aquaculture in 1984 (same figure in 1985). Much of this was probably from cichlid aquaculture, principally tilapias.
Kenya	*O. niloticus*	112	113	114	Other cultured species include *O. spilurus* and *T. zillii*. There is one highly successful, intensive tilapia farm, Baobab Farms, Mombasa.
	Other tilapias	–	490	504	

(*Continued*)

Country	*Species cultured*	*Production (t)*			*Remarks*
Liberia	Various tilapias	–	–	–	Liberia produced 35 t of cultured freshwater fishes from aquaculture in 1984 (same figure in 1985). Much of this was probably cichlid species, principally native tilapias and *O. macrochir*.
Madagascar	Various tilapias	–	–	–	Small quantities of tilapia, principally *O. niloticus*, are grown in polyculture with common carp.
Malaŵi	*O. shiranus*, *T. rendalli*	66	70	70	There is successful tilapia culture on sugar estates and in small village ponds.
Mali	*O. niloticus*	5	5	5	
Mozambique	Various tilapias	–	–	2	The 2 t (1986) is probably mainly *O. mossambicus*; culture of other native tilapias and possibly *O. niloticus* is likely to expand.
Niger	*O. niloticus*	9	7	8	*O. niloticus* is the only cichlid cultured extensively.
Nigeria	*O. niloticus*	475	534	376	Nigeria has good prospects for expansion of freshwater aquaculture (using *O. niloticus*) and coastal aquaculture using native species, such as *S. melanotheron* and hybrids.
	Various other tilapias	1 538	2 517	1 847	
Rwanda	*O. niloticus*	39	51	61	*O. niloticus* was introduced for an aquaculture development project. Rwanda also produced 646 t of freshwater fishes from aquaculture in 1985 and 1485 t in 1986. Much of this production was probably of native cichlids, especially *O. macrochir*.

Sénégal	*O. niloticus*	–	–	–	Aquaculture is just beginning in Sénégal, using *O. niloticus*. Statistics are not yet available for the small quantities produced, and the abundance of marine fish may preclude successful development.
Sierra Leone	*O. niloticus*	–	–	–	Similar situation to Sénégal, but better prospects for semi-intensive aquaculture in inland areas.
Sudan	*O. niloticus*	10	10	10	There is probably some production of other native tilapias, such as *T. zillii*. Previous failures to develop tilapia culture are being reassessed (Yousif, 1988).
Tanzania	Various tilapias	15	21	32	Tilapia culture of native species and exotics such as *O. niloticus* is likely to expand.
Togo	*O. niloticus*	12	15	20	Culture of native species such as *S. galilaeus* and *T. zillii*, and exotics such as *O. niloticus* and *O. macrochir*, is likely to expand.
Uganda	Various cichlids	–	–	–	Uganda produced 30 t of cultured freshwater fishes in 1984 (same figure in 1985). This production probably is mainly tilapias, such as *O. niloticus*.
Zaïre	*O. niloticus*	125	180	689	Aquaculture, especially *O. niloticus* culture, is expanding in Zaïre. There is probably also significant culture of other tilapias such as *O. macrochir*.

(*Continued*)

Country	*Species cultured*	*Production (t)*			*Remarks*
Zambia	*O. andersonii*	306	314	625	It is difficult to separate the statistics for culture of other native species from that of the recently introduced *O. niloticus* and hybrids. Zambia produced 975 t of freshwater fishes from aquaculture in 1985. This was principally tilapias.
Zimbabwe	*O. mossambicus*	262	9	–	The statistics mask what is at present a dynamic and expanding tilapia culture industry, changing from culture of native species such as *O. mossambicus* and *T. rendalli* to exotics such as *O. niloticus* and *O. niloticus* × *O. aureus* hybrids.
	Other *Oreochromis* spp.	–	5	–	
	T. rendalli	300	450	900	
	Other *Tilapia* and *Sarotherodon* spp.	–	362	–	
Asia					
Bangladesh	*O. niloticus*	–	–	–	*O. niloticus* culture is just beginning and will probably develop rapidly.
China (People's Republic of)	*O. niloticus*, *O. mossambicus* and hybrids	27 167	35 688	44 273	Tilapia culture is widespread in the warmer southern provinces.
Hong Kong	*O. niloticus*	2 080	1 500	1 435	Freshwater aquaculture in Hong Kong has declined as land values have risen.
India	*O. mossambicus* and red hybrids	–	–	–	Traditional attitudes are changing. There is interest in acquiring good strains of *O. niloticus*, One large intensive tilapia farm in Madras has a target production of 2–3 t d^{-1} of 'golden' (red) hybrids (Rangaswami, 1988).

Indonesia	*O. niloticus*	6 793	459	10 125	Tilapia culture, including cage culture of red hybrids, is expanding.
	O. mossambicus	23 197	33 468	36 642	
Japan	*O. niloticus* and other tilapias	3 544	4 180	4 113	Future expansion is unlikely because of the cold climate.
Kampuchea (Democratic)	*O. niloticus*	16	16	16	Culture of *O. niloticus* in rice fields and ponds is likely to expand.
	O. mossambicus	64	64	64	
Korea (Republic of)	*O. niloticus*	48	118	120	Future expansion is unlikely because of the cold climate.
Laos (People's Democratic Republic of)	*O. niloticus* (and *O. mossambicus*?)	–	–	–	Laos produced 2 500 t of freshwater fishes from aquaculture in 1984. Some of this was probably tilapias.
Malaysia	*O. niloticus*	68	2	575	There has probably been some confusion in the statistics and some cultured tilapias may be hybrids. *O. niloticus* culture is likely to expand.
	O. mossambicus	1 350	1 813	1 500	
Myanmar Burma (Union of)	*O. mossambicus*	–	–	–	No recent information is available.
Pakistan	*O. mossambicus*	–	–	–	There is a new interest in tilapia culture in Pakistan, particularly in acquiring new stocks of *O. niloticus*.
Philippines	*O. niloticus*	14 537	15 434	25 217	Culture of *O. niloticus* is expanding rapidly.
	O. mossambicus and hybrids, including red tilapias	17 578	27 206	30 602	

(*Continued*)

Country	Species cultured	Production (t)			Remarks
Taiwan, China	Tilapias; mainly *O. niloticus* × *O. aureus* and other *Oreochromis* hybrids	53 812	51 820	49 241	There is a large and dynamic tilapia culture industry using highly intensive systems. When production expands further, prices and profits may fall. The statistics here are from Taiwan Fisheries Bureau (1985, 1986, 1987).
Thailand	*O. niloticus*	7 954	15 110	15 701	*O. niloticus* is the preferred species. The Thai strain is pure. Some red hybrids are also cultured. *O. mossambicus* figured in earlier statistics but is now disappearing.
	O. mossambicus	262	63	6	
Vietnam	Various tilapias	–	–	–	Tilapia culture (mainly *O. niloticus*, *O. mossambicus* and hybrids) is likely to expand.
Carribbean					
Bahamas	*Oreochromis* hybrids	–	30	30	Freshwater and seawater systems; red tilapias are esteemed.
Cuba	*O. aureus*	14 042	14 675	14 942	Probably includes reservoir production. *O. niloticus* culture is also expanding.
Dominican Republic	*O. mossambicus*	–	67	67	The 1986 figures largely repeat 1985 data. Tilapia culture is probably now expanding. These figures may be underestimates.
	O. niloticus	–	84	82	
	Other *Oreochromis* spp.	–	5	5	
	Tilapia rendalli	–	70	70	
Jamaica	*O. niloticus*	379	834	834	Probably an underestimate for 1986. Production is expanding rapidly.

Puerto Rico	*O. niloticus*, *O. aureus* and hybrids	–	–	–	There are no clear statistics but freshwater pond and cage culture is being encouraged.
Saint Lucia	*O. niloticus*	–	–	–	Small quantities; culture is just beginning.
US Virgin Islands	Various tilapias	6	6	7	Probably mainly *O. aureus*.
Latin America					
Brazil	Various tilapias	–	–	–	Despite large research and development efforts, progress towards sustainable aquaculture has been slow.
Colombia	*O. niloticus*	210	300	300	Production is likely to expand. *T. rendalli* does well in coffee country.
	T. rendalli	–	2	2	
Costa Rica	Various tilapias	125	50	50	Probably mainly *O. niloticus*, *O. aureus* and some *T. rendalli*.
El Salvador	*O. niloticus*	41	68	61	No recent information is available.
Guatemala	*O. niloticus*	–	–	32	The other tilapias are probably *Oreochromis* hybrids.
	Other tilapias	–	–	5	
Guyana	Various tilapias	14	12	11	
Mexico	Various tilapias	1450	1611	2270	
Nicaragua	*O. aureus*	3	–	7	
Panama	*O. niloticus*	21	69	126	*O. niloticus* culture is most likely to expand.
	O. aureus	1	4	7	
	Various tilapias	23	76	139	
	Cichla ocellaris	44	49	56	
Peru	Various tilapias	8	21	42	*O. niloticus* and hybrid tilapia culture is most likely to expand in the forest regions isolated from supplies of marine fish.

(*Continued*)

Country	*Species cultured*	*Production (t)*			*Remarks*
Mediterranean/Middle East					
Cyprus	Tilapias	3	2	2	Limited culture of *Oreochromis* species and hybrids; obvious climatic restrictions.
Egypt	Various tilapias	30 000	30 000	30 000	Egypt cultures *O. niloticus*, *O. aureus* and *T. zillii*. Figures are repeated in FAO statistics for 1984–86. These statistics include a lot of tilapia production but how much falls within the new FAO definition of aquaculture is unclear. Much of Egypt is seasonally too cold for year-round tilapia culture.
Greece	Tilapias	–	–	–	As Cyprus. Only intensive commercial systems with seasonal environmental control are possible.
Israel	*O. niloticus* × *O. aureus* hybrids	3 269	4 474	3 105	A dynamic tilapia culture industry and a world leader in intensive culture technology. Expansion is limited by climatic constraints and aridity.
Kuwait	Various tilapias	–	–	–	A large research effort into saltwater culture of tilapias showed *O. spilurus* to be a suitable species for commercial aquaculture (Hopkins *et al.*, 1986a, b). Farm construction is now anticipated. Red tilapias are not well liked.

Saudi Arabia	*O. niloticus*	–	–	6	Saudi Arabia, in common with other Gulf States, seeks food security through intensive domestic production. Tilapia culture is just beginning.
	O. aureus	–	–	20	
Syria	*Oreochromis* species and hybrids	225	225	357	Climatic constraints similar to Israel's.
Other Regions					
Belgium	*Oreochromis* spp. and hybrids	–	120	160	Heated water (nuclear power station effluent).
Fiji	*O. niloticus*	1	3	3	There is a growing interest in *O. niloticus* culture.
	Other tilapias	1	–	–	
Guam	Tilapias	30	66	102	Mainly *Oreochromis* species and hybrids, including red tilapias.
USA	Various tilapias	–	5	–	There is some culture of *Oreochromis* species and hybrids in the warm southern states and in geothermal waters but the industry nationwide is small. Climatic constraints and bans on introductions or transfers of exotic species will limit expansion. The main scope for growth is in Hawaii. Red tilapias have sold well in some states.
USSR	*O. niloticus* (and other tilapias?)	–	–	–	*O. niloticus* culture at a power station in Khazakstan has just been reported (Anon., n.d.).

References

Aerts, P. (1982) Development of the musculus levator externus IV and the musculus obliquus posterior in *Haplochromis elegans* Trewavas, 1933 (Teleostei: Cichlidae): a discussion on the shift hypothesis. *J. Morph.*, **173**, 225–35.

Aerts, P. (1985) The intramandibular linkage in *Astatotilapia elegans* (Teleostei: Cichlidae): appearance and function of the meckelian cartilage. *J. Zool., Lond.*, **205**, 391–410.

Aerts, P., Osse, J. W. M. and Verraes, W. (1987) Model of jaw depression during feeding in *Astatotilapia elegans* (Teleostei, Cichlidae): mechanisms for energy storage and triggering. *J. Morph.*, **194**, 85–109.

Aerts, P., Vree, F. de and Vandewalle, P. (1986) Pharyngeal jaw movements in *Oreochromis niloticus* (Teleostei: Cichlidae): preliminary results of a cinerdiographic analysis. *Annls Sco. r. zool. Belg.*, **116**, 75–82.

Alcock, J. (1989) *Animal Behavior – An Evolutionary Approach*, 4th edn. Sinauer, Sunderland, Mass.

Andersson, M. (1982) Female choice selects for extreme tail length in a widow bird. *Nature, Lond.*, **299**, 818–20.

Anker, G. Ch. (1978) The morphology of the head muscles of a generalized *Haplochromis* species: *H. elegans* Trewavas, 1933 (Pisces, Cichlidae). *Neth. J. Zool.*, **28**, 234–71.

Anker, G. Ch. (1984) Morphology of the efferent and afferent gill-arch arteries and of the filament muscles of *Haplochromis elegans* Trewavas, 1933 (Pisces, Cichlidae). *Neth. J. Zool.*, **34**, 613–21.

Anker, G. Ch. (1986) The morphology of joints and ligaments in the head of a generalized *Haplochromis* species: *H. elegans* Trewavas, 1933 (Teleostei, Cichlidae). I. The infraorbital apparatus and the suspensorial apparatus. *Neth. J. Zool.*; **36**, 498–530.

Anker, G. Ch. (1987) The morphology of joints and ligaments in the head of a generalized *Haplochromis* species: *H. elegans* Trewavas, 1933 (Teleostei, Cichlidae). II. The jaw apparatus. *Neth. J. Zool.*, **37**, 394–426.

Anker, G. Ch. (1989) The morphology of joints and ligaments in the head of a generalized *Haplochromis* species: *H. elegans* Trewavas, 1933 (Teleostei, Cichlidae). III. The hyoid and the branchoistegal apparatus, the branchial apparatus and the shoulder girdle apparatus. *Neth. J. Zool.* **39**, 1–40.

Anon. (n.d.) Delikates iz tropikov (trans. 'Delicacies from the tropics'). *Ryb. Khoz.*, **88**, 56, (in Russian).

Anonymous (1988) Monster fish may be innocent of ecological crimes. *New Scient.*, **119** (21 July), 34.

Apfelbach, R. (1969) Vergleichend quantitative Untersuchungen des Fortpflanzungsverhaltens brutpflegemono-und-dimorpher Tilapien (Pisces, Cichlidae). *Z. Tierpsychol.*, **26**, 692–725.

Arnold, S. J. (1983) Sexual selection: the interface of theory and empiricism, in *Mate Choice* (ed. P. Bateson), Cambridge University Press, Cambridge, pp. 67–108.

Arnold, S. J. (1985) Quantitative genetic models of sexual selection. *Experientia*, **41**, 1296–310.

Aronson, L. R. (1949) An analysis of reproductive behaviour in the mouthbreeding cichlid fish, *Tilapia macrocephala* (Bleeker). *Zoologica* (N.Y.), **34**, 133–58.

Arunga, J. O. (1981) A case study of the Lake Victoria Nile Perch *Lates niloticus* (Mbuta) fishery, in *Proceedings of the Workshop of the Kenya Marine and Fisheries Research Institute on Aquatic Resources of Kenya*, July 13–19, 1981, Kenya National Academy for the Advancement of Arts and Science, pp. 165–84.

Ash, V. B. (1989) Resistance to shell breaking in two intertidal snails. *Zool. J. Linn. Soc.*, **96**, 167–84.

Avise, J. C. (1974) Systematic value of electrophoreti data. *Syst. Zool.*, **23**, 465–81.

Avise, J. C., Arnold, J., Ball, R. M., Bermingham, E., Lamb, T., Neigel, J. E., Reeb, C. A. and Saunders, N. C. (1987) Intraspecific phylogeography: the mitochondrial DNA bridge between population genetics and systematics. *A. Rev. Ecol. Syst.*, **18**, 489–522.

Avise, J. C., Helfman, G. S., Saunders, N. C. and Hales, L. S. (1986) Mitochondrial DNA differentiation in North Atlantic eels: population genetic consequences of an unusual life history pattern. *Proc. natn. Acad. Sci. U.S.A.*, **83**, 4350–4.

Avtalion, R. R. (1982) Genetic markers in *Sarotherodon* and their use for sex and species identification, in *The Biology and Culture of Tilapias* (ICLARM Conference Proceedings, Vol. 7) (eds R. S. V. Pullin and R. H. Lowe-McConnell), International Center for Living Aquatic Resources Management, Manila, pp. 269–77.

Avtalion, R. R. and Don, J. (1990) Sex-determining genes in tilapia: a model of genetic recombination emerging from sex ratio results of three generations of diploid gynogenetic *Oreochromis aureus*. *J. Fish Biol.* **37**, 167–73.

Baerends, G. P. (1984) The organization of the pre-spawning behaviour in the cichlid fish *Aequidens portalegrensis* (Hensel). *Neth. J. Zool.*, **34**, 233–66.

Baerends, G. P. (1986) The functional organisation of the reproductive behaviour in cichlid fish. *Annls Mus. r. Afr. centr. Sci. Zool.*, No. 251, 3–5.

Baerends, G. P. and Baerends-van Roon, J. M. (1950) An introduction to the study of the ethology of cichlid fishes. *Behaviour (Supp.)*·1, 1–243.

Baerends, G. P., Wanders, J. B. W. and Vodegel, R. (1986) The relationship between marking patterns and motivational state in the pre-spawning behaviour of the cichlid fish *Chromidotilapia guentheri* (Sauvage). *Neth. J. Zool.*, **36**, 88–116.

Bailey, R. G., Churchfield, S., Petr, T. and Pimm, R. (1978) The ecology of the fishes in Nyumba ya Mungu Reservoir, Tanzania. *Biol. J. Linn. Soc.*, **10**, 109–37.

Bailey, R. G., Prosser, M. V., Hickley, P. *et al.* (1983) Fish studies, in *Development Studies in the Jonglei Canal Area*. Draft Final Report for the Government of the Republic of the Sudan, vol. 3, pp. 59–186. Mefit-Babtie, Glasgow, UK.

Baker, B. S. (1989) Sex in flies: the splice of life. *Nature, Lond.*, **340**, 521–4.

Balarin, J. D. (1984) Intensive tilapia culture: a scope for the future in food production in developing countries. *Outl. Agric.*, **13**, 10–19.

Balarin, J. D. (1988) Aquaculture practices in Africa: systems in use and species cultivated, in *Village Level Aquaculture Development in Africa* (eds H. R. King and K. H. Ibrahim), Commonwealth Secretariat, London, pp. 3–24.

Balarin, J. D. and Haller, R. D. (1982) The intensive culture of tilapia in tanks, raceways and cages, in *Recent Advances in Aquaculture*. Vol. 1 (eds J. F. Muir and R. J. Roberts), Croom Helm, London, pp. 266–355.

Balarin, J. D. and Hatton, J. P. (1979) *Tilapia: a Guide to their Biology and Culture*. University of Stirling, Stirling.

Baldaccini, N. E. (1973) An ethological study of reproductive behaviour including the colour patterns of the cichlid fish *Tilapia mariae* (Boulenger). *Monitore zool. ital.*, **7**, 247–90.

Balon, E. K. (1975) Terminology of intervals of fishes development. *J. Fish. Res. Bd. Can.*, **32**, 1663–70.

Balon, E. K. (1977) Early ontogeny of *Labeotropheus* Ahl, 1927 (Mbuna, Cichlidae, Lake Malawi), with a discussion on advanced protective styles in fish reproduction and development. *Environ. Biol. Fishes*, **2**, 147–76.

Balon, E. K. (ed.) (1985) *Early Life Histories of Fishes. New Developmental, Ecological and Evolutionary Perspectives*, The Hague, Dr W. Junk.

Balon, E. K. (1986) Types of feeding in the ontogeny of fishes and the life history model. *Environ. Biol. Fishes.*, **16**, 11–24.

Barbour, C. D. (1973) A biogeographical theory of *Chirostoma* (Pisces: Atherinidae): a species flock from the Mexican plateau. *Copeia*, **1973**, 533–56.

Bardach, J. E. and Villars, T. (1974) The chemical senses of fishes, in *Chemoreception in Marine Organisms* (eds P. T. Grant and A. M. Mackie), Academic Press, London, pp. 49–104.

Barel, C. D. N. (1983) Towards a constructional morphology of cichlid fishes (Teleostei, Perciformes). *Neth. J. Zool.*, **33**, 357–424.

Barel, C. D. N. (1984) Form-relations in the context of constructional morphology: the eye and suspensorium of lacustrine Cichlidae (Pisces, Teleostei), with a discussion on the implications for phylogenetic and allometric form-interpretations. *Neth. J. Zool*, **34**, 439–502.

Barel, C. D. N. (1986) The decline of Lake Victoria's cichlid species flock. HEST Report 46. [see HEST].

Barel, C. D. N., Anker, G. Ch., Witte, F., Hoogerhoud, R. J. C. and Goldschmidt, T. (1989) Constructional constraint and its ecomorphological implications. *Acta morph. neerl.-scand.*, **27**, 83–109.

Barel, C. D. N., Dorit, P., Greenwood, P. H., Fryer, G., Hughes, N., Jackson, P. B. N., Kawanabe, H., Lowe-McConnell, R. H., Nagoshi, M., Ribbink, A. J. R., Trewavas, E., Witte, F. and Yamaoka, K. (1985) Destruction of fisheries in Africa's Lakes. *Nature, Lond.*, **315**, 19–20.

Barel, C. D. N., Oijen, M. J. P. van, Witte, F. and Witte-Maas, E. L. M. (1977) An introduction to the taxonomy and morphology of the haplochromine Cichlidae from Lake Victoria. *Neth. J. Zool.*, **27**, 333–89.

Barel, C. D. N., Witte, F. and Oijen, M. J. P. van (1976) The shape of the skeletal elements in the head of a generalized *Haplochromis* species: *H. elegans* Trewavas 1933 (Pisces, Cichlidae). *Neth. J. Zool.*, **26**, 163–265.

Barlow, G. W. (1963) Evolution of behavior. *Science, N.Y.*, **139**, 851–2.

Barlow, G. W. (1964) Ethology of the Asian teleost *Badis badis*. V. Dynamics of fanning and other parental activities, with comments on the behavior of the larvae and postlarvae. *Z. Tierpsychol.*, **21**, 99–123.

Barlow, G. W. (1970) A test of appeasement and arousal hypothese of courtship behavior in a cichlid fish, *Etroplus maculatus*. *Z. Tierpsychol.*, **27**, 779–9806.

Barlow, G. W. (1974) Contrasts in social behavior between Central American cichlid fishes and coral-reef surgeon fishes. *Am. Zool.*, **14**, 9–34.

Barlow, G. W. (1976) The Midas cichlid in Nicaragua, in *Investigations of the Ichthyofauna of Nicaraguan Lakes* (ed. T. B. Thorson), University of Nebraska, Lincoln, Nebraska, pp. 332–58.

Barlow, G. W. (1981) Patterns of parental investment, dispersal and size among coralreef fishes. *Environ. Biol. Fishes*, **6**, 65–85.

Barlow, G. W. (1983) The benefits of being gold: behavioral consequences of polychromatism in the Midas cichlid, *Cichlasoma citrinellum*. *Environ. Biol. Fishes*, **8**, 73–86 and 235–47.

Barlow, G. W. (1984) Patterns of monogamy among teleost fishes. *Arch. FischWiss.*, **35**, (Beih. 1), 75–123.

Barlow, G. W. (1986) A comparison of monogamy among freshwater and coral-reef fishes. *Proc. Int. Conf. Indo-Pac. Fish.*, pp. 767–75.

Barlow, G. W. and Ballin, P. J. (1976) Predicting and assessing dominance from size and coloration in the polychromatic Midas cichlid. *Anim. Behav.*, **24**, 793–813.

Barlow, G. W. and Green, R. F. (1969) Effect of relative size of male on color patterns in a mouthbreeding cichlid fish, *Tilapia melanotheron*. *Commun. Behav. Biol.*, **4A**, 71–8.

Barlow, G. W. and Green, R. F. (1970) The problems of appeasement and of sexual roles in the courtship behavior of the blackchin mouthbreeder, *Tilapia melanotheron* (Pisces, Cichlidae). *Behaviour*, **36**, 84–115.

Barlow, G. W. and Wallach, S. J. (1976) Colour and levels of aggression in the Midas cichlid. *Anim. Behav.*, **24**, 814–17.

Barlow, G. W., Rogers, W. and Fraley, N. (1986) Do Midas cichlids win through prowess or daring? It depends. *Behav. Ecol. Sociobiol.*, **19**, 1–8.

Barnett, C. (1977a) Aspects of chemical communication with special reference to fish. *Biosci. Commun.*, **3**, 331–92.

Barnett, C. (1977b) Chemical recognition of the mother by the young of the cichlid fish *Cichlasoma citrinellum*. *J. Chem. Ecol.*, **3**, 461–6.

Bass, T. A. (1988) This African Lake turns out to be a fine kettle of fish. *Smithsonian*, **19**, 144–55.

Bateson, P. (ed.) (1983) *Mate Choice*, Cambridge University Press, Cambridge.

Bauer, J. (1963) Lautäusserungen bei *Tilapia nilotica*. *Aquar.-u. Terarr.-Z.*, **16**, 171–2.

Baylis, J. R. (1974) The behavior and ecology of *Herotilapia multispinosa* (Teleostei, Cichlidae). *Z. Tierpsychol.*, **34**, 115–46.

Baylis, J. R. (1981) The evolution of parental care in fishes, with reference to Darwin's rule of male sexual selection. *Environ. Biol. Fishes*, **6**, 223–51.

Bayne, D. R., Dunseth, D. and Ramirios, C. G. (1976) Supplemental feeds containing coffee pulp for rearing *Tilapia* in Central America. *Aquaculture*, **7**, 133–46.

Beadle, L. C. (1981) *The Inland Waters of Tropical Africa*, 2nd edn, Longman, London.

Beer, M. de (1989) Light measurements in Lake Victoria, Tanzania. *Annls Mus. r. Afr. cent. Sci. Zool.*, No. 257, 57–60.

Bell-Cross, G. and Minshull, J. L. (1988) *The Fishes of Zimbabwe*, Nat. Mus. Monu. Zimbabwe, Harare.

Bemis, W. E. and Lauder, G. V. (1986) Morphology and function of the feeding apparatus of the lungfish, *Lepidosiren paradoxa* (Dipnoi). *J. Morph.*, **187**, 81–108.

Ben-Tuvia, A. (1959) The biology of the cichlid fishes of Lakes Tiberias and Huleh. *Bull. Res. Counc. Israel*, **8**, 153–88.

Berg, L. S. (1940) *Classification of fishes both recent and fossil.* Russian and English lithoprint. 1947, 87–517. Ann Arbor, Michigan.

Bergmann, H.-H. (1968) Eine deskriptive Verhaltensanalyse des Segelflossers (*Pterophyllum scalare* Cuv. and Val., Cichlidae. Pisces). *Z. Tierpsychol.*, **25**, 559–87.

Bernstein, J. W. (1980) Parental care benavior in the cichlid fish *Cichlasoma nigrofasciatum, Copeia, 1980*, 682–6.

Berra, T. M. (1981) *An Atlas of Distribution of the Freshwater Fish Families of the World.* University of Nebraska Press, Lincoln and London.

Bertin, L. and Arambourg, C. (1958) Superorde des Teleosteens, in *Traite de Zoologie, Anatomie, Systématique, Biologie* (ed. P. Grasse), Vol. 13, Agnathes et Poissons, Paris, pp. 2204–500.

Bertram, C. K., Borley, H. J. H. and Trewavas, E. (1942) *Report on the Fish and Fisheries of Lake Nyasa*, Crown Agents, London.

Blache, J. (1964) *Les Poissons du Bassin du Tchad et du Bassin Adjacent du Mayo Kebbi*, ORSTOM, Paris.

Blache, J., Miton, F., Stauch, A. *et al.* (1964) *Les Poissons du bassin du Tchad et du bassin adjacent du Mayo Kebbi. Etude systématique et biologique.* Mém. Orstom, Paris, 483 pp.

Blaxter, J. H. S. (ed.) (1974) *The Early Life History of Fish*, Springer-Verlag, Berlin.

Blumer, L. S. (1979) Male parental care in the bony fishes. *Q. Rev. Biol.*, **54**, 149–61.

Blumer, L. S. (1982) A bibliography and categorization of bony fishes exhibiting parental care. *Zool. J. Linn. Soc.*, **76**, 1–22.

Bock, W. J. (1977) Adaptation and the comparative method, in *Major Patterns in Vertebrate Evolution* (eds M. K. Hecht, P. C. Goody and B. M. Hecht), Plenum Press, New York, pp. 57–82.

Bock, W. J. (1980) The definition and recognition of biological adaptations. *Am. Zool.*, **20**, 217–27.

Bon, J. and Ibrahim, K. H. (1975) Small furu could boost the fish harvest from Lake Victoria. *Fish. News Int.*, July 1975, 32–3.

Bonetto, A. A. and Castello, H. P. (1985) Pesca y piscicultura en aguas continentales de America Latina (*Serie de Biologia*, Monografia No. 31), Programa Regional de Desarollo Cientifico y Tecnologico, Secretaria General de la Organizacion de los Estados Americanos, Washington, DC, pp. 74–9. (in Spanish).

Bookstein, F. L., Chernoff, B., Elder, R. L., Humphries. J. M., Jr, Smith, G. R. and Strauss, R. E. (1985) *Morphometrics in Evolutionary Biology*, Academy of Natural Sciences, Philadelphia, Peru.

Borowsky, R. (1984) The evolutionary genetics of *Xiphophorus*, in *Evolutionary Genetics of Fishes* (ed. B. J. Turner), Plenum Press, New York, pp. 235–310.

Borgia, G. (1985) Bower decorations and mating success of male satin bowerbirds. *Anim. Behav.*, **33**, 144–55.

Borgia, G., Pruett-Jones, S. G. and Pruett-Jones, M. A. (1985) The evolution of bower-building and the assessment of male quality. *Z. Tierpsychol.*, **76**, 225–36.

Boulenger, G. A. (1898) A revision of the African and Syrian fishes of the family Cichlidae. Part I. *Proc. zool. Soc. Lond.*, 1898 (IX), 132–52.

Boulenger, G. A. (1907) *Zoology of Egypt: the Fishes of the Nile.* ii. Published for the Egyptian Government by Hugh Rees, London.

Boulenger, G. A. (1915) *Catalogue of the Freshwater Fishes of Africa in the British Museum (Natural History).* Vol. III, Trustees of the British Museum, London.

Bowen, S. H. (1987) Composition and nutritional value of detritus, in *Detritus and Microbial Ecology in Aquaculture (ICLARM* Conference Proceeding. Vol 14) (eds D. J. W. Moriarty and R. S. V. Pullin), International Center for Living Aquatic Resources Management, Manila, pp. 192–216.

Bradbury, J. W. (1981) The evolution of leks, in *Natural Selection and Social Behavior: Recent Research and New Theory* (eds R. D. Alexander and D. W. Tinkle), Chiron, New York, pp. 138–69.

Brainerd, E. L. (1985) An experimental analysis of prey capture mechanisms in Teleost fishes with implications for present hydrodynamical models of suction feeding, Honours thesis, Harvard University (56 pp.)

Branson, B. A. and Moore, G. A. (1962) The lateralis components of the acoustico-lateralis system in the sunfish family Centrarchidae. *Copeia, 1962,* 1–108.

Breder, C. M. (1934) An experimental study of the reproductive habits and life history of the cichlid fish, *Aequidens latifrons* (Steindachner). *Zoologica, N.Y.,* **18**, 1–42.

Breder, C. M. and Rosen, D. E. (1966) *Modes of Reproduction in Fishes,* Natural History Press, Garden City, New York.

Brem, G., Brenig, B., Horstgen-Schwark, G. and Winnacker, E.-L. (1988) Gene transfer in tilapia. *Aquaculture,* **68**, 209–19.

Brestowsky, M. (1968) Vergleichende untersuchungen zur elternbindung von Tilapia-jungfischen (Cichlidae, Pisces). *Z. Tierpsychol.,* **25**, 761–828.

Brichard, P. (1978) *Fishes of Lake Tanganyika,* T.F.H. Publications, Neptune City, NJ.

Brinley, F. J. and Eulberg, L. (1953) Embryological head glands of the cichlid fish *Aequidens portalegrensis. Copeia, 1953,* 24–6.

Brooks, J. L. (1950) Speciation in ancient lakes. *Q. Rev. Biol.,* **23**, 30–6 and 131–76.

Brown, J. L. (1987) *Helping and Communal Breeding in Birds,* Princeton University Press, Princeton, N.J.

Buddle, C. R. (1984) Androgen-induced sex inversion of *Oreochromis* (Trewavas) hybrid fry stocked into cages standing in an earthen pond. *Aquaculture,* **40**, 233–9.

Burchard, J. E. (1965) Family structure in the dwarf cichlid *Apistogramma trifasciatum* Eigenmann and Kennedy. *Z. Tierpsychol.,* **22**, 150–62.

Burchard, J. E. (1967) The family Cichlidae, in *Fish and Fisheries of Northern Nigeria* (ed. W. Reed), Min. Agric. No. Nigeria, Zaria, pp. 123–43.

Burgess, J. W. and Coss, R. G. (1982) Effects of chronic crowding stress on midbrain development: changes in dendritic spine density and morphology in jewel fish optic tectum. *Develop. Psychobiol.,* **15**, 461–70.

Burgess, J. W., Monachello, M. P. and McGinn, M. D. (1982) Early development of spiny neurons in fish and mouse: morphometric measures of dendritic spine formation pattern. *Develop Brain Res.,* **4**, 465–72.

Busnel, R.-G. (1963) *Acoustic Behaviour of Animals,* Elsevier, Amsterdam.

Butin, R. K. and Ritchie, M. G. (1989) Genetic coupling in mate recognition systems: what is the evidence? *Biol. J. Linn. Soc.,* **37**, 237–46.

Carlstead, K. (1983) Influences of motivation on display divergences in three cichlid fish species (*Haplochromis*). *Behaviour,* **83**, 205–28.

Carlisle, T. R. (1985) Parental response to brood size in a cichlid fish. *Anim. Behav.*, **33**, 234–8.

Carpenter, J. M. (1988) Choosing among multiple equally parsimonious cladograms. *Cladistics*, **4**, 291–6.

Caryl, P. G. (1982) Animal signals: a reply to Hinde. *Anim. Behav.*, **30**, 240–4.

Caulton, M. S. (1982) Feeding, metabolism and growth of tilapias: some quantitative considerations, in *The Biology and Culture of Tilapias* (*ICLARM Conference Proceedings*, Vol. 7) (eds R. S. V. Pullin and R. H. Lowe-McConnell), International Center for Living Aquatic Resources Management, Manila, pp. 157–80.

Chandrasoma, J. (1986) Trends in reservoir fisheries in Sri Lanka with special reference to *Etroplus suratensis*, *Tilapia rendalli* and *Oreochromis niloticus*, in *The First Asian Fisheries Forum* (eds J. L. Maclean, L. B. Dizon and L. V. Hosillos), Asian Fisheries Society, Manila, pp. 353–6.

Chardon, M. and Vandewalle, P. (1971) Comparaison de la région céphalique chez cinq espèces du genre *Tilapia*, dont trois incubateurs buccaux. *Annla Soc. r. zool. Belg.*, **101**, 3–24.

Chen, M. J., Goss, R. G. and Goldthwaite, R. O. (1983) Timing of dispersal in juvenile jewel fish during development is unaffected by available space. *Develop. Psychobiol.*, **16**, 303–10.

Chien, A. K. and Salmon, M. (1972) Reproductive behavior of the angelfish, *Pterophyllum scalare*. 1. A quantitative analysis of spawning and parental behavior. *Forma Functio*, **5**, 45–74.

Chilvers, R. M. and Gee, J. M. (1974) The food of *Bagrus docmac* (Forsk.) (Pisces: Siluriformes) and its relationship with *Haplochromis* Hilgendorf (Pisces: Cichlidae) in Lake Victoria. *East Afric. J. Fish. Biol.*, **6**, 483–505.

Chimits, P. (1955) Le tilapia et son élevage: bibliographie préliminaire. *F.A.O. Fish. Bull.*, **8**, 1–38. (in French).

Chimits, P. (1957a) The tilapias and their culture: a second review and bibliography. *F.A.O. Fish. Bull.*, **10**, 1–24.

Chimits, P. (1957b) Tilapia in ancient Egypt. *F.A.O. Fish. Bull.*, **10**, 1–5.

Chourrout, D. and Itskovich, J. (1983) Three manipulations performed by artificial insemination in *Tilapia*: induced diploid gynogenesis, production of all triploid population and intergeneric hybridization, in *Int. Symp. Tilapia in Aquaculture, Nazareth*, pp. 246–55.

Cichocki, F. P. (1976) Cladistic history of cichlid fishes and reproductive strategies of the American genera *Acarichthys*, *Biotodoma* and *Geophagus*, Vol. 1, PhD thesis, University of Michigan, Ann Arbor, Mich.

Cichocki, F. P. (1977) Tidal cycling and parental behavior of the cichlid fish. *Biotodoma cupido*. *Environ. Biol. Fishes*, **1**, 159–69.

CIFA (1982) Report of the first session of the Sub-committee for the Development and Management of the Fisheries of Lake Victoria, Mwanza, Tanzania, 12–14 October 1981. *F.A.O. Fish. Rep.*, **262**, 78 pp.

CIFA (1988) Report of the fourth session of the Sub-committee for the Development and Management of the Fisheries of Lake Victoria, Kisumu, Kenya, 6–10 April 1987. *F.A.O. Fish. Rep.*, **388**, 112 pp.

Claeys, H. and Aerts, P. (1984) Note on the compound lower pharyngeal jaw operators in *Astatotilapia elegans* (Trewavas), 1933 (Teleostei: Cichlidae). *Neth. J. Zool.*, **34**, 210–14.

Coe, M. J. (1966) The biology of *Tilapia grahami* Boulenger in Lake Magadi, Kenya. *Acta tropica*, **23**, 146–77.

Cole, J. E. and Ward, J. A. (1969) The communicative function of pelvic fin-flickering in *Etroplus maculatus* (Pisces, Cichlidae). *Behaviour*, **35**, 179–99.

Colgan, P. W. (1983) *Comparative Social Recognition*, Wiley, New York.

Collias, N. E. and Collias, E. C. (1983) *Nest Building and Bird Behavior*, Princeton University Press, Princeton, N.J.

Collins, H. L. (1972) Mouth brooding behavior in the substrate spawning cichlid, *Tilapia sparrmani*. *J. Minn. Acad. Sci.*, **38**, 17–18.

Colombe, J. and Allgayer, R. (1985) Description de *Variabilochromis*, *Neolamprologus*, et *Paleolamprologus* genres nouveaux du Lac Tanganyika, avec redescription des genres *Lamprologus* Schilthuis, 1891 et *Lepidolamprologus* Pellegrin, 1904 (Pisces, Teleostei, Cichlidae). *Revue. Franc. des Cichlidophiles*, **48**, 9–28.

Cooke, S. M. and Mooyoung, R. R. (1983) History and status of commercial tilapia farming in Jamaica, in *Proc. Thirty-Fifth Annual Meeting of the Gulf and Caribbean Fisheries Institute, Nassau, Bahamas, November, 1982* (ed. J. B. Higman), Gulf and Caribbean Fisheries Institute, Miami, pp. 33–41.

Corbet, S. A., Green, J., Griffith, J. and Betney, E. (1973) Ecological studies on crater lakes in West Cameroon. Lakes Kotto and Mboandong. *J. Zool. Lond.*, **170**, 309–24.

Coss, R. G. (1979) Delayed plasticity of an instinct recognition and avoidance of 2 facing eyes by the jewel fish. *Develop. Psychobiol.*, **12**, 335–45.

Coss, R. G. and Globus, A. (1978) Spine stems on tectal interneurons in jewel fish are shortened by social stimulation. *Science, N.Y.*, **200**, 787–90.

Coss, R. G. and Globus, A. (1979) Social experience affects the development of dendritic spines and branches on tectal interneurons in the jewel fish. *Develop. Psychobiol.*, **12**, 347–58.

Costa, H. H. (1983) Biological studies of the pearl spot *Etroplus suratensis* (Pisces, Cichlidae) from three different habitats in Sri Lanka. *Int. Rev. ges. Hydrobiol.*, **68**, 565–80.

Coulter, G. W. (1967a) The deep benthic fishes at the south end of Lake Tanganyika with special reference to distribution and feeding in *Bathybates* species, *Hemibates stenosoma* and *Chrysichthys* species. *Fish. Res. Bull. Zambia*, **4** (1965–66), 33–8.

Coulter, G. W. (1967b) Low apparent oxygen requirements of deepwater fishes in Lake Tanganyika. *Nature, Lond.*, **215**, 317–18.

Coulter, G. W. (1991) *Lake Tanganyika and its Life*, Oxford University Press, Oxford and British Museum (Natural History), London.

Coulter, G. W., Allanson, B. R., Bruton, M. N., Greenwood, P. H., Hart, R. C., Jackson, P. B. N. and Ribbink, A. J. (1986) Unique qualities and special problems of the African Great Lakes, *Environ. Biol. Fishes*, **17**, 161–83.

Courtenay, S. C. and Keenleyside, M. H. A. (1983) Wriggler-hanging: a response to hypoxia by brood-rearing *Herotilapia multispinosa* (Telostei, Cichlidae). *Behaviour*, **85**, 183–97.

Couvering, J. A. H. van (1982) Fossil cichlid fishes of Africa. *Special Papers in Palaeontology: Palaeon. Assn. Lond., no. 29*, 1–103.

Crane, J. M., Jr. (1981) Feeding and growth by the sessile larvae of the teleost *Porichthys notatus*. *Copeia, 1981*, 895–7.

Crapon de Caprona, M.-D. (1974) The effect of chemical stimuli from conspecifics on the behavior of *Haplochromis burtoni* (Cichlidae, Pisces). *Experientia*, **30**, 1394–5.

Crapon de Caprona, M.-D. (1980) Olfactory communication in a cichlid fish, *Haplochromis burtoni*. *Z. Tierpsychol.*, **52**, 113–34.
Crapon de Caprona, M.-D. (1982) The influence of early experience on preferences for optical and chemical cues produced by both sexes in the cichlid fish *Haplochromis burtoni* (*Astatotilapia burtoni*, Greenwood 1979). *Z. Tierpsychol.*, **58**, 329–61.
Crapon de Crapona, M.-D. (1986) The use of fertile hybrids for the study of the accuracy of species recognition in cichlids. *Annls. Mus. r. Afr. cent. Sci. zool.*, No. 251, 117–20.
Crapon de Caprona, M.-D. and Fritzsch, B. (1984) Interspecific fertile hybrids of haplochromine Cichlidae (Teleostei) and their possible importance for speciation. *Neth. J. Zool.*, **34**, 503–38.
Cromie, W. J. (1982) Beneath the African lakes. *Mosaic*, **13**, 2–6.
Crul, R. C. M., Postma, D. J., Oijen, M. J. P. van and Bongers, G. (1989) Bibliography on Lake Victoria. HEST Report 54. [see HEST]
Cunningham, J. E. R. and Balon, E. K. (1986) Early ontogeny of *Adinia xenica* (Pisces, Cyprinodontidae). 3. Comparison and evolutionary significance of some patterns in epigenesis of egg-scattering, hiding and bearing cyprinodontiforms. *Environ. Biol. Fishes*, **15**, 91–105.
Currey, J. (1984) *The Mechanical Adaptations of Bones*, Princeton University Press, Princeton, N.J.
Daget, J. (1954) Les poissons du Niger Supérieur. *Mém. Inst. fr. Afr. noire*, No. 36, 1–391.
Daget, J. (1957) Données récentes sur la biologie des poissons dans le delta central du Niger. *Hydrobiologia*, **9**, 321–47.
Daget, J. (1988) Mutanda ichthyologica: *Thysochromis nom. nov.* en replacement de *Thysia* (Pisces, Cichlidae). *Cybium*, **12**, 97.
Daget, J. and Iltis, A. (1965) Poissons de Côte D'Ivoire (eaux douces et saumâtres). *Mém. Inst. fr. Afr. noire*, No. 74, 1–385.
Darwin, C. (1871) *The Descent of Man and Selection in Relation to Sex*, John Murray, London.
Dawkins, R. (1982) *The Extended Phenotype*, W. H. Freeman, Oxford and San Francisco.
Dawkins, R. and Krebs, J. R. (1978) Animal signals: information or manipulation? in *Behavioural Ecology* (eds J. R. Krebs and N. B. Davies), Blackwell Scientific Publications, Oxford, pp. 282–309.
Day, F. (1877) *The Fishs of India*. Vol. 1, William Dawson & Sons, London.
DeMason, L. J. (1988) Induced sex reversal in a Lake Malawi cichlid, *Pseudotropheus zebra* 'red dorsal', unpubl. MS thesis, University of Miami.
Dickman, M. C., Schliwa, M. and Barlow, G. (1988) Melanophore death and disappearance produces color metamorphosis in the polychromatic Midas cichlid (*Cichlasoma citrinellum*). *Cell Tissue Res.*, **253**, 9–14.
Dominey, W. J. (1984) Effects of sexual selection and life history on speciation: species flocks in African cichlids and Hawaiian *Drosophila*, in *Evolution of Fish Species Flocks* (eds A. A. Echelle and I. Kornfield), University of Maine at Orono Press, Orono, Maine, pp. 231–49.
Don, J. and Avtalion, R. R. (1986) The induction of triploidy in *Oreochromis aureus* by heat shock. *Theor. Appl. Genet.*, **72**, 186–92.
Don, J. and Avtalion, R. R. (1988) Comparative study on the induction of triploidy in tilapias, using cold- and heat-shock techniques. *J. Fish Biol.*, **32**, 665–72.
Dorit, R. L. (1986) Molecular and morphological variation in Lake Victoria haplo-

chromine cichlids (Perciformes: Cichlidae), unpubl. PhD diss., Harvard University, Cambridge, Mass.
Dudley, R. G. (1979) Changes in growth and size distribution of *S. macrochir* and *S. andersonii* from the Kafue floodplain, Zambia, since the construction of the Kafue Gorge dam. *J. Fish. Biol.*, **14**, 205–23.
Dullemeijer, P. and Barel, C. D. N. (1977) Functional morphology and evolution, in *Major Patterns in Vertebrate Evolution* (*NATO Advanced Study Institute*, Ser. A, Vol. 14), (eds M. Hecht, P. C. Goody and B. M. Hecht), Plenum Press, New York and London, pp. 83–117.
Dunseth, D. R. and Baye, D. R. (1978) Recruitment control and production of *Tilapia aurea* (Steindachner) with the predator, *Cichlasoma managuense* (Gunther). *Aquaculture*, **14**, 383–90.
Dupuis, H. M. C. and Keenleyside, M. H. A. (1982) Egg-care behaviour of *Aequidens paraguayensis* (Pisces, Cichlidae) in relation to predation pressure and spawning substrate. *Can. J. Zool.*, **60**, 1794–9.
Dutta, T. M. (1987) Osteology, myology and feeding mechanism of *Astronotus ocellatus* (Pisces, Perciformes). *Zoomorphologie*, **106**, 369–81.
Eccles, D. H. (1986) Is speciation of demersal fishes in Lake Tanganyika restrained by physical limnological conditions? *Biol. J. Linn. Soc.*, **29**, 115–22.
Eccles, D. H. and Lewis, D. S. C. (1976) A revision of the genus *Docimodus* Boulenger (Pisces: Cichlidae), a group of fishes with unusual feeding habits from Lake Malawi. *Zool. J. Linn. Soc.*, **58**, 165–72.
Eccles, D. H. and Lewis, D. S. C. (1977) A taxonomic study of the genus *Lethrinops* Regan (Pisces: Cichlidae) from Lake Malawi. Part I. *Ichthyol. Bull. Rhodes Univ.*, **36**, 1–12.
Eccles, D. H. and Lewis, D. S. C. (1978) A taxonomic study of the genus *Lethrinops* Regan (Pisces: Cichlidae) from Lake Malawi. Part II. *Ichthyol. Bull. Rhodes Univ.*, **37**, 1–11.
Eccles, D. H. and Lewis, D. S. C. (1979) A taxonomic study of the genus *Lethrinops* Regan (Pisces: Cichlidae) from Lake Malawi. Part III. *Ichthyol. Bull. Rhodes. Univ.*, **38**, 1–25.
Eccles, D. H. and Lewis, D. S. C. (1981) Midwater spawning in *Haplochromis chrysonotus* (Boulenger) (Teleostei, Cichlidae) in Lake Malawi. *Environ. Biol. Fishes*, **6**, 201–2.
Eccles, D. H. and Trewavas, E. (1989) Malawian cichlid fishes. The classification of some Haplochromine genera. Herten (W. Germany), Lake Fish Movies, 334 pp.
Echelle, A. A. and Kornfield, I. (eds) (1984) *Evolution of Fish Species Flocks*, University of Maine at Orono Press, Orono, Maine.
Edwards, P., Pullin, R. S. V. and Gartner, J. A. (1988) Research and education for development of integrated crop–livestock–fish farming systems in the tropics (*ICLARM Studies and Reviews*, Vol. 16), International Center for Living Aquatic Resources Management, Manila.
Eigenmann, C. H. (1912) The freshwater fishes of British Guiana, including a study of the ecological grouping of species and the relation of the fauna of the plateau to that of the lowlands. *Mem. Carnegie. Mus.*, **5**, 1–578.
Emlen, S. T. and Oring, L. W. (1977) Ecology, sexual selection, and the evolution of mating systems. *Science*, N.Y., **197**, 215–23.
Enquist, M. and Jakobsson, S. (1986) Decision making and assessment in the fighting

behaviour of *Nannacara anomala* (Cichlidae, Pisces). *Ethology*, **72**, 143–53.

Falter, U. (1987) Description des patrons de coloration chez *Oreochromis niloticus* (L.) (Teleostei, Cichlidae). *Annls. Soc. r. zool. Belg.*, **117**, 201–19.

FAO (1990) The definition of aquaculture and collection of statistics. Aquaculture Minutes No. 7, April 1990, 4 pp. Food and Agriculture Organization of the United Nations, Rome, Italy.

Farris, J. S. (1969) A successive approximations approach to character weighting. *Syst. Zool.*, **18**, 374–85.

Feldberg, E. and Bertollo, L. A. C. (1985) Karyotypes of 10 species of Neotropical cichlids (Pisces: Perciformes). *Caryologia*, **38**, 257–68.

Felsenstein, J. (1988) Phylogenies from molecular sequences: inference and reliability. *A. Rev. Genet.*, **22**, 521–65.

Fernald, R. D. (1976) The effect of testosterone on the behavior and coloration of adult male cichlid fish (*Haplochromis burtoni* Gunther). *Hormone Res.*, **7**, 172–8.

Fernald, R. D. (1984) Vision and behavior in an African cichlid fish. *Am. Scient.*, **72**, 58–65.

Fernald, R. D. and Hirata, N. R. (1977) Field study of *Haplochromis burtoni*: habits and co-habitant. *Environ. Biol. Fishes.*, **2**, 299–308.

Fernö, A. and Sjölander, S. (1976) Influence of previous experience on the mate selection of two colour morphs of the convict cichlid, *Cichlasoma nigrofasciatum* (Pisces, Cichlidae). *Behav. Process.*, **1**, 3–14.

Fischer, E. A. (1988) Simultaneous hermaphroditism, tit-for-tat and the evolutionary stability of social systems. *Ethol. Sociobiol.*, **9**, 119–36.

Fishelson, L. (1966) Untersuchungen zur vergleichenden Entwicklungsgeschichte der Gattung *Tilapia* (Cichlidae, Teleostei). *Zool. Jb., Abt. Anat. Ontog. Tiere*, **83**, 571–656.

Fishelson, L. and Heinrich, W. (1963) Some observations on the mouthbreeding *Tilapia galilaea* L. (Pisces, Cichlidae). *Ann. Mag. nat. Hist., Ser. 13*, **6**, 507–8.

Fishelson, L. and Yaron, Z. (compilers) (1983) *International Symposium on Tilapia in Aquaculture*, Tel Aviv University, Tel Aviv.

Fisher, R. A. (1930) *The Genetical Theory of Natural Selection*, Dover, New York.

Flegler-Balon, C. (1989) Direct and indirect development in fishes – examples of alternative life-history styles, in *Alternative Life History Styles in Animals* (ed. M. N. Bruton), Kluwver Scientific Publ., Dordrecht, pp. 71–100.

Foerster, W. and Schartl, M. (1987) Karyotype and isozyme patterns of five species of *Aulonocara* Regan, 1922. *Cour. Forsch.-inst. Senckenberg*, **94**, 55–62.

Fraley, N. B. and Fernald, R. D. (1982) Social control of developmental rate in the African cichlid, *Haplochromis burtoni*. *Z. Tierpsychol.*, **60**, 66–82.

Francis, R. C. (1988) Socially mediated variation in growth rate of the Midas cichlid: the primacy of early size differences. *Anim. Behav.*, **36**, 1844–5.

Fryer, G. (1956) Biological notes on some cichlid fishes of Lake Nyasa. *Revue Zool. Bot. afr.*, **54**, 1–7.

Fryer, G. (1959) The trophic interrelationships and ecology of some littoral communities of Lake Nyasa with especial reference to the fishes, and a discussion of the evolution of a group of rock-frequenting Cichlidae. *Proc. Zool. Soc. Lond.*, **132**, 153–281.

Fryer, G. (1972) Conservation of the Great Lakes of East Africa: a lesson and a warning. *Biol. Conserv.*, **4**, 256–62.

Fryer, G. (1973) The Lake Victoria fisheries: some facts and fallacies. *Biol. Conserv.*, **5**, 304–8.

Fryer, G. (1977) Evolution of species of cichlid fishes in African lakes. *Z. Zool. Syst. Evol.-forsch.*, **15**, 141–65.

Fryer, G. and Iles, T. D. (1972) *The Cichlid Fishes of the Great Lakes of Africa: their Biology and Evolution*, Oliver and Boyd, Edinburgh.

Fryer, G., Greenwood, P. H. and Trewavas, E. (1955) Scale-eating habits of African cichlid fishes. *Nature, Lond.*, **175**, 1089–90.

Fujita, K. (1981) Oviphagous embryos of the pseudo-carchariid shark *Pseudocarcharius kamoharai* from the central Pacific. *Jap. J. Ichthyol.*, **28**, 37–44.

Futuyma, D. J. (1986) *Evolutionary Biology*, Sinauer Associates, Sunderland, Mass.

Futuyma, D. J. and Moreno, G. (1988) The evolution of ecological specialization. *A. Rev. Ecol. Syst.*, **19**, 207–33.

Gaemers, P. A. M. (1984) Taxonomic position of the Cichlidae as demonstrated by the morphology of their otoliths. *Neth. J. Zool.*, **34**, 566–95.

Galis, F. and Barel, C. D. N. (1980) Comparative functional morphology of the gills of African lacustrine Cichlidae (Pisces, Teleostei). *Neth. J. Zool.*, **30**, 392–430.

Galis, F. and Jong, P. W. de (1988) Optimal foraging and ontogeny; food selection by *Haplochromis piceatus. Oecologia*, **75**, 175–84.

Gans, C. (1969) Functional components versus mechanical units in descriptive morphology. *J. Morph.*, **128**, 365–8.

Gashagaza, M. M. (1988) Feeding activity of a Tanganyikan cichlid fish *Lamprologus brichardi. Afr. Stud. Monogr. Kyoto Univ.*, **9**, 1–9.

Gashagaza, M. M. and Nagoshi, M. (1986) Comparative study on the food habits of six species of *Lamprologus* (Osteichthyes, Cichlidae). *Afr. Stud. Monogr. Kyoto Univ.*, **6**, 37–44.

Gee, J. M. (1969a) A comparison of certain aspects of the biology of *Lates niloticus* (Linné) in some East African lakes. *Revue Zool. Bot. afr.*, **80**, 244–62.

Gee, J. M. (1969b) A comparison of certain aspects of the biology of *Lates niloticus* (Linné) in endemic and introduced environments in East Africa, in *Man-made Lakes: the Accra Symposium* (ed. L. E. Obeng), Ghana University Press, Accra, Ghana, pp. 251–9.

Geisler, R. (1969) Untersuchungen über der Sauerstoffgehalt, den biochemischen Sauerstoffbedarf und der Sauerstoffverbrauch von Fischen in einem Schwarzwasser (Rio Negro, Amazonien, Brasilien), *Arch. Hydrobiol.*, **66**, 307–25.

Gittleman, J. L. (1981) The phylogeny of parental care in fishes. *Anim. Behav.*, **29**, 936–41.

Gjedrem, T. (1985) Improvement of productivity through breeding schemes. *Geojournal*, **10**, 233–41.

Goedel, W. (1974a) Beiträge zur vergleichenden und funktionellen Anatomie des Kopfes von *Tilapia* (Cichlidae, Teleostei). Teil I, Vol I, *Zool. Jb. Abt. Anat. Ontog. Tiere*, **92**, 220–74

Goedel, W. (1974b) Beiträge zur vergleichenden und funktionellen Anatomie des Kopfes von *Tilapia* (Cichlidae, Teleostei). Teil 2, Vol II, *Zool. Jb., Abt. Anat. Ontog. Tiere*, **92**, 321–83.

Goldschmidt, T. (1986) Niche differentiation in planktivorous haplochromines in the Mwanza Gulf of Lake Victoria and a hypothesis on the speciation of haplochromines in general. *Ann. Kon. Mus. Mid. Afr. Zool. Wetensch.*, **251**, 45–52.

Goldschmidt, T. (1989a) An ecological and morphological field study of the haplochromine cichlid fishes of Lake Victoria, thesis, Rijksuniversiteit, Leiden, The Netherlands.

Goldschmidt, T. (1989b) Ecomorphology of anal spots in haplochromine cichlids (Pisces, Perciformes) from Lake Victoria, in Goldschmidt (1989a), pp. 123–49.

Goldschmidt, T. (1989c) Reproductive strategies, subtrophic niche differentiation and the role of competition for food in haplochromine cichlids (Pisces) from Lake Victoria, Tanzania. *Annls. Mus. r. Afr. cent. Sci. zool.*, No. 257, 119–32.

Goldschmidt, T. and Goudswaard, P C. (1989) Reproductive strategies in haplochromines (Pisces, Cichlidae) from Lake Victoria, in Goldschmidt (1989a), pp. 93–119.

Goldschmidt, T. and Visser, J. de (1990) On the possible role of egg mimics in speciation. *Acta. Biotheor.*, **38**, 125–34.

Goldschmidt, T. and Witte, F. (1990) Reproductive strategies of zooplanktivorous haplochromine cichlids (Pisces) from Lake Victoria before the Nile perch boom.*Oikos*, **58**, 356–68.

Goldschmidt, T., Witte, F. and Visser, J. de (1990) Ecological segregation in zooplanktivorous haplochromine species (Pisces, Cichlidae) from Lake Victoria. *Oikos*, **58**, 343–355.

Goldstein, R. J. (1973) *Cichlids of the World*, T. F. H. Publs Inc., Neptune City, N.J.

Gosse, J.-P. (1963) Le milieu aquatique et l'écologie des poissons dans la région de Yangambi. *Annls Mus. r. Afr. cent. Sér. 8vo sci. zool.*, No. 116, 1–270.

Gosse, J.-P. (1975) Révision du genre *Geophagus* (Pisces, Cichlidae). *Mém. Acad. r. Sci. d'outre-mer Cl. Sci. nat. méd.* **19**, 1–172.

Gottfried, M. D. (1986) Developmental transition in feeding morphology of the Midas cichlid. *Copeia, 1986*, 1028–30.

Goude, G., Edlund, B., Engqvist-Edlund, U. and Andersson, M. (1972) Approach and withdrawal in young of *Tilapia mossambica* (Cichlidae, Pisces) as a function of age and onset of stimulation. *Z. Tierpsychol.*, **31**, 60–77.

Goudswaard, P.C. (1988) A comparison of trawl surveys in 1969/70 and 1984/85 in the Tanzanian part of Lake Victoria. *F.A.O. Fish. Rep.*, **388**, 86–100.

Goudswaard, P. C. and Ligtvoet, W. (1988) Recent developments in the fishery for haplochromines (Pisces, Cichlidae) and Nile perch (*Lates niloticus* L.) (Pisces, Centropomidae) in Lake Victoria. *F.A.O. Fish. Rep.*, **388**, 101–12.

Gould, S. J. and Lewontin, R. C. (1979) The spandrels of San Marcos and the panglossian paradigm: a critique of the adaptationist programme. *Proc. R. Soc.*, **205B**, 581–98.

Goulding, M., Carvalho, M. L. and Ferreira, E. G. (1988) *Rio Negro; Rich Life in Poor Water*, SPB Academic Publishers, The Hague.

Govani, J. J., Boehlert, G. W. and Watanabe, Y. (1986) The physiology of digestion in fish larvae. *Environ. Biol. Fishes.*, **16**, 59–77.

Graham, M. (1929) *The Victoria Nyanza and its Fisheries. A Report on the Fishing Surveys of Lake Victoria*, Crown Agents for the Colonies, London.

Grant, P. R. (1986) *Ecology and Evolution of Darwin's Finches*, Princeton University Press, Princeton, N. J.

Greenwood, P. H. (1954) On two species of cichlid fishes from the Malagarazi River (Tanganyika), with notes on the pharyngeal hypophesis in species of the *Haplochromis* group. *Ann. Mag. nat. Hist. (12)*, **7**, 401–14.

Greenwood, P. H. (1956) A revision of the Lake Victoria *Haplochromis* species (Pisces, Cichlidae), Part I. *Bull. Br. Mus. nat. Hist. (Zool.)*, **4**, 223–44.

Greenwood, P. H. (1959a) A revision of the Lake Victoria *Haplochromis* species (Pisces, Cichlidae), Part III. *Bull. Br. Mus. nat. Hist. (Zool.)*, **5**, 179–218.

Greenwood, P. H. (1959b) Evolution and speciation in the *Haplochromis* (Pisces, Cichlidae) of Lake Victoria, *Proc. 15th Int. Congr. Zool., 1958*, 147–50.

Greenwood, P. H. (1959c) The monotypic genera of cichlid fishes in Lake Victoria. Part II. *Bull. Br. Mus. nat. Hist. (Zool.)*, **5** (7), 163–77.

Greenwood, P. H. (1965a) The cichlid fishes of Lake Nabugabo, Uganda. *Bull. Br. Mus. nat. Hist. (Zool.)*, **12**, 315–57.

Greenwood, P. H. (1965b) Environmental effects on the pharyngeal mill of a cichlid fish, *Astatoreochromis alluaudi*, and their taxonomic implications. *Proc. Linn. Soc. Lond.*, **176**, 1–10.

Greenwood, P. H. (1967) A revision of the Lake Victoria *Haplochromis* species (Pisces, Cichlidae), Part VI. *Bull. Br. Mus. nat. Hist. (Zool.)*, **15**, 29–119.

Greenwood, P. H. (1973a) Morphology, endemism and speciation in African cichlid fishes. *Verh. dt. zool. Ges. Mainz. 1973*, **66**, 115–24.

Greenwood, P. H. (1973b) A revision of the *Haplochromis* and related species Pisces, Cichlidae) from Lake George, Uganda. *Bull. Br. Mus. nat. Hist.* (*Zool.*), **25**, 139–242.

Greenwood, P. H. (1974a) The cichlid fishes of Lake Victoria, East Africa: the biology and evolution of a species flock. *Bull. Br. Mus. nat. Hist. (Zool.)*, Supp.6, 1–134.

Greenwood, P. H. (1974b) The *Haplochromis* species (Pisces, Cichlidae) of Lake Rudolf, East Africa. *Bull. Br. Mus. nat. Hist. (Zool.)*, **27**, 141–65.

Greenwood, P. H. (1978) A review of the pharyngeal apophysis and its significance in the classification of African cichlid fishes. *Bull. Br. Mus. nat. Hist. (Zool.)*, **33**, 297–323.

Greenwood, P. H. (1979) Towards a phyletic classification of the 'genus' *Haplochromis* (Pisces, Cichlidae) and related taxa. Part I. *Bull. Br. Mus. nat. Hist. (Zool.)*, **35**, 265–322.

Greenwood, P. H. (1980) Towards a phyletic classification of the 'genus' *Haplochromis* (Pisces, Cichlidae) and related taxa. Part II; the species from Lakes Victoria, Nabugabo, Edward, George and Kivu. *Bull. Br. Mus. nat. Hist. (Zool.)*, **39**, 1–101.

Greenwood, P. H. (1981a) *The Haplochromine Fishes of the East African Lakes*, British Museum (Natural History), London, and Kraus International Publications, Munich.

Greenwood, P. H. (1981b) Species flocks and explosive evolution, in *Chance, Change and Challenge – The Evolving Biosphere* (eds P. H. Greenwood and P. L. Forey), Cambridge University Press and British Museum (Natural History), London, pp. 61–74.

Greenwood, P. H. (1983a) On *Macropleurodus*, *Chilotilapia* (Teleostei, Cichlidae) and the interrelationships of African cichlid species in flocks. *Bull. Br. Mus. nat. Hist. (Zool.)*, **45**, 209–31.

Greenwood, P. H. (1983b) The *Ophthalmotilapia* assemblage of cichlid fishes reconsidered. *Bull. Br. Mus. nat. Hist. (Zool.)*, **44**, 249–90.

Greenwood, P. H. (1984) African cichlids and evolutionary theories, in *Evolution of Fish Species Flocks* (eds A. A. Echelle and I. Kornfield), University of Maine at Orono Press, Orono, Maine, pp. 141–54.

Greenwood, P. H. (1985a) Notes on the anatomy and phyletic relationships of

Hemichromis Peters, 1858. *Bull. Br. Mus. nat. Hist. (Zool.)*, **48**, 131–71.

Greenwood, P. H. (1985b) The generic status and affinities of *Paratilapia thomasi* Blgr 1915 (Teleostei, Cichlidae). *Bull. Br. Mus. nat. Hist. (Zool.)*, **49**, 257–72.

Greenwood, P. H. (1986a) The pars jugularis and the intrarelationships of cichlid fishes (Labroidei, Teleostei). *J. nat. Hist.*, **20**, 949–74. [*Note*: All citations 'Greenwood (1986)' refer to this paper and not to the following entry.]

Greenwood, P. H. (1986b) The pharyngeal apophysis on the base of the skull in cichlid fishes. A reply to Trewavas (1985). *Neth. J. Zool.*, **36**, 562–4.

Greenwood, P. H. (1987a) The genera of pelmatochromine fishes (Teleostei, Cichlidae). A phylogenetic review. *Bull. Br. Mus. nat. Hist. (Zool.)*, **53**, 139–203.

Greenwood, P. H. (1987b) The pharyngeal apophysis on the base of the skull in cichlid fishes. A reply to Trewavas (1985). *Neth. J. Zool.*, **36**, 562–4.

Greenwood, P. H., Rosen, D. E., Weitzman, S. H. and Myers, G. S. (1966) Phyletic studies of teleostean fishes, with a provisional classification of living forms. *Bull. Am. Mus. nat. Hist.*, **131**, 339–445.

Gross, M. R. (1984) Sunfish, salman, and the evolution of alternative reproductive strategies and tactics in fishes, in *Fish Reproduction: Strategies and Tactics* (eds G. W. Potts and R. J. Wootton), Academic Press, London, pp. 55–73.

Gross, M. R. (1984) Sneakers, satellites and parentals: polymorphic mating strategies in North American sunfishes. *Z. Tierpsych.*, **60**, 1–26.

Gross, M. R. and Sargent, R. C. (1985) The evolution of male and female parental care in fishes. *Am. Zool.*, **25**, 807–22.

Guerrero, R. D. III (1982) Control of tilapia reproduction, in *The Biology and Culture of Tilapias* (*ICLARM Conference Proceedings*, Vol. 7) (eds R. S. V. Pullin and R. H. Lowe-McConnell), International Center for Living Aquatic Resources Management, Manila, pp. 309–16.

Guerrero, R. D. III (1987) *Tilapia Farming in the Philippines*, National Book Store, Manila.

Hailman, J. P. (1977) *Optical Signals*, Indiana University Press, Bloomington, Ind.

Hamblyn, E. L. (1966) The food and feeding habits of Nile perch: *Lates niloticus* (Linné) (Pisces, Centropomidae). *Revue Zool. Bot. afr.*, **74**, 1–28.

Hamilton, W. D. and Zuk, M. (1982) Heritable true fitness and bright birds: a role for parasites? *Science, N.Y.*, **218**, 384–7.

Hanon, L. (1975) Adaptations morphologiques et comportementales à l'incubation buccale chez les poissons cichlides: oeufs et alevins. *Annls Soc. r. zool. Belg.*, **105**, 169–92.

Hay, T. F. (1978) Filial imprinting in the convict cichlid fish *Cichlasoma nigrofasciatum*. *Behaviour*, **65**, 129–60.

Heijne, G. von (1987) *Sequence Analysis in Molecular Biology*, Academic Press, NY.

Heiligenberg, W., Kramer, U. and Schulz, V. (1972) The angular orientation of the black eye-bar in *Haplochromis burtoni* (Cichlidae, Pisces) and its relevance to aggressivity. *Z. vergl. Physiol.*, **76**, 168–76.

Heisenberg, W. (1971) *Physics and Beyond*, Harper and Row, New York.

Heisler, I. L. (1985) Quantitative genetic models of female choice based on 'arbitrary' male characters. *Heredity, Lond.*, **55**, 187–98.

Henderson, H. F. (1982) A programme for evaluating the fishery potential of Cuban reservoirs and for monitoring the state of their fish resources. *UNDP/FAO Report to the Government of the Republic of Cuba*. Food and Agriculture Organization of the United Nations, Rome.

Herke, S. W., Kornfield, I., Moran, P. and Morling, J. (1990) Molecular confirmation of hybridization between northern pike (*Esox lucius*) and chain pickerel (*E. niger*). *Copeia*, 842–6.

Hert, E. (1985) Individual recognition of helpers by the breeders in the cichlid fish *Lamprologus brichardi* (Poll, 1974). *Z. Tierpsychol.*, **68**, 313–25.

Hert, E. (1986) On the significance of the egg-spots in an African mouth brooder (*Haplochromis elegans*, Trewavas 1933). *Annls Mus. r. Afr. cent., Sci. zool.*, No. 251, 25–6.

Hert, E. (1987) On the significance of the egg-spots in an African mouth brooder (*Haplochromis elegans*, Trewavas 1933). *Abstr. XXth Int. Ethol. Conf., Madison, WI*, 85.

HEST: Haplochromis Ecology Survey Team: P.O. Box 903, Mwanza, Tanzania; and c/o Research Group in Ecological Morphology, Zoölogische Laboratorium, Rijksuniversiteit Leiden, Postbus 9516, 2300 RA Leiden, The Netherlands.

Heywood, V. H. (1988) The structure of systematics, in *Prospects in Systematics* (ed. D. L. Hawksworth), Clarendon Press, Oxford, pp. 44–56.

Hickley, P. and Bailey, R. G. (1987) Food and feeding relationships of fish in the Sudd swamps (River Nile, southern Sudan). *J. Fish Biol.*, **30**, 147–59.

Hildemann, W. H. (1959) A cichlid fish, *Symphysodon discus*, with unique nurture habits. *Am. Nat.*, **93**, 27–34.

Hilgendorf, E. (1888) Fische aus dem Victoria-Nyanza (Ukerewe See). *Sber. Ges. naturf. Freunde Berl., 1888*, 75–9.

Hillis, D. M. (1987) Molecular versus morphological approaches to systematics. *A. Rev. Ecol. Syst.*, **18**, 23–42.

Hillis, D. M. and Moritz, C. (eds) (1990) *Molecular Systematics*. Sinauer Publishers, Sunderland, MA.

Hinde, R. A. (1981) Animal signals: ethological and games-theory approaches are not incompatible. *Anim. Behav.*, **29**, 535–42.

Hoffman, S. G. (1983) Sex-related foraging behavior in sequentially hermaphroditic hogfishes (*Bodianus* spp.). *Ecology*, **64**, 798–808.

Holzberg, S. (1978) A field and laboratory study of the behaviour and ecology of *Pseudotropheus zebra* (Boulenger), an endemic cichlid of Lake Malawi (Pisces: Cichlidae). *Z. Zool. Syst. Evol.-forsch.*, **16**, 171–87.

Hoogerhoud, R. J. C. (1984) A taxonomic reconsideration of the haplochromine genera *Gaurochromis* Greenwood, 1980 and *Labrochromis* Regan, 1920 (Pisces, Cichlidae). *Neth. J. Zool.*, **34**, 539–65.

Hoogerhoud, R. J. C. (1986a) Ecological morphology of some cichlid fishes, doctoral thesis, Rijksuniversitiet Leiden, The Netherlands.

Hoogerhoud, R. J. C. (1986b) Taxonomic and ecological aspects of morphological plasticity in molluscivorous haplochromines (Pisces, Cichlidae). *Annls Mus. r. Afr. cent. Sci. zool.*, No. 251, 131–4.

Hoogerhoud, R. J. C. (1987) The adverse effects of shell ingestion for molluscivorous cichlids, a constructional morphological approach. *Neth. J. Zool.*, **37**, 277–300.

Hoogerhoud, R. J. C. and Barel, C. D. N. (1978) Integrated morphological adaptations in piscivorous and mllusc-crushing *Haplochromis* species, in *Proc. Zodiac Symposium on Adaptation*, 24–26 May 1978, Pudoc, Wageningen, The Netherlands, pp. 52–6.

Hoogerhoud, R. J. C., Witte, F. and Barel, C. D. N. (1983) The ecological differentiation of two closely resembling *Haplochromis* species from Lake Victoria (*H. iris* and *H. hiatus*; Pisces, Cichlidae). *Neth. J. Zool.*, **33**, 283–305.

Hopkins, K. D., Hopkins, M. L., LeClerq, D. and Al-Ameeri, A. (1986a) Tilapia culture in Kuwait: a preliminary economic analysis of production systems. *Kuwait Bull. Mar. Sci.*, **7**, 45–64.

Hopkins, K. D., Pauly, D. L., Cruz, E. M. and Weerd, J. H. van (1982) An alternative to predator–prey ratios in predicting recruitment. *Meeresforschung*, **29**, 125–34.

Hopkins, K. D., Riha, M., LeClerq, D. and Al-Ahmad, T. C. (1986b) Tilapia culture in Kuwait: screening tilapia for seawater culture. *Tech. Rep. Kuwait Inst. Scient. Res.* No. KISR 2070. Kuwait Institute of Scientific Research, Safat, Kuwait.

Hori, M. (1983) Feeding ecology of thirteen species of *Lamprologus* (Teleostei: Cichlidae) coexisting at a rocky shore of Lake Tanganyika. *Physiol. Ecol. Japan*, **20**, 129–49.

Hori, M. (1987) Mutualism and commensalism in a fish community in Lake Tanganyika, in *Evolution and Coadaptation in Biotic Communities* (eds. S. Kawano, J. H. Connell and T. Hidaka), University of Tokyo Press, Tokyo, pp. 219–39.

Hori, M., Yamaoka, K. and Takamura, K. (1983) Abundance and microdistribution of cichlids on a rocky shore in Lake Tanganyika. *Afr. Stud. Monogr. Kyoto Univ.*, **3**, 25–38.

Huang, C. M., Chang, S. L., Cheng, H. J. and Liao, I. C. (1988) Single gene inheritance of red body coloration in Taiwanese red tilapia. *Aquaculture*, **74**, 227–32.

Hughes, N. F. (1986) Changes in the feeding biology of the Nile perch, *Lates niloticus* (L.) (Pisces, Centropomidae), in Lake Victoria, East Africa since its introduction in 1960, and its impact on native fish community of the Nyanza Gulf. *J. Fish. Biol.*, **29**, 541–8.

Huisman, E. A. (ed.) (1986) *Aquaculture Research in the African Region* (*Proc. International Foundation for Science (IFS, Stockholm, Sweden) African Seminar on Aquaculture, Kisumu, Kenya, 7–11 October 1985*) Pudoc, Wageningen, The Netherlands.

Hulscher-Emeis, T. M. (1986) Are colour markings in *Tilapia zillii* associated with motivational systems? *Annls Mus. r. Afr. cent. Sci. zool.*, **251**, 35–8.

Hunter, J. R. (1981) Feeding ecology and predation of marine fish larvae, in *Marine Fish Larvae: Morphology, Ecology and Relation to Fisheries* (ed. R. Lasker), University of Washington Press, Seattle, Wash., pp. 33–77.

Huntingford, F. A. (1986) Development of behaviour in fishes, in *The Behaviour of Teleost Fishes* (ed. T. J. Pitcher), Croom Helm, London, pp. 47–68.

Huxley, J. (ed.) (1940) *The New Systematics*, Oxford University Press, Oxford.

ICLARM (1988) Research for the development of tropical aquaculture: a proposed new venture for the Consultative Group on International Agricultural Research. International Center for Living Aquatic Resources Management, Manila.

Iles, T. D. (1960) A group of zooplankton feeders of the genus *Haplochromis* (Cichlidae) in Lake Nyasa. *Ann. Mag. nat. Hist.*, **2**, 257–80.

Iles, T. D. and Holden, M. J. (1969) Bi-parental mouth brooding in *Tilapia galilaea* (Pisces, Cichlidae). *J. Zool., Lond.*, **158**, 327–33.

Ismail, M. H. (1979) The ontogeny of the head parts in *Haplochromis elegans* Trewavas, 1933 (Teleostei, Cichlidae), thesis, University of Ghent, Belgium.

Ismail, M. H., Verraes, W. and Huysseune, A. (1982) Developmental aspects of the pharyngeal jaws in *Astatotilapia elegans*. *Neth. J. Zool.*, **32**, 513–43.

Itzkovich, J., Routhbard, S. and Hulata, G. (1981) Inheritance of pink body coloration in *Cichlasoma nigrofasciatum* Günther (Pisces, Cichlidae). *Genetica*, **55**, 15–16.

Itzkowitz, M. (1984) Parental division of labor in a monogamous fish. *Behavior*, **89**, 251–60.

Itzkowitz, M. and Nyby, J. (1982) Field observations of parental behaviour of the Texas cichlid *Cichlasoma cyanoguttatum*. *Am. Mildl. Nat.*, **108**, 364–8.

Jakobsson, S., Radesäter, T. and Järvi, T. (1979) On the fighting behaviour of *Nannacara anomala* (Pisces, Cichlidae) males. *Z. Tierpsychol.*, **49**, 210–20.

Janetos, A. C. (1980) Strategies of female mate choice: a theoretical analysis. *Behav. Ecol. Sociobiol.*, **7**, 107–12.

Jarman, P. J. (1974) The social organization of antelope in relation to their ecology. *Behaviour*, **48**, 215–67.

Jauncey, K. and Ross, B. (1982) *A Guide to Tilapia Feeds and Feeding*, University of Stirling, Stirling.

Jayaprakas, V. and Phil, M. (1980) Culture possibilities of pearl spot (*Etroplus suratensis*) in Kerala. *Seafood Export J.*, **12**, 13–15.

Jeffreys, A. J., Wilson, V. and Thein, S. L. (1985) Hypervariable 'minisatellite' regions in human DNA. *Nature, Lond.*, **314**, 67–73.

Jobling, M. and Reinsnes, T. G. (1987) Effect of sorting on size-frequency distributions and growth of arctic charr, *Salvelinus alpinus* L. *Aquaculture*, **60**, 27–31.

Johnson, D. J. (1980) The limits and relationships of the Lutjanidae and associated families. *Bull. Scripps Instn. Oceanog.*, **24**, 1–114.

Johnston, J. W., Moulton, D. G. and Turk, A. (1970) *Communication by Chemical Signals*, Appleton-Century-Crofts, New York.

Johnstone, R., Macintosh, D. J. and Wright, R. S. (1983) Elimination of orally administered 17-methyltestosterone by *Oreochromis mossambicus* (Tilapia) and *Salmo gairdneri* (Rainbow trout) juveniles. *Aquaculture*, **35**, 249–57.

Jones, A. J. (1972) The early development of substrate-brooding cichlids (Telostei, Cichlidae) with a discussion of a new system of staging. *J. Morph.*, **136**, 255–72.

Jones, R. S. (1968) Ecological relationships in Hawaiian and Johnston Island Acanthuridae (Surgeonfishes). *Micronesica*, **4**, 309–61.

Jordan, D. S. (1905) *A Guide to the Study of Fishes*. Vol. II, Henry Holt and Co., New York.

Jubb, R. A. (1967) *Freshwater Fishes of Southern Africa*. A. A. Balkema, Cape Town.

Junk, W. J., Soares, G. M. and Carvalho, F. M. (1983) Distribution of fish species in a lake of the Amazon River floodplain near Manaus (Lago Camaleão), with special reference to extreme oxygen conditions. *Amazoniana*, **7**, 397–431.

Kapetsky, J. M. (1974) The Kafue River floodplain: an example of preimpoundment potential for fish production, in *Lake Kariba: a Man-made Tropical Ecosystem in Central Africa* (eds E. K. Balon and A. G. Coche), Dr W. Junk, The Hague, pp. 497–523.

Karlan, U. (1985) Superintensive fish farming. *Caribbean Farming*, July 1985, p. 8.

Karlin, S. and Raper, J. (1979) Sexual selection encounter models. *Theor. Popul. Biol.*, **15**, 246–56.

Kaufman, L. and Liem, K. F. (1982) Fishes of the suborder Labroidei (Pisces: Perciformes): phylogeny, ecology and evolutionary significance. *Breviora*, **472**, 1–19.

Kawabata, M. and Mihigo, N. Y. K. (1982) Littoral fish fauna near Uvira, northwestern end of Lake Tanganyika. *Afr. Stud. Monogr., Kyoto Univ.*, **2**, 133–43.

Kawanabe, H. (1981) Territorial behaviour of *Tropheus moorei* (Osteichthyes: Cichlidae) with a preliminary consideration on the territorial forms in animals. *Afr. Stud. Monogr., Kyoto Univ.*, **1**, 101–8.

Kawanabe, H. (1983) Ecological and limnological study on Lake Tanganyika and its adjacent regions II. Report of the work supported by Grant-in-Aid for Overseas Scientific Survey of the Ministry of Education, Science and Culture of Japan, during 1981 and 1982. 68 pp.

Keenleyside, M. H. A. (1972) Intraspecific intrusions into nests of spawning longer sunfish (Pisces, Centrarchidae). *Copeia*, **1972**, 272–8.

Keenleyside, M. H. A. (1978) Parental care behavior in fishes and birds, in *Contrasts in Behavior* (eds E. S. Reese and F. J. Lighter), John Wiley and Sons, New York, pp. 1–19.

Keenleyside, M. H. A. (1979) *Diversity and Adaptation in Fish Behaviour*, Springer-Verlag, Berlin.

Keenleyside, M. H. A. (1983) Mate desertion in relation to adult sex ratio in the biparental cichlid fish *Herotilapia multispinosa*. *Anim. Behav.*, **31**, 683–8.

Keenleyside, M. H. A. (1985) Bigamy and mate choice in the biparental cichlid fish *Cichlasoma nigrofasciatum*. *Behav. Ecol. Sociobiol*, **17**, 285–90.

Keenleyside, M. H. A. and Bietz, B. F. (1981) The reproductive behaviour of *Aequidens vittatus* (Pisces, Cichlidae) in Surinam, South America. *Environ. Biol. Fishes*, **6**, 87–94.

Keenleyside, M. H. A. and Prince, C. E. (1976) Spawning-site selection in relation to parental care of eggs in *Aequidens paraguayensis* (Pisces, Cichlidae). *Can. J. Zool.*, **54**, 2135–9.

Keenleyside, M. H. A., Bailey, R. C. and Young, V. H. (1990) Variation in the mating system and associated parental behaviour of captive and free-living *Cichlasoma nigrofasciatum* (Pisces, Cichlidae). *Behaviour*, **112**, 202–21.

Kessler, L. G. and Avise, J. C. (1985) A comparative description of mitochondrial DNA differentiation in selected avian and other vertebrate genera. *Mol. Biol. Evol.*, **2**, 109–25.

Kiener, A. and Maugé, M. (1966) Contributions à l'étude systématique et écologique des poissons Cichlidae endémiques de Madagascar. *Mém. Mus. Nat. Hist. Paris, Sér. A., Zool.*, **40**, 51–99.

Kirkpatrick, M. (1987) Sexual selection by female choice in polygynous animals. *A. Rev. Ecol. Syst.*, **18**, 43–70.

Kirpichnikov, V. S. (1981) *Genetic Bases of Fish Selection*. Springer-Verlag, NY.

Knöppel, H.-A. (1970) Food of Central Amazonian fishes. *Amazoniana*, **2**, 257–352.

Knowlton, N. (1979) Reproductive synchrony, parental investment, and the evolutionary dynamics of sexual selection. *Anim. Behav.*, **27**, 1022–33.

Kocher, T. and McKaye, K. R. (1983) Defense of heterospecific cichlids by *Cyrtocara moorii* in Lake Malawi, Africa, *Copeia* 1983, 544–7.

Kocher, T. D. and White, T. J. (1989) Evolutionary analysis via PCR, in *PCR Technology* (ed. H. A. Erlich), M. Stockton Press, NY, pp. 137–47.

Kocher, T., Thomas, W. K., Meyer, A., Edwards, S. V., Paabo, S., Villablanca, F. X. and Wilson, A. C. (1989) Dynamics of mitochondrial DNA evolution in animals: amplification and sequencing with conserved primers. *Proc. natn. Acad. Sci. U.S.A.*, **86**, 6196–6200.

Kondo, T. (1986) Feeding habits of *Lamprologus savoryi* (Teleostei: Cichlidae) with

special reference to its social behaviour. *Physiol. Ecol. Japan*, **23**, 1–15.
Konings, A. (1988) *Tanganyika Cichlids*, Raket, B. V., Pijnacker, The Netherlands.
Kop, P. P. A. M. and Heuts, B. A. (1973) An experiment on sibling imprinting in the jewel fish *Hemichromis bimaculatus* (Gill 1862, Cichlidae). *Revue Comp. Animal*, **7**, 63–76.
Kornfield, I. (1978) Evidence for rapid speciation in African cichlid fishes. *Experientia*, **34**, 335–6.
Kornfield, I. (1984) Descriptive genetics of cichlid fishes, in *Evolutionary Genetics of Fishes* (ed. B. J. Turner), Plenum Press, NY, pp. 591–615.
Kornfield, I. and Koehn, R. K. (1975) Genetic variation and evolution in New World cichlids. *Evolution*, **29**, 427–37.
Kornfield, I. and Taylor, J. N. (1983) A new species of polymorphic fish, *Cichlasoma minckleyi* from Cuatro Cienegas, Mexico, (Teleostei: Cichlidae). *Prob. biol. Soc. Wash.*, **96**, 253–69.
Kornfield, I., McKaye, K. R. and Kocher, T. (1985) Evidence for the immigration hypothesis in the endemic cichlid fauna of Lake Tanganyika. *Isozyme Bull.*, **18**, 76.
Kornfield, I., Ritte, U., Richler, C. and Wahrman, J. (1979) Biochemical and cytologcial differentiation among cichlid fishes of the Sea of Galilee. *Evolution*, **33**, 1–14.
Kramer, D. L. and McClure, M. (1982) Aquatic surface respiration, a widespread adaptation to hypoxia in tropical freshwater fishes. *Environ. Biol. Fishes*, **7**, 47–55.
Kramer, D. L., Lindsey, C. C., Moodie, G. E. E. and Stevens, E. D. (1978) The fishes and the aquatic environment of the central Amazon basin, with particular reference to respiratory patterns. *Can. J. Zool.*, **56**, 717–29.
Krawetz, S. A. (1989) Sequence errors described in GenBank: a means to determine the accuracy of DNA sequence interpretation. *Nucleic Acids Res.*, **17**, 3951–7.
Krebs, J. R. and Davies, N. B. (1987) *An Introduction to Behavioural Ecology*, 2nd edn, Blackwell Scientific Publications, Oxford.
Krischik, V. A. and Weber, P. G. (1975) Induced parental care in male convict cichlid fish. *Develop. Psychobiol.*, **8**, 1–11.
Kudhongania, A. W. and Cordone, A. J. (1974) Batho-spatial distribution pattern and biomass estimate of the major demersal fishes in Lake Victoria. *Afr. J. Trop. Hydrobiol. Fish.*, **3**, 15–31.
Kuenzer, P. (1975) Analyse der auslosenden Reizsituationen fur die Anschwimm-, Eindring- und Fluchtreaktion junger *Hemihaplochromis multicolor* (Cichlidae). *Z. Tierpsychol.*, **38**, 505–45.
Kuenzer, P. and Peters, I. (1974) Versuche zur Auslosung und Umstellung der Brutpflegephasen bei Weibchen von *Nannacara anomala* (Cichlidae). *Z. Tierpsychol.*, **35**, 425–36.
Kühme, W. (1963) Chemisch ausgelöste Brutpflege- und Schwarmreaktion bei *Hemichromis bimaculatus* (Pisces). *Z. Tierpsychol.*, **20**, 688–704.
Kühme, W. (1964) Eine chemisch ausgelöste Schwarmreaktion bei jungen Cichliden (Pisces). *Naturwissenschaften*, **51**, 120–1.
Kullander, S. O. (1980) A taxonomical study of the genus *Apistogramma* Regan, with a revision of the Brazilian and Peruvian species. *Bonn. zool. Monogr.*, **14**, 1–152.
Kullander, S. O. (1981) A cichlid from Patagonia. *Buntbarsche Bull.*, **85**, 13–23.
Kullander, S. O. (1982) Cichlid fishes from the La Plata basin. Part III. The *Crenicichla lepidota* species group. *Revue. suisse Zool.*, **89**, 627–61.

Kullander, S. O. (1983a) *A Revision of the South American Cichlid Genus Cichlasoma*, Swedish Museum of Natural History, Stockholm.
Kullander, S. O. (1983b) Review of the South American Cichlidae, PhD thesis, University of Stockholm.
Kullander, S. O. (1985) Ökologie sudamerikanische Zwergbuntbarsche, in *Die Buntbarsche der Neuen Welt. Zwergcichlidae* (ed. I. Koslowski), edn Kernen, Osnabrück, Federal Republic of Germany, pp. 27–36.
Kullander, S. O. (1986) *Cichlid Fishes of the Amazon River Drainage of Peru*, Swedish Museum of Natural History, Stockholm.
Kullander, S. O. (1988) *Teleocichla*, a new genus of South American rheophilic cichlid fishes with six new species (Teleostei: Cichlidae). *Copeia, 1988*, 196–230.
Kullander, S. O. and Nijssen, H. (1989) *The Cichlids of Surinam-Teleostei: Labroidei*. E. J. Brill, Leiden, The Netherlands.
Kuwamura, T. (1986a) Parental care and mating systems of cichlid fishes in Lake Tanganyika: a preliminary field survey. *J. Ethol.*, **4**, 129–46.
Kuwamura, T. (1986b) Feeding habits of *Lamprologus savoryi* (Teleostei, Cichlidae) with special reference to its social behaviour. *Physiol. Ecol. Japan*, **23**, 1–15.
Kuwamura, T. (1986c) Substratum spawning and biparental guarding of the Tanganyikan cichlid *Boulengerochromis microlepis*, with notes on its life history. *Physiol. Ecol. Japan*, **23**, 31–43.
Kuwamura, T. (1987) Distribution of fishes in relation to the depth and substrate at Myako, east-middle coast of Lake Tanganyika. *Afr. Stud. Monogr., Kyoto Univ.*, **7**, 1–14.
Lagomarsino, I. V., Francis, R. C. and Barlow, G. W. (1988) The lack of correlation between size of egg and size of fry in the Midas cichlid, *Cichlasoma citrinellum*. *Copeia, 1988*, 1086–9.
Lai, C. F. and Huang, L. C. (1981) A bibliography of *Tilapia* (family *Cichlidae*) in Taiwan. *Aquaculture*, **22**, 389–94.
Lalande, M., Noolandi J., Turmel, C., Brousseau, R., Rousseau, J. and Slater, G. S. (1988) Scrambling of bands in gel electrophoresis of DNA. *Nucleic Acids Res.*, **16**, 5427–37.
Lamon, M. S. and Ward, J. A. (1983) Measurements of reproductive effort from successive reproductive cycles for the Asian cichlid *Etroplus maculatus*. *Environ. Biol. Fishes*, **8**, 311–20.
Lande, R. (1980) Sexual dimorphism, sexual selection, and adaptation in polygenic characters. *Evolution*, **34**, 292–305.
Lande, R. (1981) Models of speciation by sexual selection on polygenic traits. *Proc. Nat. Acad. Sci. U.S.A.*, **78**, 3721–5.
Lande, R. (1982) Rapid origin of sexual isolation and character divergence in a cline. *Evolution*, **36**, 213–23.
Lanzing, W. J. R. and Bower, C. C. (1974) Development of colour patterns in relation to behaviour in *Tilapia mossambica* (Peters). *J. Fish Biol.*, **6**, 29–41.
Lauder, G. V. (1980) The suction feeding mechanism in sunfishes (*Lepomis*): an experimental analysis. *J. exp. Biol.*, **88**, 49–72.
Lauder, G. V. (1983a) Functional and morphological bases of trophic specialization in sunfishes (Teleostei; Centrarchidae). *J. Morph.*, **178**, 1–21.
Lauder, G. V. (1983b) Prey capture hydrodynamics in fishes: experimental tests of two models. *J. exp. Biol.*, **104**, 1–13.

Lauder, G. V. and Clark, B. D. (1984) Water flow patterns during prey capture in teleost fishes. *J. exp. Biol.*, **113**, 143–50.

Leeuwen, J. L. van and Müller, M. (1984) Optimum prey sucking techniques for predatory fish. *Trans. zool. Soc. Lond.*, **37**, 137–69.

Leibel, W. S. and Peairs, K. (1990) Molecular homology of eye- and liver-specific LDH isozymes from the basketmouth cichlid assessed by oxamate-sepharose and blue dextran-sepharose affinity chromatography. *J. exp. Zool.*, **253**, 107–14.

Lester, L. J., Lawson, K. S., Abella, T. A. *et al.* (1989) Estimated heritability of sex ratio and sexual dimorphism in tilapia. Aquaculture and Fisheries Management, **20**, 369–80.

Levine, J. S. and MacNichol, E. F. (1979) Visual pigments in teleost fishes: effects of habitat, microhabitat, and behavior in visual system evolution. *Sens. Process.*, **3**, 95–131.

Levine, J. S. and MacNicol, E. F., Jr. (1982) Color vision in fishes. *Scient. Am.*, **246**(2), 108–17.

Levine, J. S., Lobel, P. S. and MacNichol, E. F. (1981) Visual communication in fishes, in *Environmental Physiology of Fishes* (ed. M. A. Ali), Plenum Press, NY, pp. 447–75.

Lewis, D. B. and Gower, D. M. (1980) *Biology of Communication*, Blackie and Son Ltd, Glasgow.

Lewis, D. S. C. (1980) Mixed species broods in Lake Malawi cichlids: an alternative to the cuckoo theory. *Copeia, 1980*, 874–5.

Lewis, D. S. C. (1981) Preliminary comparisons between the ecology of haplochromine cichlid fishes of Lake Victoria and Lake Malawi. *Neth. J. Zool.*, **31**, 746–61.

Lewis, D. S. C. (1982) A revision of the genus *Labidochromis* (Teleostei: Cichlidae) from Lake Malawi. *Zool. J. Linn. Soc.*, **75**, 189–265.

Lewis, D., Reinthal, P. and Trendall, J. (1986) *A Guide to the Fishes of Lake Malawi National Park.* Gland, Switzerland, World Wildlife Fund.

Lewontin, R. C. (1989) Inferring the number of evolutionary events from DNA coding sequence differences. *Mol. Biol. Evol.*, **6**, 15–32.

Liem, K. F. (1973) Evolutionary strategies and morphological innovations: cichlid pharyngeal jaws. *Syst. Zool.*, **22**, 425–41.

Liem, K. F. (1978) Modulatory multiplicity in the functional repertoire of the feeding mechanism in cichlid fishes. Part I: Piscivores. *J. Morph.*, **158**, 323–60.

Liem, K. F. (1979) Modulatory multiplicity in the feeding mechanism in cichlid fishes, as exemplified by the invertebrate pickers of Lake Tanganyika. *J. Zool., Lond.*, **189**, 93–125.

Liem, K. F. (1980a) Adaptive significance of intra- and interspecific differences in the feeding repertoires of cichlid fishes. *Am. Zool.*, **20**, 295–314.

Liem, K. F. (1980b) Acquisition of energy by teleosts: adaptive mechanisms and evolutionary patterns, in *Environmental Physiology of Fishes* (ed. M. A. Ali), Plenum Publishing Corporation, New York, pp. 299–334.

Liem, K. F. (1981a) A phyletic study of the Lake Tanganyika cichlid genera *Asprotilapia, Ectodus, Lestradea, Cunningtonia, Ophthalmochromis,* and *Ophthalmotilapia. Bull. Mus. Comp. Zool.*, **149**, 191–214.

Liem, K. F. (1981b) Larvae of air-breathing fishes as countercurrent flow devices in hypoxic environments. *Science, N.Y.*, **211**, 1177–9.

Liem, K. F. (1986) The pharyngeal jaw apparatus of the Embiotocidae (Teleostei): a functional and evolutionary perspective. *Copeia, 1986*, 311–23.

Liem, K. F. and Greenwood, P. H. (1981) A functional approach to the phylogeny of pharyngognath teleosts. *Am. Zool.*, **21**, 83–101.

Liem, K. F. and Kaufman, L. S. (1984) Intraspecific macroevolution: functional biology of the polymorphic cichlid species *Cichlasoma minckleyi*, in *Evolution of Fish Species Flocks* (eds A. A. Echelle and I. Kornfield), University of Maine at Orono Press, Orono, Maine, pp. 203–15.

Liem, K. F. and Osse, J. W.M. (1975) Biological versatility, evolution and food resource exploitation in African cichlid fishes. *Am. Zool.*, **15**, 427–54.

Liem, K. F. and Sanderson, L. (1986) The pharyngeal jaw apparatus of labrid fishes: a functional morphological perspective. *J. Morph.*, **187**, 143–58.

Liem, K. F. and Stewart, D. J. (1976) Evolution of the scale-eating cichlid fishes of Lake Tanganyika: a generic revision with a description of a new species. *Bull. Mus. comp. Zool. Harv.*, **147**, 319–50.

Ligtvoet, W. (1988) The Nile perch in Lake Victoria: a blessing or a disaster. HEST Report 42. [see HEST].

Ligtvoet, W. and Mkumbo, O. C. (1990) Stock assessment of Nile perch in Lake Victoria. FAO Fish Report (in press).

Ligtvoet, W. and Witte, F. (in press). Perturbation by predator introduction: effects on trophic structure and fisheries yields in Lake Victoria (East Africa), in *Perturbation and Recovery of Terrestrial and Aquatic Ecosystems* (ed. O. Ravera), Ellis Horwood Ltd, Chichester.

Ligtvoet, W., Chande, A. I. and Mosille, O. I. I. W. (1988) A preliminary description of the artisanal Nile perch (*Lates niloticus*) fishery in southern Lake Victoria. FAO Fish Rep., **388**, 72–85.

Limberger, D. (1983) Pairs and harems in a cichlid fish, *Lamprologus brichardi*. *Z. Tierpsychol.*, **62**, 115–44.

Linke, H. and Staeck, W. (1981a) *Afrikanische cichliden I–Buntbarsche aus Westafrika*, Tetra-Verlag, Melle, Federal Republic of Germany.

Linke, H. and Staeck,.W. (1981b) *African Cichlids. I. Cichlids of West Africa*, Tetra Press, American Cichlid Association.

Livingstone, D. A. (1980) Environmental changes in the Nile headwaters, in *The Sahara and the Nile* (eds M. A. J. Williams and H. Faure), Balkema, Rotterdam, pp. 339–59.

Livingstone, D. A. and Melack, J. M. (1984) Some lakes of subsaharan Africa, in *Lakes and Reservoirs Ecosystems of the World* (ed. F. B. Taub), Elsevier Publications, Amsterdam, pp. 467–97.

Lofstedt, C. Hansson, B. S., Roelofs, W. and Bengtsson, B. O. (1989) No linkage between genes controlling female pheromone production and male pheromone response in the European corn borer. *Ostinia nubilalis Hubner* (*Lepidoptera: Pyralidae*). *Genetics*, **123**, 553–6.

Loiselle, P. V. (1977) Colonial breeding by an African substratum-spawning cichlid fish, *Tilapia zillii* (Gervais). *Biol. Behav.*, **2**, 129–42.

Loiselle, P. V. (1979) A revision of the genus *Hemichromis* Peters (Teleostei, Cichlidae). *Annls. Mus. r. Afr. cent. Sér., 8vo, Sci. zool.*, no. 228, 1–124.

Loiselle, P. V. (1985a) *The Cichlid Aquarium*, Tetra-Press, Melle, Federal Republic of Germany.

Loiselle, P. V. (1985b) The care and breeding of *Sarotherodon occidentalis*. *Buntbarsche Bull.*, (109), 14–24.

Loiselle, P. V. and Barlow, G. W. (1979) Do fishes lek like birds? in *Contrasts in Behavior:*

Adaptations in the Aquatic and Terrestrial Environment (eds E. Reese and F. J. Lighter), John Wiley and Sons, New York, pp. 31–75.

Loiselle, P. V. and Welcomme, R. L. (1972) Description of a new genus of cichlid fish from West Africa. *Revue Zool. Bot. afr.*, **85**, 37–58.

Lovshin, L. L. (1982) Tilapia hybridization, in *The Biology and Culture of Tilapias* (eds R. S. V. Pullin and R. H. Lowe-McConnell), ICLARM Conf. Proc. 7, International Center of Living Aquatic Resources Management, Manila, pp. 279–308.

Lowe, R. H. (1952) Report on the *Tilapia* and other fish and fisheries of Lake Nyasa. *Fishery Publs Colon. Off. Lond.*, **1**, 1–126.

Lowe (McConnell), R. H. (1956) The breeding behaviour of *Tilapia* species (Pisces; Cichlidae) in natural waters: observations on *T. karomo* Poll and *T. variabilis* Boulenger. *Behaviour*, **9**, 140–63.

Lowe-McConnell, R. H. (1959) Breeding behaviour patterns and ecological differences between *Tilapia* species and their significance for evolution within the genus *Tilapia* (Pisces: Cichlidae). *Proc. zool. Soc. Lond.*, **132**, 1–30.

Lowe-McConnell, R. H. (1964) The fishes of the Rupununi savanna district of British Guiana, South America. Part 1. Ecological groupings of fish species and effects of the seasonal cycle on the fish. *J. Linn. Soc., zool.*, **45**, 103–44.

Lowe-McConnell, R. H. (1967) Tilapias in fish communities, in *The Biology and Culture of Tilapias* (eds R. S. V. Pullin and R. H. Lowe-McConnell) ICLARM Conf. Proc. 7, Manila, pp. 83–7.

Lowe-McConnell, R. H. (1969a) Speciation in tropical freshwater fishes. *Biol. J. Linn. Soc.*, **1**, 51–75.

Lowe-McConnell, R. H. (1969b) The cichlid fishes of Guyana, South America, with notes on their ecology and breeding behaviour. *Zool. J. Linn. Soc.*, **48**, 255–302.

Lowe-McConnell, R. H. (1975) *Fish Communities in Tropical Freshwaters: Their Distribution, Ecology and Evolution*, Longman, London, New York.

Lowe-McConnell, R. H. (1982) Tilapias in fish communities, in *The Biology and Culture of Tilapias* (*ICLARM Conference Proceedings*, Vol. 7) (eds R. S. V. Pullin and R. H. Lowe-McConnell), International Center for Living Aquatic Resources Management, Manila, pp. 83–113.

Lowe-McConnell, R. H. (1987) *Ecological Studies in Tropical Fish Communities*, Cambridge University Press, Cambridge.

Lowe-McConnell, R. H. (1988) Ecology and distribution of tilapias in Africa that are important for aquaculture, in *Tilapia Genetic Resources for Aquaculture*, (*ICLARM Conference Proceedings*, Vol. 16) (ed. R. S. V. Pullin), International Center for Living Aquatic Resources Management, Manila, pp. 12–18.

Lüling, K. H. (1971) *Aequidens vittata* (Heckel) und andere Fische des Rio Huallaga im Ubergangs bereich zur Hylea. *Zool. Beitr.*, **17**, 193–226.

Lüling, K. H. (1975) Ichthyologische und gewasserkundliche Beobachtungen und Untersuchungen an der Yarina Cocha in der Umgebung von Pucallpa und am Rio Pacaya (mittlerer und unterer Ucayali, Ostperu.) *Zool. Beitr.* (*N.F.*), **21**, 29–96.

Lynch, M. (1988) Estimation of relatness by DNA fingerprinting. *Mol. Biol. Evol.*, **5**, 584–99.

McAndrew, B. J., Roubal, F. R., Roberts, R. J., Bullock, A. M. and McEwan, I. M. (1988) The genetics and history of red, blond, and associated color variants in *Oreochromis niloticus*. *Genetica*, **76**, 172.

Macaranas, J. M., Taniguchi, N., Pante, M. J. R., Capili, J. B. and Pullin, R. S. V. (1986) Electrophoretic evidence for extensive hybrid gene introgression into commercial *Oreochromis niloticus* (L.) stocks in the Philippines. *Aquacult. Fish. Magmt.* **17**, 249–58.

McElroy, D. M. and Kornfield, I. (1990) Sexual selection, reproductive behavior, and speciation in the mbuna species flock of Lake Malawi (Piaces: Cichlidae). *Env. Biol. Fishes*, **28**, 285–94.

Macintosh, D. J. (1987) The potential of cichlid fishes for aquaculture in the inland waters of Southern Asia. *Arch. Hydrobiol. Beih. Ergebn. Limnol.*, **28**, 349–53.

Macintosh, D. J., Singh, T. B., Little, D. C. and Edwards, P. (1988) Growth and sexual development of 17α-methyltestosterone- and progesterone-treated Nile tilapia (*Oreochromis niloticus*) reared in eastern ponds, in The Second International Symposium on Tilapia in Aquaculture (eds R. S. V. Pullin, T. Bhukaswan, K. Tonguthai and J. L. Maclean), ICLARM Conference Proceedings 15, 623 pp. Department of Fisheries, Bangkok, Thailand and International Center for Living Aquatic Resources Management, Manila, Philippines.

Macintosh, D. J., Varghese, T. J. and Rao, G. P. S. (1985) Hormonal sex reversal of wild spawned tilapia in India. *J. Fish. Biol.*, **26**, 87–94.

McKaye, K. R. (1977) Competition for breeding sites between the cichlid fishes of Lake Jiloá, Nicaragua, *Ecology*, **58**, 291–302.

McKaye, K. R. (1980) Seasonality in habitat selection by the gold color morph of *Cichlasoma citrinellum* and its relevance to sympatric speciation in the family Cichlidae. *Environ. Biol. Fishes*, **5**, 75–8.

McKaye, K. R. (1981a) Field observations on death feigning: a unique hunting behavior by the predatory cichlid, *Haplochromis livingstoni*, of Lake Malawi, Africa. *Environ. Biol. Fishes*, **6**, 361–5.

McKaye, K. R. (1981b) Natural selection and the evolution of interspecific brood care in fishes, in *Natural Selection and Social Behaviour* (eds R. D. Alexander and D. W. Tinkle), Blackwell Scientific Publications, Oxford, pp. 173–83.

McKaye, K. R. (1983) Ecology and breeding behavior of a cichlid fish, *Cyrtocara eucinostomus*, on a large lek in Lake Malawi, Africa. *Environ. Biol. Fishes*, **8**, 81–96.

McKaye, K. R. (1984) Behavioural aspects of cichlid reproductive strategies: patterns of territoriality and brood defence in Central American substratum spawners and African mouth brooders, in *Fish Reproduction: Strategies and Tactics* (eds G. W. Potts and R. J. Wootton), Academic Press, London, pp. 245–73.

McKaye, K. R. (1986) Mate choice and size assortative pairing by the cichlid fishes of Lake Jiloá, Nicaragua. *J. Fish. Biol.*, **29** (Supp. A), 135–50.

McKaye, K. R. and Barlow, G. W. (1976a) Competition between color morphs of the Midas cichlid, *Cichlasoma citrinellum*, in Lake Jiloá, Nicaragua, in *Investigations of the Ichthyofauna of Nicaraguan Lakes* (ed. T. B. Thorson), School of Life Sciences, University of Nebraska-Lincoln, Lincoln, Nebr., pp. 465–75.

McKaye, K. R. and Barlow, G. W. (1976b) Chemical recognition of young by the Midas cichlid, *Cichlasoma citrinellum*. *Copeia*, *1976*, 276–82.

McKaye, K. R. and Gray, W. M. (1984) Extrinsic barriers to gene flow in rock-dwelling cichlids of Lake Malawi: macrohabitat heterogeneity and reef colonization, in *Evolution of Fish Species Flocks* (eds A. A. Echelle and I. Kornfield), University of Maine at Orono Press, Orono, Maine, pp. 169–83.

McKaye, K. R. and Kocher, T. (1983) Head ramming behaviour by three paedophagous cichlids in Lake Malawi, Africa. *Anim. Behav.*, **31**, 206–10.
McKaye, K. R. and McKaye, N. M. (1977) Communal care and kidnapping of young by parental cichlids. *Evolution*, **31**, 674–81.
McKaye, K. R. and Mackenzie, C. (1982) *Cyrtocara liemi*, a previously undescribed paedophagous cichlid fish (Teleostei: Cichlidae) from Lake Malawi, Africa. *Proc. biol. Soc. Wash.*, **95**, 398–402.
McKaye, K. R. and Marsh, A. C. (1983) Food switching by two specialized algae-scraping cichlid fishes in Lake Malawi, Africa. *Oecologia*, **56**, 245–8.
McKaye, K. R. and Stauffer, J. R., Jr. (1986) Description of a gold cichlid (Teleostei, Cichlidae) from Lake Malawi, Africa. *Copeia*, *1986*, 870–5.
McKaye, K. R. and Stauffer, J. R., Jr. (1988) Seasonality, depth and habitat distribution of breeding males of *Oreochromis* spp. 'Chambo' in Lake Malawi National Park. *J. Fish. Biol.*, **33**, 825–34.
McKaye, K. R., Kocher, T., Reinthal, P., Harrison, R. and Kornfield, I. (1984) Genetic evidence for allopatric and sympatric differentiation among color morphs of a Lake Malawi cichlid fish. *Evolution*, **38**, 215–19.
McKaye, K. R., Kocher, T., Reinthal, P. and Kornfield, I. (1982) Genetic analysis of a sympatric sibling species complex of *Petrotilapia* Trewavas (Cichlidae, Lake Malawi). *Zool. J. Linn. Soc.*, **76**, 91–6.
McKaye, K. R., Louda, S. M. and Stauffer, J. R., Jr. (1990) Bower size and male reproductive success in a cichlid fish lek. *Am. Nat.*, **135**, 597–613.
McRae, A. F. and Anderson, W. W. (1988) Evidence for non-neutrality of mitochondrial DNA haplotypes in *Drosophila pseudoobscura*. *Genetics*, **120**, 485–94.
Maddison, W. P., Donoghue, M. J. and Maddison, D. R. (1984) Outgroup analysis and parsimony. *Syst. Zool.*, **33**, 83–103.
Maitland, P. S. (1977) *The Hamlyn Guide to Freshwater Fishes*, Hamlyn Publ. Group Ltd.
Majerus, M. E. N., O'Donald, P., Kearns, P. W. E. and Ireland, H. (1986) Genetics and evolution of female choice. *Nature, Lond.*, **321**, 164–7.
Majerus, M. E. N., O'Donald, P. and Weir, J. (1982) Female mating preference is genetic. *Nature, Lond.*, **300**, 521–3.
Majumdar, K. C. and McAndrew, B. J. (1986) Relative DNA content of somatic nuclei and chromosomal studies in three genera, *Talapia*, *Sarotherodon*, and *Oreochromis* of the tribe Tilapiini (Pisces, Cichlidae). *Genetica*, **68**, 175–88.
Marathe, V. B. and Khosla, R. (1958) The cranial osteology of *Etroplus suratensis* Bloch. *J. Univ. Bombay, New Ser. B*, **27**, 37–45.
Marlier, G. (1959) Observations sur la biologie littorale du Lac Tanganyika. *Revue Zool. Bot. afr.*, **59**, 164–83.
Marlier, G. and Leleup, N. (1954) A curious ecological 'niche' among the fishes of Lake Tanganyika. *Nature, Lond.*, **174**, 935–6.
Marsh, A. C. (1981) A contribution to the ecology and systematics of the genus *Petrotilapia* (Pisces: Cichlidae) in Lake Malawi, MSc thesis, Rhodes University, Republic of South Africa.
Marsh, A. C. and Ribbink, A. J. (1985) Feeding-site utilization in three sympatric species of *Petrotilapia* (Pisces, Cichlidae) from Lake Malawi. *Biol. J. Linn. Soc.*, **25**, 331–8.
Marsh, A. C. and Ribbink, A. J. (1986) Feeding schools among Lake Malawi cichlid fishs. *Environ. Biol. Fishes.*, **15**, 75–9.

Marsh, A. C., Ribbink, A. J. and Marsh, B. A. (1981) Sibling species complexes in sympatric populations of *Petrotilapia* Trewavas (Cichlidae, Lake Malawi). *Zool. J. Linn. Soc.*, **71**, 253–64.

Marsh, B. A., Marsh, A. C. and Ribbink, A. J. (1986) Reproductive seasonality in a group of rock-frequenting cichlid fishes in Lake Malawi. *J. Zool., Lond. (A)*, **209**, 9–20.

Marten, G. G. (1979) Impact of fishing on the inshore fishery of Lake Victoria (East Africa). *J. Fish. Res. Bd Can.*, **36**, 891–900.

Martinez-Palacios, C. A. and Ross, L. G. (1986) The effects of temperature, body weight and hypoxia on the oxygen consumption of the Mexican mojarra, *Cichlasoma urophthalmus* (Günther). *Aquacult. Fish. Manage.*, **17**, 243–8.

Masters, J. C., Rayner, R. J., McKay, I. J., Potts, A. D., Nails, D., Ferguson, J. W., Weissenbacher, B. K., Allsopp, M. and Anderson, M. L. (1987) The concept of species: recognition versus isolation. *S. Afr. J. Sci.*, **83**, 534–7.

Matsuno, T., Katsuyama, M., Iwahashi, M., Koike, T. and Okada, M. (1980) Intensification of color of red *Tilapia* with Lutein, Rhodoxanthin and Spirulina. *Bull. Jap. Soc. scient. Fish.*, **46**, 479–82.

Matthes, H. (1961) *Boulengerochromis microlepis*, a Lake Tanganyika fish of economical importance. *Bull. aquat. Biol.*, **3**, 1–15.

Matthes, H. (1962) Poissons nouveaux ou intéressants du Lac Tanganika et du Ruanda. *Annls. Mus. r. Afr. cent. Sér.. 8vo. Sci. zool.*, No. 111, 27–88.

Matthes, H. (1964) Les poissons du lac Tumba et de la région d'Ikela Etude systématique et écologique, *Annals. Mus. r. Afr. cent. Sér. 8vo. Sci. zool.*, No. 126, 1–204.

Maynard Smith, J. (1977) Parental investment: a prospective analysis. *Anim. Behav.*, **25**, 1–9.

Maynard Smith, J. (1982) *Evolution and the Theory of Games*, Cambridge University Press, Cambridge.

Maynard Smith, J. (1985) Mini review: sexual selection, handicaps and true fitness. *J. theor. Biol.*, **115**, 1–8.

Mayr, E. (1942) *Systematics and the Origin of Species*, Columbia University Press, New York.

Mayr, E. (1963) *Animal Species and Evolution*, Belknap Press, Cambridge, Mass.

Mayr, E. (1964) *Systematics and the Origin of Species*, Dover Publications Inc., New York.

Mayr, E. (1976) Sympatric speciation, in *Evolution and Diversity of Life: Selected Essays* (ed. E. Mayr) Belknap Press, Cambridge, Mass., pp. 144–75.

Mayr, E. (1982) *The Growth of Biological Thought*, Belknap Press, Cambridge, Mass.

Mbomba, N. B. (1986) Comparative feeding ecology of Aufwuchs eating cichlid fishes in Lake Tanganyika with reference to their developmental changes. *Physiol. Ecol. Japan*, **23**, 79–108.

Meer, H. J. van der (1989) Ecological significance of retinal receptor patterns in four sympatric haplochromine cichlids. *Annls. Mus. r. Afr. cent. Sci. zool.*, **257**, 51–6.

Meer, H. J. van der and Anker, G. Ch. (1984) Retinal resolving power and sensitivity of the photopic system in seven haplochromine species (Cichlidae, Teleostei). *Neth. J. Zool.*, **34**, 137–209.

Meer, H. J. van der and Anker, G. Ch. (1986) The influence of light deprivation on the development of the eye and retina in the cichlid *Sarotherodon mossambicus* (Teleostei). *Neth. J. Zool.*, **36**, 480–98.

Meral, G. H. (1973) The adaptive significance of territoriality in New World Cichlidae, PhD thesis. University of California, Berkeley.

Meyer, A. (1986) Changes in behavior with increasing experience with a novel prey in fry of the Central American cichlid, *Cichlasoma managuense* (Teleostei, Cichlidae). *Behaviour*, **98**, 145–67.

Meyer, A. (1987) Phenotypic plasticity and heterochrony in *Cichlasoma managuense* (Pisces, Cichlidae) and their implications for speciation in cichlid fishes. *Evolution*, **41**, 1357–69.

Meyer, A. (1988) Influence of age and size on the response to novel prey by fry of the cichlid fish *Cichlasoma managuense* (Pisces, Cichlidae). *Ethology*, **78**, 199–208.

Meyer, A., Kocher, T. D., Basasibwaki, P. and Wilson, A. C. (1990) Monophyletic origin of Lake Victoria cichlid fishes suggested by mitochondrial DNA sequences. *Nature*, **347**, 550–3.

Meyer, A. (1989) Trophic polymorphisms in cichlid fishes: do they represent intermediate steps during sympatric speciation and explain their rapid adaptive radiation? in *New Trends in Ichthyology* (ed. J. H. Schroder), pp. 000–000. (in press).

Meyer, M. K. and Schartl, M. (1984) *Pseudotropheus* (*Maylandia*) *hajomaylandi* n. sp. a new taxon from Lake Malawi. *Revue fr. Aquariol.*, **11**, 51–6.

Miller, R. R. (1966) Geographical distribution of Central American freshwater fishes. *Copeia*, *1966*, 773–802.

Miller, R.R. (1976) Geographical distribution of Central American freshwater fishes, with Addendum, in *Investigations of the Ichthyofauna of Nicaraguan Lakes* (ed. T. B. Thorson), Univ. of Nebraska, Lincoln, Nebraska, pp. 125–56.

Miller, R. R. and Nelson, B. C. (1961) Variation, life colors and ecology of *Cichlasoma callolepis*, a cichlid fish from Southern Mexico, with a discussion of the *Thorichthys* species group. *Occ. Pap. Mus. Zool. Univ. Mich.*, No. 622, 1–9.

Minnier, D. E. and Cole, J. E. (1973) Parental influence on the larval and early post-larval development of young *Etroplus maculatus* (Osteichthyes, Cichlidae). *Proc. Penn. Acad. Sci.*, **47**, 117–21.

Moller, A. P. (1988) Female choice selects for male sexual tail ornaments in the monogamous swallow. *Nature, Lond.*, **332**, 640–2.

Moreau, J., Bambino, C. and Pauly, D. (1986) Indices of overall growth performance of 100 tilapia (Cichlidae) populations, in *The First Asian Fisheries Forum* (eds J. L. Maclean, L. B. Dizon, and L. V. Hosillos), Asian Fisheries Society, Manila, pp. 201–6.

Moriarty, D. J. W. and Pullin, R. S. V. (eds) (1987) *Detritus and Microbial Ecology in Aquaculture* (*ICLARM Conference Proceedings*, Vol. 14), International Center for Living Aquatic Resources Management, Manila.

Moritz, C., Dowling, T. E. and Brown, W. M. (1987) Evolution of animal mitochondrial DNA: relevance for population biology and systematics. *A. Rev. Ecol. Syst.*, **18**, 269–92.

Motta, P. J. (1985) Functional morphology of the head of Hawaiian and mid-Pacific butterflyfishes (Perciformes, Chaetodonatidae). *Environ. Biol. Fishes.*, **13**, 253–76.

Moyle, P. B. and Cech, J. J. (1988) *Fishes: An introduction to Ichthyology*. Prentice Hall, Englewood Cliffs, NJ.

Mrowka, W. (1984) Is the parental-care behaviour of *Aequidens paraguayensis* (Cichlidae) optimal? *Behaviour*, **89**, 128–46.

Mrowka, W. (1987a) Brood adoption in a mouthbrooding cichlid fish: experiments and a hypothesis. *Anim. Behav.*, **35**, 922–3.

Mrowka, W. (1987b) Oral fertilization in a mouthbrooding cichlid fish. *Ethology*, **74**, 293–6.
Müller, J. (1843) Nachtrage zu der Abhandlung uber die naturlich Familien der Fische. *Arch. Naturgesch.*, **9**, 381–4.
Müller, M. and Osse, J. W. M. (1984) Hydrodynamics of suction feeding in fish. *Trans. zool. Soc. Lond.*, **37**, 1–136.
Müller, M., Osse, J. W. M. and Verhagen, J. H. G. (1982) A quantitative hydrodynamical model of suction feeding in fish. *J. theor. Biol.*, **95**, 49–79.
Müller, R. G. and Benda, R. S. (1981) Comparison of bottom trawl stock densities in the inner Kavirondo Gulf of Lake Victoria. *J. Fish. Biol.*, **19**, 399–401.
Myers, G. S. (1939) A possible method of evolution of oral brooding habits in cichlid fishes. *Stanford Ichthyol. Bull.*, **3**, 85–7.
Myers, G. S. (1966) Derivation of the freshwater fish fauna of Central America. *Copeia*, *1966*, 766–73.
Myers, J. M. (1986) Tetraploid induction in *Oreochromis* spp. *Aquaculture*, **57**, 281–7.
Myrberg, A. A., Jr. (1964) An analysis of the preferential care of eggs and young by adult cichlid fishes. *Z. Tierpsychol.*, **21**, 53–98.
Myrberg, A. A., Jr. (1965) A descriptive analysis of the behaviour of the African cichlid fish, *Pelmatochromis guentheri* (Sauvage). *Anim. Behav.*, **13**, 312–29.
Myrberg, A. A., Jr. (1966) Parental recognition of young in cichlid fishes. *Anim. Behav.*, **14**, 565–71.
Myrberg, A. A., Jr. (1975) The role of chemical and visual stimuli in the preferential discrimination of young by the cichlid fish, *Cichlasoma nigrofasciatum* (Günther). *Z. Tierpsychol.*, **37**, 274–97.
Myrberg, A. A., Jr., Kramer, E. and Heinecke, P. (1965) Sound production by cichlid fishes. *Science, N.Y.*, **149**, 555–8.
Nagoshi, M. (1983) Distribution, abundance and parental care of the genus *Lamprologus* (Cichlidae) in Lake Tanganyika. *Afr. Stud. Monogr., Kyoto Univ.*, **3**, 39–47.
Nagoshi, M. (1985) Growth and survival in larval stage of the genus *Lamprologus* (Cichlidae) in Lake Tanganyika. *Verh. Int. Verein. Limnol.*, **22**, 2663–70.
Nagoshi, M. (1987) Survival of broods under parental care and parental roles of the cichlid fish, *Lamprologus toae*, in Lake Tanganyika. *Jap. J. Ichthyol.*, **34**, 71–5.
Naik, I. U. (1973) Studies on *Tilapia mossambica* Peters in Pakistan. *Agric. Pakist.*, **24**, 47–74.
Naish, K. A. and Riffink, A. J. (1990) A preliminary investigation of sex change in *Pseudotropheus lombardoi* (Pisces Cichlidae). *Envir. Bio. Fishes* **28**, 285–94.
Nei, M. (1987) *Molecular Evolutionary Genetics*, Columbia Univ. Press, NY.
Neil, S. J. (1984) Field studies of the behavioral ecology and agonistic behavior of *Cichlasoma meeki* (Pisces: Cichlidae). *Environ. Biol. Fishes.*, **10**, 59–68.
Nelissen, M. (1975) Contribution to the ethology of *Simochromis diagramma* (Günther) (Pisces, Cichlidae). *Acta Zool. Path. Antv.*, **61**, 31–46.
Nelissen, M. (1977a) Sound production by *Haplochromis burtoni* (Günther) and *Tropheus moorii* Boulenger (Pisces: Cichlidae). *Annls. Soc. r. zool. Belg.*, **106**. 155–66.
Nelissen, M. (1977b) Contribution to the ethology of *Tropheus moorii* Boulenger (Pisces, Cichlidae) and a discussion of the significance of its colour patterns. *Revue. Zool. afr.*, **90**, 17–29.
Nelissen, M. (1978) Sound production by some Tanganyikan cichlid fishes and a

hypothesis for the evolution of their communication mechanisms. *Behaviour*, **64**, 137–47.

Nelissen, M. (1985) Structure of the dominance hierarchy and dominance determining 'group factors' in *Melanochromis auratus* (Pisces, Cichlidae). *Behaviour*, **94**, 85–107.

Nelson, J. S. (1984) *Fishes of the World*, John Wiley and Sons, Toronto.

Nishida, M. (1985) Genetic difference and interrelation among cichlid fishes of Lake Tanganyika, in *Ecological and Limnological Study on Lake Tanganyika and its Adjacent Regions*, II (ed. H. Kawanabe), Kyoto University, Kyoto, p. 11.

Noakes, D. L. G. (1973) Parental behaviour and some histological features of scales in *Cichlasoma citrinellum* (Pisces, Cichlidae), *Can. J. Zool.*, **51**, 619–22.

Noakes, D. L. G. (1978a) Ontogeny of behavior in fishes: a survey and suggestions, in *The Development of Behavior: Comparative and Evolutionary Aspects* (eds G. M. Burghardt and M. Bekoff), Garland, New York, pp. 103–25.

Noakes, D. L. G. (1978b) Early behavior in fishes. *Environ. Biol. Fishes*, **3**, 321–6.

Noakes, D. L. G. (1979) Parent-touching behavior by young fishes: incidence, function and causation. *Environ. Biol. Fishes*, **4**, 389–400.

Noakes, D. L. G. (1981) Comparative aspects of behavioural development: a philosophy from fishes, in *Behavioral Development* (eds K. Immelman, G. W. Barlow, L. Petrinovitch and M. Main), Cambridge University Press, Cambridge, pp. 491–508.

Noakes, D. L. G. and Balon, E. K. (1982) Life histories of tilapias: an evolutionary perspective, in *The Biology and Culture of Tilapias (ICLARM Conference Proceedings*, Vol. 7) (eds R. J. V. Pullin and R. H. Lowe-McConnell), International Center for Living Aquatic Resources Management, Manila, pp. 61–82.

Noakes, D. L. G. and Barlow, G. W. (1973a) Ontogeny of parent-contacting in young *Cichlasoma citrinellum* (Pisces, Cichlidae). *Behaviour*, **46**, 221–55.

Noakes, D. L. G. and Barlow, G. W. (1973b) Cross-fostering and parent–offspring responses in *Cichlasoma citrinellum* (Pisces, Cichlidae). *Z. Tierpsychol.*, **33**, 147–52.

Noakes, D. L. G. and Godin, J. -G. J. (1988) Ontogeny of behavior and concurrent developmental changes in sensory systems in teleost fishes, in *Fish Physiology*, (Vol. 11b) (eds W. S. Hoar and D. J. Randall). Academic Press, New York. pp. 345–95.

Noakes, D. L. G. and Leatherland, J. F. (1977) Social dominance and interrenal cell activity in rainbow trout, *Salmo gairdneri* (Pisces, Salmonidae). *Environ. Biol. Fishes*, **2**, 131–6.

Noble, G. K. and Curtis, B. (1939) The social behavior of the jewel fish *Hemichromis bimaculatus* Gill. *Bull. Am. Mus. nat. Hist.*, **76**, 1–46.

Norton, B. G. (ed.) (1986) *The Preservation of species. The Value of Biological Diversity*, Princeton University Press, Princeton, N.J.

Nshombo, M., Yanagisawa, Y. and Nagoshi, M. (1985) Scale-eating in *Perissodus microlepis* (Cichlidae) and change of its food habits with growth. *Jap. J. Ichthyol.*, **32**, 66–73.

Nur, N. and Hasson, O. (1984) Phenotypic plasticity and the handicap principle. *J. theor. Biol.*, **110**, 275–97.

O'Connell, C. P. (1981) Development of organ systems in the northern anchovy, *Engraulis mordax*, and other teleosts. *Am. Zool.*, **21**, 429–46.

O'Donald, P. (1980) *Genetic Models of Sexual Selection*, Cambridge University Press, Cambridge.

O'Donald, P. and Majerus, M. E. N. (1985) Sexual selection and the evolution of preferential mating in ladybirds. I. Selection for high and low lines of female preference. *Heredity, Lond.*, **55**, 401–12.

Ochumba, P. B. O. (1988) Periodic massive fish kills in the Kenyan part of Lake Victoria. *F.A.O. Fish. Rep.*, **388**, 47–60.

Ogari, J. (1985) Distribution, food and feeding habits of *Lates niloticus* in the Nyanza Gulf of Lake Victoria (Kenya). *F.A.O. Fish. Rep.*, **335**, 68–80.

Ogari, J. (1988) Predator-prey relationship in Lake Victoria with special reference to the Nile perch *Lates niloticus* (Linné), in *Predator–prey Relationships, Population Dynamics and Fisheries Productivities of Large African Lakes* (ed. D. Lewis), *CIFA FAO, Rome, Occ. Pap.*, **15**, 53–68.

Ogari, J. and Dadzie, S. (1988) The food of the Nile perch, *Lates niloticus* (L.) after the disappearance of the haplochromine cichlids in the Nyanza Gulf of Lake Victoria (Kenya) *J. Fish Biol.*, **32**, 570–1.

Ogutu-Ohwayo, R. (1985) The effects of predation by Nile perch introduced into Lake Kyoga (Uganda) in relation to the fisheries of Lake Kyoga and Lake Victoria. *F.A.O. Fish. Rep.*, **335**, 18–39.

Ogutu-Ohwayo, R. (1990) The decline of the native fishes of Lakes Victoria and Kyoga (East Africa) and the impact of introduced species, especially the Nile perch, *Lates niloticus*, and the Nile tilapia, *Oreochromis niloticus*. *Envir. Biol. Fish.*, **27**, 81–96.

Ohm, D. (1958) Die ontogenetische Entwicklung des Kampverhaltens bei *Aequidens portalegrensis* Hensel und *Ae. latifrons* Steindachner (Cichlidae). *Verh. Deutsch. Zool. Gesel. Frankfurt a. Main.*, *1958*, 182–94.

Oijen, M. J. P. van (1982) Ecological differentiation among the piscivorous haplochromine cichlids of Lake Victoria (East Africa), *Neth. J. Zool.*, **32**(3), 336–63.

Oijen, M. J. P. van, Witte, F. and Witte-Maas, E. L. M. (1981) An introduction to ecological and taxonomic investigations on the haplochromine cichlids from the Mwanza Gulf of Lake Victoria. *Neth. J. Zool.*, **31**, 149–74.

Okaronon, J., Acere, T. and Ocenodongo, D. (1985) The current state of the fisheries in the northern portion of Lake Victoria. *FAO Fish. Rep.*, **335**, 89–98.

Okedi, J. (1971) Further observations on the ecology of the Nile perch (*Lates niloticus.* Linné) in Lake Victoria and Lake Kyoga. *East Afr. Freshw. Fish Res. Org. Ann. Rep.*, 1970, 42–55.

Okedi, J. (1981) The Engraulicypris 'Dagaa' fishery of Lake Victoria with special reference to the southern waters of the Lake, in *Proc. Workshop Kenya Marine and Fisheries Research Institute on Aquatic Resources of Kenya*, July 13–19, 1981, pp. 445–84, Kenya National Academy for the Advancement of Arts and Science.

Okemwa, E. N. (1981) Changes in fish species composition of Nyanza Gulf of Lake Victoria, in *Proc. Workshop Kenya Marine and Fisheries Research Institute on Aquatic Resources of Kenya*, July 13–19, 1981, pp. 138–56, Kenya National Academy for the Advancement of Arts and Science.

Okemwa, E. N. (1984) Potential fishery of Nile perch *Lates niloticus* Linné (Pisces: Centropomidae) in Nyanza Gulf of Lake Victoria, East Africa. *Hydrobiologia*, **108**, 121–6.

Oliver, M. K. (1984) Systematics of African cichlid fishes; determination of the most primitive taxon, and studies on the haplochromines of Lake Malawi (Teleostei: Cichlidae), PhD thesis, Yale University.

Oppenheimer, J. (1970) Mouthbreeding in fishes. *Anim. Behav.*, **18**, 493–503.

Oppenheimer, J. R. and Barlow, G. W. (1968) Dynamics of parental behavior in the black-chinned mouthbreeder, *Tilapia melanotheron* (Pisces: Cichlidae). *Z. Tierpsychol.*, **25**, 889–914.

Oppenheimer, S. A., Boynton, S. C. and Leibel, W. S. (1989) Lactate dehydrogenase polymorphism in natural and cultivated populations of the angelfish cichlid, *Pterophyllum scalare. Comp. Biochem. Physiol.*, **93B**, 99–106.

Otten, E. (1983) The jaw mechanism during growth of a generalized *Haplochromis* species: *H. elegans* Trewavas, 1933 (Pisces, Cichlidae). *Neth. J. Zool.*, **33**, 55–98.

Owen, R. B., Crossley, R., Johnson, T. C. *et al.* (1990) Major low levels of Lake Malawi and implications for speciation rates in cichlid fishes. *Proc. Roy. Soc. Lond.* **240B**, 519–53.

Parker, G. A. (1974) Assessment strategy and the evolution of fighting behaviour. *J. theor. Biol.*, **47**, 223–43.

Paterson, H. E. H. (1980) A comment on 'mate recognition systems'. *Evolution*, **34**, 330–1.

Paterson, H. E. H. (1985) The recognition concept of species, in *Species and Speciation* (ed. E. S. Vrba), Transvaal Museum, Pretoria, Republic of South Africa, pp. 21–9.

Patterson, C. (1982) Morphological characters and homology, in *Problems of Phylogenetic Reconstruction* (eds K. A. Joysey and E. A. Friday), Academic Press, London, pp. 21–74.

Patterson, C. (1987) Introduction, in *Molecules and Morphology in Evolution* (ed. C. Patterson), Cambridge Univ. Press, London, pp. 1–22.

Patterson, C. (1988) The impact of evolutionary theories on systematics, in *Prospects in Systematics* (ed. D. L. Hawksworth), Clarendon Press, Oxford, pp. 60–91.

Pauly, D. (1976) The biology, fishery and potential for aquaculture of *Tilapia melanotheron* in a small West African lagoon. *Aquaculture*, **7**, 33–49.

Pauly, D., Moreau, J. and Prein, M. (1988) A comparison of overall growth performance of tilapia in open waters and aquaculture, in *The Second International Symposium on Tilapia in Aquaculture* (ICLARM Conference Proceedings, Vol. 15) (eds R. S. V. Pullin, T. Bhukasawan, K. Tonguthai and J. L. Maclean), Department of Fisheries, Bangkok and International Center for Living Aquatic Resources Management, Manila, pp. 469–79.

Pellegrin, J. (1904) Contribution à l'étude anatomique, biologique et taxonomique des poissons de la famille Cichlides. *Mém. Soc. Zool. Fr.*, **16**, 41–399.

Perrone, M. (1978a) Mate size and breeding success in a monogamous cichlid fish. *Environ Biol. Fishes*, **3**, 193–201.

Perrone, M. (1978b) The economy of brood defence by parental cichlid fishes, *Cichlasoma maculicauda, Oikos*, **31**, 137–41.

Peters, H. M. (1937) Experimentelle Untersuchungen uber die Bruptflege von *Haplochromis multicolor*, eine maulbrutenden Knochenfische. *Z. Tierpsychol.*, **1**, 201–18.

Peters, H. M. and Berns, S. (1982) Die Maulbrutpflege der Cichliden. Untersuchungen zur Evolution eines Verhaltensmusters. *Z. Zool. Syst. Evol-forsch.*, **20**, 18–52.

Philippart, J.-Cl. and Ruwet, J.-Cl. (1982) Ecology and distribution of tilapias, in *The Biology and Culture of Tilapias* (*ICLARM Conference Proceedings*. Vol. 7) (eds R. S. V. Pullin and R. H. Lowe-McConnell), International Center for Living Aquatic Resources Management, Manila, pp. 15–59.

Pitcher, T. J. (1986) Functions of shoaling behaviour in teleosts, in *The Behaviour of Teleost Fishes* (ed. T. J. Pitcher), Croom Helm, London, pp. 294–337.

Plate, M. (1913) *Selektionsprinzip und Probleme der Artbildung*, Leipzig and Berlin.

Poll, M. (1952) Ségrégation géographique et formation des espèces. *Annls. Soc. zool. Belg.*, **83**, 211–24.

Poll, M. (1956) Poissons Cichlidae. *Résult, scient. Explor. hydrobiol. Lac Tanganyika (1946–1947)*, Vol. 3, Fasc. 5b, Institut Royal des Sciences Naturelles de Belgique, Bruxelles, pp. 1–619.

Poll, M. (1957) Les genres des poissons d'eau douce de l'Afrique. *Annls Mus. r. Congo Belg., Sér. 8vo., Sci. zool.*, No. 54, 1–191; later *Publ. Dir. del'Agriculture, des Forêts et de l'Elevage, Bruxelles*. 5–191.

Poll, M. (1959) Recherches sur la faune ichthyologique de la région du Stanley Pool. *Annls. Mus. r. Congo Belge. Sér. 8vo. Sci. zool.*, No. 71, 75–174.

Poll, M. (1978) Contribution à la connaissance du genre *Lamprologus* SCH description de quatre espèces nouvelles, réhabilitation de *L. mondabu* et synopsis remanié des esp. du lac Tanganyika. *Bul. Acad. r. Belg. Cl. Sci.*, **64**, 725–58.

Poll, M. (1986) Classification des Cichlidae du lac Tanganyika: Tribus, genres et espèces. *Mém. Acad. r. Belg. Cl. Sci.*, **45**, 5–163.

Poll, M. and Matthes, H. (1962) Trois poissons remarquables du lac Tanganyika. *Annls Mus. r. Afr. cent. Sér. 8vo. Sci. zool.*, No. 111, 1–26.

Powles, H. (ed.) (1987) *Research Priorities for African Aquaculture* (*Report of a Workshop held in Dakar, Senegal, 13–16 October 1986.* IDRC-MR149e), International Development Research Center, Ottawa.

Pretto Malca, R. (1979) Pasos a seguir en la production de hibridos de tilapia *Tilapia nilotica* hembra × *Tilapia hornorum* macho. Latinoam. Acuicult. *Rev.* (Trans. 'Procedures for production of tilapia hybrids, *Tilapia nilotic* female × *Tilapia hornorum* male'). *Lima-Peru*, No. 1, pp. 29–32 (in Spanish).

Pullin, R. S.V. (1982) *Tilapia, Sarotherodon* or *Oreochromis? ICLARM Newsl.*, **5**, 19.

Pullin, R. S. V. (1983) Choice of tilapia species for aquaculture, in *International Symposium on Tilapia in Aquaculture* (Compilers L. Fishelson and Z. Yaron), Tel Aviv University, Tel Aviv, pp. 64–76.

Pullin, R. S. V. (1984) Tilapia – potentially an international food community. *INFOFISH Mark. Dig.*, No. 3/84, 35–6.

Pullin, R. S. V. (1985) Tilapias: 'everyman's fish', *Biologist*, **32**, 84–8.

Pullin, R. S. V. (1986) Culture of herbivorous tilapias, in *Development and Management of Tropical Living Aquatic Resources* (eds H. H. Chan, K. J. Ang. A. T. Law, M. H. M. Ibrahim and H. O. Ishak), Penerbit Universiti Pertanian Malaysia, Selangor, Malaysia, pp. 145–9.

Pullin, R. S. V. (ed.) (1988) *Tilapia Genetic Resources for Aquaculture* (*ICLARM Conference Proceedings*, Vol. 16), International Center for Living Aquatic Resources Management, Manila.

Pullin, R. S. V. and Capili, J. B. (1988) Genetic improvement of tilapias: problems and prospects, in Pullin *et al.* (1988), pp. 259–66.

Pullin, R. S. V. and Lowe-McConnell, R. H. (eds) (1982) *The Biology and Culture of Tilapias* (*ICLARM Conference Proceedings*, Vol. 7), International Center for Living Aquatic Resources Management, Manila.

Pullin, R. S. V. Bhukasawan, T., Tonguthai, K. and Maclean, J. L. (eds) (1988) *The Second International Symposium on Tilapia in Aquaculture* (ICLARM Conference Proceedings, Vol. 15), Department of Fisheries, Bangkok, and International Center for Living Aquatic Resources Management, Manila.

Quertermus, C. J. and Ward, J. A. (1969) Development and significance of two motor patterns used in contacting parents by young orange chromides (*Etroplus maculatus*). *Anim. Behav.*, **17**, 624–35.

Ramshaw, J. A. M., Coyne, J. A. and Lewontin, R. C. (1979) The sensitivity of gel electrophoresis as a detector of genetic variation. *Genetics*, **93**, 1019–37.

Rangaswami, G. (1988) 'Golden fish' culture in India. *Naga, The ICLARM Quarterly*, **11**, 24–6.

Rechten, C. (1980) A note on the reproductive coloration of the cichlid fish *Etroplus maculatus*. *Z. Tierpsychol.*, **52**, 135–40.

Regan, C. T. (1906) A revision of the fishes of the South American cichlid genera *Cichla, Chaetobranchus* and *Chaetobranchopsis*, with notes on the genera of the American Cichlidae. *Ann. Mag. nat. Hist.*, **7** (17), 230–9.

Regan, C. T. (1906–08) Pisces. *Biologia cent.-am.*, **8**, 1–203.

Regan, C. T. (1913) The classification of percoid fishes. *Ann. Mag. nat. Hist.*, **8** (12), 111–45.

Regan, C. T. (1920) The classification of the fishes of the family Cichlidae. I. The Tanganyikan genera. *Ann. Mag. nat. Hist.*, **9** (5), 33–53.

Regan, C. T. (1921) The cichlid fishes of Lake Nyasa. *Proc. zool. Soc. Lond.*, *1921*, 675–727.

Regan, C. T. (1922a) The cichlid fishes of Lake Victoria. *Proc. zool. Soc. Lond.*, *1922*, 157–91.

Regan, C. T. (1922b) The classification of the fishes of the family Cichlidae. II. On African and Syrian genera not restricted to Great Lakes. *Ann. Mag. nat. Hist.*, **9** (10), 249–64.

Regnier, F. E. (1971) Semiochemicals – structure and function. *Biol. Reprod.*, **4**, 309–26.

Reid, M. J. and Atz, J. W. (1958) Oral incubation in the cichlid fish *Geophagus jurupari* Heckel. *Zoologica*, **43**, 77–88.

Reinthal, P. (1987) Morphology, ecology and behavior of groups of the rock-dwelling cichlid fishes (Cichlidae: Perciformes) from Lake Malawi, Africa, PhD thesis, Yale University.

Reynolds, J. E. and Greboval, D. F. (1988) Socio-economic effects of the evolution of Nile perch fisheries in Lake Victoria: a review. *CIFA Tech. Pap.* No. 17. (CIFA = Committee for Inland Fisheries of Africa; Food and Agriculture Organization of the United Nations)

Ribbink, A. J. (1977) Cuckoo among Lake Malawi cichlid fish. *Nature, Lond.*, **267**, 243–4.

Ribbink, A. J. (1986) The species concept, sibling species and speciation. *Annls. Mus. r. Afr. cent. Sci. zool.*, No. 251, 109–16.

Ribbink, A. J. (1987) African Lakes and their fishes: conservation scenarios and suggestions. *Environ. Biol. Fishes*, **19**, 3–26.

Ribbink, A. J. (1988) Evolution and speciation of African cichlids, in *Biology and Ecology of African Freshwater Fishes* (eds C. Lévêque, M. N. Bruton and G. W. Ssentongo), Editions de l'ORSTOM, Travaux et Documents no. 216, Paris, pp. 35–51.

Ribbink, A. J., Marsh, A. C. and Marsh, B. A. (1981) Nest-building and communal care of young by *Tilapia rendalli* Dumeril (Pisces, Cichlidae) in Lake Malawi. *Environ. Biol. Fishes.*, **6**, 219–22.

Ribbink, A. J., Marsh, A. C., Marsh, B. and Sharp, B. J. (1980) Parental behaviour and mixed broods among cichlid fish of Lake Malawi. *S. Afr. J. Zool.*, **15**, 1–6.

Ribbink, A. J., Marsh, B. A., Marsh, A. C., Ribbink, A. C. and Sharp, B. J. (1983a) A preliminary survey of the cichlid fishes of rocky habitats in Lake Malawi. *S. Afr. J. Zool.*, **18**, 149–310.

Ribbink, A. J., Marsh, A. C., Marsh, B. A. and Sharp, B. J. (1983b) The zoogeography, ecology and taxonomy of the genus *Labeotropheus* Ahl, 1927, of Lake Malawi (Pisces: Cichlidae). *Zool. J. Linn. Soc.*, **79**, 223–43.

Ridley, M. (1978) Paternal care. *Anim. Behav.*, **26**, 904–32.

Ringler, N. H. (1983) Variation in foraging tactics of fishes, in *Predators and Prey in Fishes* (eds D. L. G. Noakes, D. G. Lindquist, G. S. Helfman and J. A. Ward), Dr W. Junk, The Hague, pp. 159–71.

Roberts, T. R. (1972) Ecology of fishes in the Amazon and Congo basins. *Bull. Mus. comp. Zool. Harv.*, **143**, 117–47.

Roberts, T. R. (1982) *Gobiocichla ethelwynnae*, a new species of goby-like cichlid fish from rapids in the Cross River, Cameroon. *Proc. K. ned. Acad. Wet. Ser. C. Biol. Med. Sci.*, **85**, 575–87.

Roberts, T. R. and Stewart, D. J. (1976) An ecological and systematic survey of fishes in the rapids of the Lower Zaire or Congo River. *Bull. Mus. comp. Zool. Harv.*, **147**, 239–317.

Robins, C. R., Bailey, R. M., Bond, C. E., Brooker, J. R., Lachner, E. A., Lea, R. N. and Scott, W. B. (1980) *A List of Common and Scientific Names of Fishes from the United States and Canada.* American Fisheries Society, Maryland.

Rodman, D. T. (1966) Sound production by the African cichlid *Tilapia mossambica. Ichthyologica*, **38**, 279–80.

Rogers, W. (1987) Sex ratio, monogamy and breeding success in the Midas cichlid (*Cichlasoma citrinellum*). *Behav. Ecol. Sociobiol.*, **21**, 47–51.

Rogers, W. (1988) Parental investment and division of labor in the Midas cichlid (*Cichlasoma citrinellum*). *Ethology*, **79**, 126–42.

Rompaey, J. van, Ahmadyar, S. Z. and Verheyen, E. (1988) General protein patterns of muscle homogenates of some Lake Victoria haplochromines (Pisces: Cichlidae). *J. Zool., Lond.*, **214**, 505–18.

Rosen, D. E. (1973) Interrelationships of higher euteleostean fishes, in *Interrelationships of Fishes* (eds P. H. Greenwood, R. S. Miles and C. Patterson), Academic Press, London, pp. 397–513.

Rosen, D. E. and Patterson, C. (1977) Review of ichthyodectiform and other Mesozoic fishes and the theory and practice of classifying fossils. *Bull. Am. Mus. nat. Hist.*, **158**, 85–172.

Rosenzweig, M. L. (1978) Competitive speciation. *Biol. J. Linn. Soc.*, **10**, 275–89.

Rowland, W. J. (1978) Sound production and associated behavior in the jewel fish, *Hemichromis bimaculatus. Behaviour*, **64**, 125–36.

Russock, H. I. (1986) Preferential behaviour of *Sarotherodon* (*Oreochromis*) *mossambicus* (Pisces: Cichlidae) fry to maternal models and its relevance to the concept of imprinting. *Behaviour*, **96**, 304–21.

Russock, H. I. and Schein, M. W. (1977) Effect of age and experience on the filial behaviour of *Tilapia mossambica* fry (Pisces: Cichlidae). *Behaviour*, **61**, 276–303.

Russock, H. I. and Schein, M. W. (1978) Effect of socialization on adult social preferences in *Tilapia mossambica* (*Sarotherodon mossambicus*); Pisces: Cichlidae. *Anim. Behav.*, **26**, 148–59.

Sage, R. D. and Selander, R. K. (1975) Trophic radiation through polymorphism in cichlid fishes. *Proc. natu. Acad. Sci.·U.S.A.*, **72**, 4669–73.

Sage, R. D., Loiselle, P. V., Basasibwaki, P. and Wilson, A. C. (1984) Molecular versus morphological change among cichlid fishes of Lake Victoria, in *Evolution of Fish Species Flocks* (eds A. A. Echelle and I. Kornfield), University of Maine at Orono Press, Orono, Maine, pp. 185–201.

Saint-Paul, U. (1986) Potential for aquaculture of South American freshwater fishes: a review. *Aquaculture*, **54**, 205–40.

Saint-Paul, U. (1987) Recent developments and prospects of cultured indigenous freshwater fish in Latin America, in *Aquaculture Research in Latin America* (eds J. A. J. Verreth, M. Carillo, S. Zanuy and E. A. Huisman), Pudoc, Wageningen, The Netherlands, pp. 9–22.

Samarakoon, J. I. (1985) Experimental seed production in *Etroplus suratensis* by induced pair formation and breeding through environmental manipulation. *Aquaculture*, **46**, 323–32.

Satia, B. P. N. (1987) Situation of aquaculture in Cameroon, in Powles (1987), pp. 70–82.

Saul, W. G. (1975) An ecological study of fishes at a site in Upper Amazonian Ecuador. *Proc. Acad. nat. Sci. Philad.*, **127**, 93–134.

Schmidt-Focke, E. and Homburg, B. (1977) The red discus: heredity or environment? *Trop. Fish Hobby.*, **26**, 4–14.

Schoenen, P. (1982) *A Bibliography of Important Tilapias (Pisces: Cichlidae) for Aquaculture (ICLARM Bibliographies*, Vol. 3), International Center for Living Aquatic Resonance Management, Manila.

Schoenen, P. (1984) *A Bibliography of Important Tilapias (Pisces: Cichlidae) for Aquaculture. Oreochromis macrochir, O. aureus, O. hornorum, O. mossambicus, O. niloticus, Sarotherodon galilaeus, Tilapia rendalli* and *T. zillii (ICLARM Bibliographies*, Vol. 3, Suppl. 1), International Center for Living Aquatic Resources Management, Manila.

Schoenen, P. (1985) *A Bibliography of Important Tilapias (Pisces: Cichlidae) for Aquaculture. Oreochromis variabilis, O. andersonii, O. esculentus, O. leucostictus, O. mortimeri, O. spilurus niger, Sarotherodon melanotheron* and *Tilapia sparrmanii (ICLARM Bibliographies*, Vol. 6), International Center for Living Aquatic Resources Management, Manila.

Schoener, T. W. (1971) Theory of feeding strategies. *A. Rev. Ecol. Syst.*, **11**, 369–404.

Scholl, A. and Holzberg, S. (1972) Zone electrophoretic studies on lactate dehydrogenase isoenzymes in South American cichlids (Teleostei, Percomorphi). *Experientia*, **28**, 489–91.

Scholz, C. A. and Rosendahl, B. R. (1988) Low lake stands in Lakes Tanganyika and Malawi, East Africa, delineated with multifold seismic data. *Science, Wash., DC*, **240**, 1645–8.

Schröder, J. H. (1980) Morphological and behavioural differences between the BB/OB and B/W colour morphs of *Pseudotropheus zebra* Boulenger (Pisces, Cichlidae). *Z. Zool. Syst. Evol. Forsch.*, **18**, 69–76.

Schröder, J. H. (1986) Speciation within *Pseudotropheus zebra* Boulenger. *Annls. Mus. r. Afr. Cent. Sci. zool.*, No. 251, 121–8.

Schröder, S. L. and Zaret, T. M. (1979) The adaptive significance of color patterns in *Cichla ocellaris*. *Copeia*, *1979*, 43–7.

Schumm, J. W., Knowlton, R. G., Braman, J. C., Barker, D. F., Botstein, D., Akots, G., Brown, V., Gravius, T., Helms, C., Hsiao, K., Rediker, K., Thurston, J. and Donis-Keller, H. (1988) Identification of more than 500 restriction fragment length polymorphisms by screening random genomic clones. *Am. J. hum. Genet.*, **42**, 143–59.

Schuster, W. H. (1952) *Fish culture in brackishwater ponds of Java* (*Spec. Publ. Indo-Pac. Fish. Counc.*, No. 1). FAO, Bangkok, Thailand.

Schwanck, E. (1987a) Lunar periodicity in the spawning of *Tilapia mariae* in the Ethiop River, Nigeria. *J. Fish. Biol.*, **30**, 533–7.

Schwanck, E. (1987b) Reproductive behaviour of a monogamous cichlid fish *Tilapia mariae*, PhD dissertation, University of Stockholm.

Schwarz, A. (1974a) Sound production and associated behavior in a cichlid fish, *Cichlasoma centrarchus*. I. Male–male interactions. *Z. Tierpsychol.*, **35**, 147–56.

Schwarz, A. (1974b) The inhibition of aggressive behavior by sound in the cichlid fish, *Cichlasoma centrarchus*. *Z. Tierpsychol.*, **35**, 508–17.

Scott, A. G., Penman, D. J., Beardmore, J. A. and Skibinski, D. O. F. (1989) The 'YY' supermale in *Oreochromis niloticus* (L.) and its potential in aquaculture. *Aquaculture*, **78**, 237–51.

Sebeok, T. A. (1968) *Animal Communication*, Indiana University Press, Bloomington, Ind. and London.

Sebeok, T. A. (1977) *How Animals Communicate*, Indiana University Press, Bloomington, Ind. and London.

Seegers, L. (1987) Ein neuer Haplochromisverwandter aus Ost Afrika sowie einige Bemerkungen zur Fischfauna des Victoria Sees. *DATZ* (*Dt. Aquar.-Terrar.-Z.*), **401**, 263–7.

Seitz, A. (1940) Die Paarbildung bei einigen Cichliden. I. Die Paarbildung bei *Astatotilapia strigigena*. *Z. Tierpsychol.*, **4**, 50–84.

Seitz, A. (1942) Die Paarbildung bei einigen Cichliden. II. Die Paarbildung bei *Hemichromis bimaculatus*. *Z. Tierpsychol.*, **5**, 74–101.

Seyoum, S. (1989) Stock identification and the evolutionary relationship of the genera *Oreochromis*, *Sarotherodon* and *Tilapia* (Pisces: Cichlidae) using alozyme analysis and restriction endonuclease analysis of mitochondrial DNA, PhD thesis, University of Waterloo, Waterloo, Ontario, Canada.

Seyoum, S. (1990) Allozyme variation in subspecies of *Oreochromis niloticus*. *Isozyme Bull.*, **23**, p. 97.

Sharp, B. J. (1981) An ecological study of territoriality in four cichlid species resident on rocky shores near Monkey Bay, Lake Malawi, MSc thesis, Rhodes University, Republic of South Africa.

Shaw, E. S. and Aronson, L. R. (1954) Oral incubation in *Tilapia macrocephala*. *Bull. Am. Mus. nat. Hist.*, **103**, 379–415.

Shaw, E. S. and Innes, K. (1980) The 'infantilization' of a cichlid fish. *Develop. Psychobiol.*, **13**, 131–9.

Shine, R. (1978) Propagule size and parental care: the 'safe harbor' hypothesis. *J. theor. Biol.*, **75**, 417–24.

Sibly, R. M. and Calow, P. (1987) *Physiological Ecology of Animals*, Blackwell Scientific Publications, Oxford.

Siepen, G. and Crapon de Caprona, M.-D. (1986) The influence of parental color morph on mate choice in the cichlid fish *Cichlasoma nigrofasciatum*. *Ethology*, **71**, 187–200.

Silva, S. S., de and Perera, M. K. (1983) Digestibility of an aquatic macrophyte by cichlids *Etroplus suratensis* (Bloch) with observations on the relative merits of three indigenous components as markers and daily changes in protein digestibility. *J. Fish Biol.*, **23**, 675–84.

Sioli, H. (ed.) (1984) *The Amazon: Limnology and Landscape Ecology of a Mighty River and its Basin* (*Monographs in Biology*, Vol. 56), Dr W. Junk, Dordrecht, The Netherlands.

Sjölander, S. (1972) Feldbeobachtungen an einigen westafrikanischen Cichliden. *Aquar. Terr.*, **19**, 42–5.

Sjölander, S. and Fernö, A. (1973) Sexual imprinting on another species in a cichlid fish, *Haplochromis burtoni*. *Revue. Comp. Animal*, **7**, 77–81.

Slootweg, R. (1987) Prey selection by molluscivorous cichlids foraging on a schistosomiasis vector snail, *Biomphalaria glabrata*. *Oecologia*, **74**, 193–202.

Smith, C. L. and Tyler, J. C. (1973) Direct observations of resource sharing in coral reef fish. *Helgoländer wiss. Meeresunters.*, **24**, 264–75.

Smith, I. R., Torres, E. B. and Tan, E. O. (eds) (1985) *Philippine Tilapia Economics* (*ICLARM Conference Proceedings*, Vol. 12), Philippine Council for Agriculture and Resources Research Development, Laguna, and International Center for Living Aquatic Resources Management, Manila.

Smith, W. J. (1977) *The Behavior of Communicating. An Ethological Approach*, Harvard University Press, Cambridge, Mass.

Smith-Grayton, P. K. and Keenleyside, M. H. A. (1978) Male–female parental roles in *Herotilapia multispinosa* (Pisces, Cichlidae). *Anim. Behav.*, **26**, 520–6.

Smithe, F. B. (1975) *Naturalist's Color Guide*, Am. Mus. Nat. Hist., NY.

Smithe, F. B. (1981) *Naturalist's Color Guide*, Part III, Am. Mus. Nat. Hist., NY.

Sneath, P. H. and Sokal, R. R. (1973) *Numerical Taxonomy*, W. H. Freeman, San Franscisco.

Soares, M. G. M., Almeida, R. G. and Junk, W. J. (1986) The trophic status of the fish fauna in Lago Camaleâo, a macrophyte dominated floodplain lake in the middle Amazon. *Amazoniana*, **9**, 511–26.

Spiers, E. B. (1982) Do female flies choose their mates? *Am. Nat.*, **119**, 675–93.

Sprenger, K. (1971) The red hump *Geophagus*. *Buntb. Bull.*, No. 26, 24–5.

Ssentongo, G. W. and Welcomme, R. L. (1985) Past history and current trends in the fisheries of Lake Victoria. *F.A.O. Fish. Rep.*, **335**, 123–38.

Staeck, W. (1974, 1977, 1983) *Cichliden, Verbreitung, Verhalten, Arten* (3 vols), E. Pfriem Verlag, Wuppertal, Federal Republic of Germany.

Staeck, W. and Linke, H. (1984) *Amerikanische Cichliden. I. Kleine Buntbarsche*, Tetra-Verlag, Melle, Federal Republic of Germany.

Staeck, W. and Linke, H. (1985) *Amerikanische Cichliden. II. Grosse Buntbarsche*, Tetra-Verlag, Melle, Federal Republic of Germany.

Stager, J. C., Reinthal, P. N. and Livingstone, D. A. (1986) A 25 000-year history for Lake Victoria, East Africa, and some comments on its significance for the evolution of cichlid fishes. *Freshwat. Biol.*, **16**, 15–19.

Stiassny, M. L. J. (1980) The anatomy and phylogeny of two genera of African cichlid fishes, PhD thesis, London University.

Stiassny, M. L. J. (1981a) The phyletic status of the family Cichlidae (Pisces: Perciformes): a comparative anatomical investigation. *Neth. J. Zool.*, **31**, 275–314.

Stiassny, M. L. J. (1981b) Phylogenetic versus convergent relationship between piscivorous cichlid fishes from Lakes Malawi and Tanganyika. *Bull. Br. Mus. nat. Hist. (Zool.)*, **40**, 67–101.

Stiassny, M. L. J. (1982) The relationships of the neotropical genus *Cichla* (Perciformes, Cichlidae): a phyletic analysis including some functional considerations. *J. Zool., Lond.*, **197**, 427–53.

Stiassny, M. L. J. (1986) The limits and relationships of the acanthomorph teleosts. *J. Zool., Lond. (B)*, **1**, 411–60.

Stiassny, M. L. J. (1987) Cichlid familial intrarelationships and the placement of the neotropical genus *Cichla* (Perciformes, Labroidei). *J. nat. Hist.*, **21**, 1311–31.

Stiassny, M. L. J. (1989) A taxonomic revision of the African genus *Tylochromis* (Labroidei, Cichlidae); with notes on the anatomy and relationships of the group. *Annls. Mus. R. Afs. Cent. Sci. zool.* **258**, 1–161.

Stiassny, M. L. J. and Jensen, J. S. (1987) Labroid intrarelationships revisited: morphological complexity, 'key innovations', and the study of comparative diversity. *Bull. Mus. comp. Zool.*, **151**, 269–319.

Stiassny, M. L. J. (in press) *Tylochromis*, relationship and the phylogenetic status of the African Cichlidae. *Ann. Mus. Novit.*

Strauss, R. E. (1984) Allometry and functional feeding morphology in haplochromine cichlids, in *Evolution of Fish Species Flocks* (eds A. A. Echelle and I. Kornfield), University of Maine at Orono Press, Orono, Maine, pp. 217–29.

Sultan, S. E. (1987) Evolutionary implications of phenotypic plasticity in plants. *Evol. Biol.*, **21**, 127–8.

Swofford, D. L. (1985) *PAUP, Phylogenetic Analysis Using Parsimony, Version 2.3.* Illinois Nat. Hist. Survey, Urbana. Ill.

Taborsky, M. (1984) Broodcare helpers in the cichlid fish *Lamprologus brichardi*: their costs and benefits. *Anim. Behav.*, **32**, 1236–52.

Taborsky, M. (1985) Breeder–helper conflict in a cichlid fish with broodcare helpers: an experimental analysis. *Behaviour*, **95**, 45–75.

Taborsky, M. and Limberger, D. (1981) Helpers in fish. *Behav. Ecol. Sociobiol.*, **8**, 143–5.

Taborsky, M., Hert, E., Siemens, M. and Stoerig, P. (1986) Social behaviour of *Lamprologus* species: functions and mechanisms. *Annls. Mus. r. Afr. Cent., Sci. zool,* No. 251, 7–11.

Taiwan Fisheries Bureau (1985, 1986, 1987) *Fisheries Yearbook(s), Taiwan Area 1984, 1985, 1986*, Taiwan Fisheries Bureau, Department of Agriculture and Forestry, Provincial Government of Taiwan, Taipei.

Takamura, K. (1983) Interspecific relationship between two Aufwuchs eaters *Petrochromis polyodon* and *Tropheus moorei* (Pisces: Cichlidae) of Lake Tanganyika, with a discussion on the evolution and functions of a symbiotic relationship. *Physiol. Ecol. Japan*, **20**, 59–69.

Takamura, K. (1984) Interspecific relationship between two aufwuchs-eating fishes in Lake Tanganyika. *Environ. Biol. Fishes*, **10**, 225–41.

Tave, D. (1987) All-male hybrid tilapia. *Aquaculture Mag.*, (Nov.–Dec.), 55–6.

Tave, D. (1988) Genetics and breeding of tilapia: a review, in Pullin *et al.* (1988), pp. 285–93.

Tave, D., Rezk, M. and Smitherman, R. O. (1989) Gold-colored tilapia now possible. *Highlts. agric. Res.*, **36**, 4.

Tavolga, W. N. (1977) *Sound Production in Fishes* (*Benchmark Papers in Animal Behaviour*, Vol. 9), Dowden, Hutchinson and Ross Inc., Stroudsbury.

Thorpe, J. E. (1983) Inheritance of developmental rates in Atlantic salmon, *Salmo salar L. Aquaculture*, **33**, 119–28.

Thorpe, J. E. (1987) Environmental regulation of growth patterns in juvenile Atlantic salmon, in *Age and Growth of Fish* (eds R. C. Summerfelt and G. E. Hall), Iowa State University Press, Ames, Iowa, pp. 463–74.

Thys, van den Audenaerde, D. F. E. (1963) Description d'une espèce nouvelle d'*Haplochromis* (Pisces, Cichlidae) avec observations sur les *Haplochromis* rhéophiles du Congo orintal, *Revue Zool. Bot. afr.*, **68**, 140–52.

Thys, van den Audenaerde, D. F. E. (1964) Révision systématique des espèces Congolaises du genre *Tilapia* (Pisces, Cichlidae). *Annls. Mus. r. Afr. cent. Sér. 8vo. Sci. zool.*, No. 124, 1–155.

Thys, van den Audenaerde, D. F. E. (1968a) A preliminary contribution to a systematic revision of the genus *Pelmatochromis* Hubrecht *sensu lato* (Pisces, Cichlidae), *Revue Zool. Bot. afr.*, **77**, 349–91.

Thys, van den Audenaerde, D. F. E. (1968b) *An annotated bibliography of Tilapia (Pisces, Cichlidae)* (*Documentation Zoologique* No. 14), Musée Royale de l'Afrique Centrale, Tervuren, Belgium.

Timms, A. M. and Keenleyside, M. H. A. (1975) The reproductive behaviour of *Aequidens paraguayensis* (Pisces, Cichlidae). *Z. Tierpsychol.*, **39**, 8–23.

Townshend, T. J. and Wootton, R. J. (1984) Effects of food supply on the reproduction of the convict cichlid, *Cichlasoma nigrofasciatum. J. Fish Biol.*, **24**, 91–104.

Townshend, T. J. and Wootton, R. J. (1985) Variation in the mating system of a biparental cichlid fish, *Cichlasoma panamense. Behaviour*, **95**, 181–97.

Travers, R. A. (1981) The interarcual cartilage: a review of its development, distribution and value as an indicator of phylogenetic relationships in euteleostean fishes. *J. nat. Hist.*, **15**, 853–71.

Trendall, J. (1988a) The distribution and dispersal of introduced fish at Thumbi West Island in Lake Malawi, Africa. *J. Fish Biol.*, **33**, 357–69.

Trendall, J. (1988b) Recruitment of juvenile mbuna (Pisces: Cichlidae) to experimental rock shelters in Lake Malawi, Africa. *Eniv. Biol. Fishes*, **22**, 117–31.

Trewavas, E. (1931) A revision of the cichlid fishes of the genus *Lethrinops* Regan. *Ann. Mag. nat. Hist.*, **10**, 133–52.

Trewavas, E. (1933) Scientific results of the Cambridge Expedition to the East African lakes 1930–31. II. The cichlid fishes. *J. Linn. Soc.* (*Zool.*), **38**, 309–41.

Trewavas, E. (1935) A synopsis of the cichlid fishes of Lake Nyasa. *Ann. Mag. nat. Hist., ser.* 10, **16**, 65–118.

Trewavas, E. (1938) Lake Albert fishes of the genus *Haplochromis. Ann. Mag. nat. Hist.*, **11**(1), 435–49.

Trewavas, E. (1947) An example of 'mimicry' in fishes. *Nature, Lond.*, **160**, 120.

Trewavas, E. (1949) The origin and evolution of the cichlid fishes of the great African lakes, with special reference to Lake Nyasa. *13th Int. Congr. Zool.*, 365–8.

Trewavas, E. (1973) On the cichlid fishes of the genus *Pelmatochromis* with a proposal of a new genus for *P. congicus*; on the relationship between *Pelmatochromis* and *Tilapia* and the recognition of *Sarotherodon* as a distinct genus. *Bull. Br. Mus. nat. Hist.* (*Zool.*), **25**, 2–26.

Trewavas, E. (1974) The freshwater fishes of the Rivers Mungo and Meme and Lakes Kotto, Mboandong and Soden, West Cameroon. *Bull. Br. Mus. nat. Hist.* (*Zool.*), **26**, 331–419.

Trewavas, E. (1982) Generic groupings of Tilapiini used in aquaculture. *Aquaculture*, **27**, 79–81.

Trewavas, E. (1983) *Tilapiine Fishes of the Genera Sarotherodon, Oreochromis and Danakilia*, British Museum (Natural History), London.

Trewavas, E., Green, J. and Corbet, S. A. (1972) Ecological studies on crater lakes in West Cameroon. Fishes of Barombi Mbo. *J. Zool., Lond.*, **167**, 41–95.

Trivers, R. L. (1972) Parental investment and sexual selection, in *Sexual Selection and the Descent of Man* (ed. B. Campbell), Aldine, Chicago, pp. 136–79.

Turner, G. F. (1988) Environmental influences on male breeding coloration in *Oreochromis niloticus*. *J. Fish. Biol.*, **32**, 155–6.

Turner, G. F. and Huntingford, F. A. (1986) A problem for game theory analysis: assessment and intention in male mouthbrooder contests. *Anim. Behav.*, **34**, 961–70.

Turner, J. L. (1977a) Changes in the size structure of cichlid populations of Lake Malawi resulting from bottom trawling. *J. Fish. Res. Bd. Can.*, **34** (2), 232–8.

Turner, J. L. (1977b) Some effects of demersal trawling in Lake Malawi (Lake Nyasa) from 1968 to 1974. *J. Fish Biol.*, **10**, 261–73.

Ufermann, A., Alligayer, R. and Geerts, M. (1987) *Cichlid Catalogue*, Martin Brumath, France.

Valerio, M. and Barlow, G. W. (1986) Ontogeny of young Midas cichlids: a study of feeding, filial cannibalism and agonism in relation to differences in size. *Biol. Behav.*, **11**, 16–35.

Vandewalle, P. (1971) Comparasion ostéologique et myologique de cinq Cichlidae Africains et Sud-Americains. *Annls. Soc. r. zool. Belg.*, **101**, 259–92.

Vandewalle, P. (1976) Ostéologie et myologie de *Tilapia guineensis*. *Annls. Mus. r. Afr. cent.*, No. 196, 1–50.

Varadaraj, K. and Pandian, T. J. (1990) Production of all-female, sterile-triploid *Oreochromis mossambicus*. *Aquaculture*, **84**, 117–23.

Verheyen, E. and Rompaey, J. van (1986) Genetics and speciation in African lacustrine cichlids. *Annls. Mus. r. Afr. cent. Sci. zool.*, **251**, 95–101.

Verheyen, E., Rompaey, J. van and Selens, M. (1985) Enzyme variations in haplochromine cichlid fishes from Lake Victoria. *Neth. J. Zool.*, **35**, 469–78.

Vermeij, G. J. and Currey, J. D. (1980) Geographical variation in the strength of thaidid snail shells. *Biol. Bull. mar. biol. Lab., Woods Hole*, **158**, 383–9.

Voss, J. (1977) Les livrées ou patrons de coloration chez les poissons Cichlides africains. *Rev. fr. Aquariol.*, **4**, 34–80.

Wachtel, S. S. (1988) *Evolutionary Mechanisms in Sex Determination*, CRC Press, Boca Raton, Florida.

Waddington, C. H. (1975) *The Evolution of an Evolutionist*, Edinburgh University Press, Edinburgh.

Wanink, J. H. (1988) Recent changes in the zooplanktivorous/insectivorous fish community of the Mwanza Gulf. HEST Report 45. [see HEST]

Ward, J. A. (1973) The significance of contacting in cichlids. *Buntbarsche Bull.*, **36**, 9–12.

Ward, J. A. (1978) The red chromide. *Trop. Fish Hobby.*, **26**, 64–7.

Ward, J. A. and Barlow, G. W. (1967) The maturation and regulation of glancing off the parents by young orange chromides (*Etroplus maculatus*; Pisces, Cichlidae). *Behaviour*, **29**, 1–56.

Ward, J. A. and Samarakoon, J. I. (1981) Reproductive tactics of the Asian cichlids of the genus *Etroplus* in Sri Lanka: preliminary findings. *Environ. Biol. Fishes*, **6**, 95–103.

Ward, J. A. and Wyman, R. L. (1975) The cichlids of the resplendent isle. *Oceans*, **8**, 42–7.
Ward, J. A. and Wyman, R. L. (1977) Ethology and ecology of cichlid fishes of the genus *Etroplus* in Sri Lanka: preliminary findings. *Environ. Biol. Fishes*, **2**, 137–45.
Watrous, L. E. and Wheeler, Q. D. (1981) The out-group comparison method of character analysis. *Syst. Zool.*, **30**, 1–11.
Watt, W. B., Carter, P. A. and Donohue, K. (1986) Females' choice of 'good genotypes' as mates is promoted by an insect mating system. *Science*, **233**, 1187–90.
Weber, P. G. and Weber, S. P. (1976) The effect of female color, size, dominance and early experience upon mate selection in male convict cichlids, *Cichlasoma nigrofasciatum* Günther (Pisces, Cichlidae). *Behaviour*, **56**, 116–35.
Welcomme, R. L. (1967a) Observations on the biology of the introduced species of *Tilapia* in Lake Victoria. *Revue. Zool. Bot. afr.*, **76**, 249–79.
Welcomme, R. L. (1967b) The relationship between fecundity and fertility in the mouthbrooding cichlid fish *Tilapia leucosticta*. *J. Zool., Lond.*, **151**, 453–68.
Welcomme, R. L. (1984) International transfer of inland fish species, in *Distribution, Biology and Management of Exotic Fishes* (eds W. R. Courtenay, and J. Stauffer), The Johns Hopkins University Press, Baltimore, Maryland, and London, pp. 21–40.
West-Eberhard, M. J. (1983) Sexual selection, social competition, and evolution. *Q. Rev. Biol.*, **58**, 155–83.
Whyte, S. A. (1975) Distribution, trophic relationships and breeding habits of the fish populations in a tropical lake basin (Lake Bosumtwi–Ghana). *J. Zool., Lond.*, **177**, 25–56.
Wickler, W. (1962a) Ei-Attrappen und Maulbruten bei afrikanischen Cichliden. *Z. Tierpsychol.*, **19**, 129–64.
Wickler, W. (1962b) 'Egg-dummies' as natural releasers in mouth-brooding cichlids. *Nature, Lond.*, **194**, 1092–4.
Wickler, W. (1963) Zur Klassification der Cichlidae, am Beispiel der Gattungen *Tropheus, Petrochromis, Haplochromis* und *Hemihaplochromis n. gen.* (Pisces, Perciformes). *Senckenberg. biol.*, **44**, 83–96.
Wickler, W. (1965a) Neue Varianten des Fortpflanzungsverhaltens afrikanischer Cichliden (Pisces, Perciformes). *Naturwissenschaften*, **52**, 219.
Wickler, W. (1965b) Signal value of the genital tassel in the male *Tilapia macrochir* Blgr. (Pisces, Cichlidae). *Nature, Lond.*, **208**, 595–6.
Wickler, W. (1966) Sexualdimorphismus, Paarbildung und Versteckbruten bei Cichliden (Pisces: Perciformes). *Zool. Jb. Abt. Syst. Oekol. Geogr. Tiere.*, **93**, 127–38.
Wiley, E. O. (1981) *Phylogenetics: The Theory and Practice of Phylogenetic Systematics*, John Wiley and Sons, New York.
Wiley, E. O. (1987) *Phylogenetics*, John Wiley and Sons, N.Y.
Wilhelm, W. (1980) The disputed feeding behaviour of a paedophagous haplochromine cichlid (Pisces) observed and discussed. *Behaviour*, **74**, 310–23.
Williams, G. C. (1966) *Adaptation and Natural Selection*, Princeton University Press, Princeton, NJ.
Williams, G. C. (1975) *Sex and Evolution*, Princeton University Press, Princeton, NJ.
Williams, G. C. (1972) On the ontogeny of behaviour of the cichlid fish *Cichlasoma nigrofasciatum* (Günther), PhD thesis, Univ. Groningen, Holland.
Wilson, E. O. (1975) *Sociobiology – The New Synthesis*, Belknap Press, Cambridge, Mass.
Wilson, E. O. (ed.) (1988) *Biodiversity*, National Academy Press, Washington, DC.
Witte, F. (1981) Initial results of the ecological survey of the haplochromine cichlid

fishes from the Mwanza Gulf of Lake Victoria (Tanzania): breeding patterns, trophic and species distribution, with recommendations for commercial trawl-fisher, *Neth. J. Zool.*, **31**, 175–202.

Witte, F. (1984a) Ecological differentiation in Lake Victoria haplochromines: comparison of cichlid species flocks in African lakes, in *Evolution of Fish Species Flocks* (eds A. A. Echelle and I. Kornfield), University of Maine at Orono Press, Orono, Maine, pp. 155–67.

Witte, F. (1984b) Consistency and functional significance of morphological differences between wild-caught and domestic *Haplochromis squamipinnis* (Pisces, Cichlidae). *Neth. J. Zool.*, **34**, 596–612.

Witte, F. (1987) From form to fishery. An ecological and taxonomical contribution to morphology and fishery of Lake Victoria cichlids, thesis, Rijksuniversiteit Leiden.

Witte, F. and Barel, C. D. N. (1976) The comparative functional morphology of the pharyngeal jaw apparatus of piscivorous and intrapharyngeal mollusc-crushing *Haplochromis species. Revue. Trav. Inst. Pêch marit.*, **40**, 793–6.

Witte, F., Barel, C. D. N. and Hoogerhoud, R. J. C. (1990) Phenotypic plasticity of anatomical structures and its ecomorphological significance. *Neth. J. Zool.*, **40**, 278–98.

Witte, F. and Goudswaard, P. C. (1984) Prospects of the haplochromine fishery in Southern Lake Victoria. Reports from the *Haplochromis* Ecology Survey Team (HEST) operating in the Mwanza area of Lake Victoria, No. 33, 1–13. [see HEST]

Witte, F. and Goudswaard, P. C. (1985) Aspects of the haplochromine fishery in Southern Lake Victoria. *F.A.O. Fish. Rep.*, **335**, 81–8.

Witte, F. and Goudswaard, P. C. (1987) Prospects of the haplochromine fishery in Southern Lake Victoria, in Witte (1987), Chap. IV.

Witte, F. and Oijen, M. J. P. van (in press) Taxonomy, ecology and fishery of haplochromine trophic groups. *Zoöl. Verhandelingen Leiden*.

Witte, F. and Witte-Maas, E. L. M. (1981) Haplochromine cleaner fishes: a taxonomic and eco-morphological description of two new species. Revision of the haplochromine species (Teleostei, Cichlidae) from Lake Victoria. Part I. *Neth. J. Zool.*, **31**, 203–31.

Witte, F. and Witte-Maas, E. L. M. (1987) Implications for taxonomy and functional morphology of intra-specific variation in haplochromine cichlids of Lake Victoria, in Witte (1987), Chap. II.

Witte, F., Barel, C. D. N. and Hoogerhoud, R. J. C. (1990) Phenotypic plasticity of anatomical structures and its ecomorphological significance. *Neth. J. Zool.*, **70**, FAO Fish Report (in press).

Witte, F. Wanink, J. H., Goldschmidt, T., van Oÿen, M. J. P., Goudswaard, P. C. and Bouton, N. (submitted) Quantitative data on the dechine of haplochromine species from the Mwanza Gulf (Lake Victoria). *Env. Biol. Fishes.*

Witte-Maas, E. L. M. (1981) Egg snatching: an observation on the feeding behaviour of *Haplochromis barbarae* Greenwood, 1967 (Pisces, Cichlidae). *Neth. J. Zool.*, **31**, 786–9.

Wittenberger, J. F. (1981) *Animal Social Behavior*, Duxbury Press, Boston, Mass.

Wholfarth, G. W. and Hulata, G. (1983) *Applied Genetics of Tilapias* (*ICLARM Studies and Reviews*, Vol. 6, 2nd edn, revised), International Center for Living Aquatic Resources Management, Manila.

Wolters, W. R., Libey, G. S. and Chrisman, C. L. (1982) Effect of triploid on growth and gonad development of channel catfish. *Trans. Am. Fish. Soc.*, **111**, 102–5.

Woodward, A. S. (1939) Tertiary fossil fishes from Maranhao, Brasil. *Ann. Mag. nat. Hist.*, **2**(3), 450–3.

Worthington, E. B. (1937) On the evolution of fish in the Great Lakes of Africa. *Int. Rev. ges. Hydrobiol. Hydrog.*, **35**, 305–17.

Wourms, J. P. (1981) Viviparity: the maternal–fetal relationship in fishes. *Am. Zool.*, **21**, 473–515.

Wrede, W. L. (1932) Versuche über den Artduft der Elritzen. *Z. vergl. Physiol.*, **17**, 510–19.

Wright, J. M. (1989) Nucleotide sequence, genomia organization and evolution of a major repetitive DNA family in tilapia (*Oreochromis mossambicus/hornorum*). *Nucleic Acids Res.*, **17**, 5071–9.

Wyles, J. S., Kunkel, J. G. and Wilson, A. C. (1983) Birds, behaviour, and anatomical evolution. *Proc. Natn. Acad. Sci. U.S.A.*, **80**, 4394–7.

Wyman, R. L. and Ward, J. A. (1972) A cleaning symbiosis between the cichlid fishes *Etroplus maculatus* and *Etroplus suratensis*. I. Description and possible evolution. *Copeia, 1972*, 834–8.

Wyman, R. L. and Ward, J. A. (1973) The development of behavior in the cichlid fish *Etroplus maculatus* (Bloch). *Z. Tierpsychol.*, **33**, 461–91.

Wynne-Edwards, V. C. (1962) *Animal Dispersion in Relation to Social Behaviour*, Oliver and Boyd, Edinburgh.

Yamagami, K. (1988) Mechanisms of hatching in fish, *Fish Physiology*, Vol. 11A (eds W. S. Hoar and D. J. Randall), Academic Press, New York, pp. 447–99.

Yamaoka, K. (1982) Morphology and feeding behaviour of five species of genus *Petrochromis* (Teleostei: Cichlidae). *Physiol. Ecol. Japan*, **19**, 57–75.

Yamaoka, K. (1983a) Feeding behaviour and dental morphology of algae scraping cichlids (Pisces: Teleostei) in Lake Tanganyika. *Afr. Stud. Monogr., Kyoto Univ.*, **4**, 77–89.

Yamaoka, K. (1983b) A revision of the cichlid fish genus *Petrochromis* from Lake Tanganyika, with description of a new species. *Jap. J. Ichthyol.*, **30**, 129–41.

Yamaoka, K. (1985) Intestinal coiling pattern in the epilithic algal-feeding cichlids (Pisces, Teleostei) of Lake Tanganyika, and its phylogenetic significance. *Zool. J. Linn. Soc.*, **84**, 235–61.

Yamaoka, K. (1987) Comparative osteology of the jaw of algal-feeding cichlids (Pisces, Teleostei) from Lake Tanganyika. *Rep. Usa mar. biol. Inst., Kochi Univ.*, **9**, 87–137.

Yamaoka, K. (1990) Feeding behaviour of *Asprotilapia leptura*, an epilithic algal feeding cichlid fish in Lake Tanganyika. *Jap. J. Ichthyol.*, **37**, 80–2.

Yamaoka, K., Hori, M. and Kuratani, S. (1986) Eco-morphology of feeding in 'goby-like' cichlid fishes in Lake Tanganyika. *Physiol. Ecol. Japan*, **23**, 17–29.

Yanagisawa, Y. (1985) Parental strategy of the cichlid fish *Perissodus microlepis*, with particular reference to intraspecific brood 'framing out'. *Environ. Biol. Fishes*, **12**, 241–9.

Yanagisawa, Y. (1986) Parental care in a monogamous mouthbrooding cichlid *Xenotilapia flavipinnis* in Lake Tanganyika. *Jap. J. Ichthyol.*, **33**, 249–61.

Yanagisawa, Y. (1987) Social organization of a polygynous cichlid *Lamprologus furcifer* in Lake Tanganyika. *Jap. J. Ichthyol.*, **34**, 82–90.

Yanagisawa, Y. and Nshombo, M. (1983) Reproduction and parental care of the scale-eating cichlid fish *Perissodus microlepis* in Lake Tanganyika. *Physiol. Ecol. Japan*, **20**, 23–31.

Yousif, O. M. (1988) Tilapia culture in Sudan. *Naga, The ICLARM Quarterly*, **11**, 13.

Youson, J. H. (1988) First metamorphosis, in *Fish Physiology*, Vol. 11B (eds W. S. Hoar and D. J. Randall), Academic Press, New York, pp. 135–96.

Zahavi, A. (1975) Mate selection – a selection of a handicap. *J. theor. Biol.*, **53**, 205–14.

Zahavi, A. (1979) Ritualization and the evolution of movement signals. *Behaviour*, **72**, 77–81.

Zaret, T. M. (1977) Inhibition of cannibalism in *Cichla ocellaris* and hypothesis of predator mimicry among South American fishes. *Evolution*, **31**, 421–37.

Zaret, T. M. (1980) Life history and growth relationships of *Cichla ocellaris*, a predatory South American cichlid. *Biotropica*, **12**, 144–57.

Zaret, T. M. (1984) Fish/zooplankton interactions in Amazon floodplain lakes. *Verh. Int. Verein. Limnol.*, **22**, 1305–9.

Zaret, T. M. and Paine, R. T. (1973) Species introduction in a tropical lake. *Science, N.Y.*, **182**, 449–55.

Zihler, F. (1982) Gross morphology and configuration of digestive tracts of Cichlidae (Teleostei, Perciformes): phylogenetic and functional significance. *Neth. J. Zool.*, **32**, 544–71.

Zoran, M. J. and Ward, J. A. (1983) Parental egg care behavior and fanning activity for the orange chromide, *Etroplus maculatus*, in *Predators and Prey in Fishes* (eds D. L. G. Noakes, D. G. Lindquist, G. S. Helfman and J. A. Ward), Dr. W. Junk, The Hague, pp. 139–48.

Author index

Fish name index

River name index

Lake name index

Subject index

For fish, river and lake names see the separate indexes.